Microwave Devices and Circuits for Advanced Wireless Communication

This book offers a comprehensive overview of design and analysis of microwave devices and circuits for 5G and beyond wireless communication systems. It focuses on modern microwave antennas, filters, metamaterials, and MIMO systems. It includes a design approach based on Artificial Intelligence and the practical use of microwave devices and circuits in commercial, medical, and military applications.

Microwave Devices and Circuits for Advanced Wireless Communications: Design and Analysis explores the performance of microwave devices and circuits by highlighting the difficulties encountered by researchers and designers such as latency, interoperability, wireless coexistence, data streaming, safety, security, and privacy. The book explores the most important aspects of antenna design, including radiation pattern control, impedance matching with bandwidth improvement, and gain enhancement. It also examines different categories of metasurfaces, including frequency-selective surfaces (FSS) and electromagnetic bandgap (EBG) structures, and their distinct roles in antenna design. Additionally, the book examines concepts such as ultra-wideband (UWB) radar for 5G millimeter wave applications, and advanced techniques such as synthetic aperture radar (SAR), beam-forming, compressed sensing, and diffraction tomography for enabling high-resolution imaging across wider application areas. The authors also present an overview on applying machine learning (ML) techniques to advanced wireless communication for signal-processing tasks such as signal denoising, equalization, and modulation recognition. They then discuss the potential significance of UAV communication systems in achieving seamless connection, quality of service (QoS), as well as the difficulties and potential remedies involved in building dependable networks using UAVs. Throughout the book the authors offer a critical assessment of the strengths and limitations of each topic and approach presented, thus providing valuable guidance for future research in this exciting field.

This book will be helpful for graduate students, researchers, and engineers working in the area of design and reliability of circuits for microwave and communication systems.

Materials, Devices, and Circuits: Design and Reliability

Series Editor: Shubham Tayal, K. K. Paliwal, Amit Kumar Jain

This series is designed to illustrate the many new and exciting challenges in the expansive and interdisciplinary field of materials science, device engineering, reliability, device-circuit co-design and their applications. The scope of this series is broad, with reference works and textbooks offering insight into all aspects of materials development, device fabrication, circuit analysis and their reliability. The titles in this series deliver authoritative information to professionals, researchers, and students across material and device engineering as well as other scientific disciplines. Each volume offers a comprehensive approach covering fundamentals, technologies, their evolution advancements and applications and includes real-world examples where appropriate.

Device Circuit Co-Design Issues in FETs
Edited by Shubham Tayal, Billel Smaani, Shiromani Balmukund Rahi, Samir Labiod, Zeinab Ramezani

Human-Machine Interface Technology Advancements and Applications
Edited by Edited By Ravichander Janapati, Usha Desai, Shrirang Ambaji Kulkarni, Shubham Tayal

5G and Beyond Wireless Networks: Technology, Network Deployments and Materials for Antenna Design
Edited by Indrasen Singh, Shubham Tayal, Niraj Pratap Singh, Vijay Shankar Tripathi, and Ghanshyam Singh

Negative Capacitance Field Effect Transistors: Physics, Design, Modeling and Applications
Edited by Young Suh Song, Shubham Tayal, Shiromani Balmukund Rahi, Abhishek Kumar Upadhyay

Laser Applications in Manufacturing
Edited by Pankaj Kumar, Manowar Hussain, Amit Kumar Jain, Sunil Pathak

Microwave Devices and Circuits for Advanced Wireless Communication: Design and Analysis
Edited by Dilip Kumar Choudhary, Naveen Mishra, Indrasen Singh, Naser Ojaroudi Parchin, and Ghanshyam Singh

For more information about this series, please visit: www.routledge.com/Materials-Devices-and-Circuits/book-series/MDCDR

Microwave Devices and Circuits for Advanced Wireless Communication

Design and Analysis

Edited by

Dilip Kumar Choudhary, Naveen Mishra, Indrasen Singh, Naser Ojaroudi Parchin, and Ghanshyam Singh

CRC Press
Taylor & Francis Group
Boca Raton London New York

CRC Press is an imprint of the Taylor & Francis Group, an **informa** business

Designed cover image: Shutterstock

First edition published 2025
by CRC Press
2385 NW Executive Center Drive, Suite 320, Boca Raton FL 33431

and by CRC Press
4 Park Square, Milton Park, Abingdon, Oxon, OX14 4RN

CRC Press is an imprint of Taylor & Francis Group, LLC

ISBN: 978-1-032-65600-7 (hbk)
ISBN: 978-1-032-65601-4 (pbk)
ISBN: 978-1-032-65602-1 (ebk)

DOI: 10.1201/9781032656021

Typeset in Times LT Std
by Apex CoVantage, LLC

Contents

About the Editors

Dilip Kumar Choudhary received a Ph.D. in electronics engineering from Indian Institute of Technology (ISM) Dhanbad, India in 2020. Currently, he is working as Assistant Professor in the School of Electronics Engineering, Vellore Institute of Technology, Vellore, India. His main expertise comprises the electromagnetic simulation and characterization of microwave filtering circuits using metamaterial structures for wireless communication. He has authored more than 34 referred journal and conference papers including 16 SCI journals. He has authored one copyright and two book chapters. He was the recipient of the International Travel Grant from DST, India in 2017. He was the recipient of the Best Paper at PIERS, Singapore in 2017. He served as a session chair "Poster Session 2-P" at the 2017 IEEE Asia Pacific Microwave Conference (APMC), Kuala Lumpur, Malaysia. He is a potential reviewer of many journals and conferences such as *IEEE Access*, *IEEE TCAS-II*, IEEE TEMC, *Electronics Letters*, *MOTL*, RFCAD, *Radioengineering*, PIER, *Wireless Personal Communications*, and EuCAP conferences. He is a senior member of IEEE, MTT, APS, and AESS societies. His current research interests include microwave filtering circuits, metamaterial, and their applications, microwave filters, antenna, MIMO antenna, absorber, power dividers, and coupler for wireless communication.

Naveen Mishra was born in Deoria, India, in 1990. He received a B.Tech degree in electronics and communication engineering from Uttar Pradesh Technical University, Lucknow, in 2010. He completed his Ph.D. degree with the Department of Electronics Engineering, IIT (Indian School of Mines) Dhanbad, India. He joined the Department of Electronics Engineering, IIT (Indian School of Mines) Dhanbad, in January 2014, as a Junior Research Fellow, where he worked as a Senior Research Fellow, from January 2016 to April 2019. He is currently working as Assistant Professor with the Department of Electronics Engineering, School of Electronics Engineering, Vellore Institute of Technology, Vellore, Tamil Nadu, India. He joined VIT University in May 2019. He has authored or co-authored more than 30 peer-reviewed international/national journals/conference research articles. His current research interests involve metamaterial structures and their various applications. He is a potential reviewer of many journals and conferences, such as the *IEEE Access*, *IEEE Open Journal of Antennas and Propagation*, *Progress in Electromagnetic Research*, and *International Journal of RF and Microwave Computer-Aided Engineering.*

Indrasen Singh is working as Assistant Professor (Sr. Grade-2) in the Department of Embedded Technology under the School of Electronics Engineering (SENSE), Vellore Institute of Technology (VIT), Vellore, Tamil Nadu, India. He received his B.Tech and M.Tech degrees in electronics and communication engineering from Uttar Pradesh Technical University, Lucknow, India in 2006, and 2010, respectively. He obtained his Ph.D. degree in electronics and communication engineering from National Institute of Technology Kurukshetra, Haryana, India in 2019. He has qualified GATE examination many times. He has more than 15 years of teaching/research experience in the various reputed technical institutes/universities. He is supervising B.Tech, M.Tech, and Ph.D. students. He is the editorial board member of *AJECE*, Science Publishing Group, USA. He is the reviewer of many SCI/SCOPUS indexed international journals and served as TPC member and reviewer in different conferences. He has published more than 25 research papers in national/international journals/conferences of repute and many are under review. His research interests are in the area of cooperative communication, stochastic geometry, modeling of wireless networks, heterogeneous networks, millimeter wave communications, device-to-device communication, Machine Learning, and 5G/6G communication.

Naser Ojaroudi Parchin is a senior member of IEEE. He received a Ph.D. degree in electrical engineering from the University of Bradford, U.K. He is currently Assistant Professor at Edinburgh Napier University, U.K. He was a postdoctoral research assistant at the Faculty of Engineering and Informatics, University of Bradford. He worked as a Research Fellow in the SATNNEX V Project, funded by the European Space Agency. From 2018 to 2020, he was a Marie Curie Research Fellow in the H2020-ITN-SECRET Project, funded by EU Commission, targeting 5G mobile small cells. From 2014 to 2018, he worked with the APMS Section, Aalborg University, Denmark. In 2016, he was a Visiting Researcher at Ankara University, Turkey. He has over 12 years of research experience in antennas and microwave engineering. His research interests include phased arrays, MIMO systems, smartphone antennas, SAR/user-impact, full-duplex diversity, 5G antennas, implementable and biomedical sensors, RFID tag antennas, millimeter-wave and terahertz components, fractal structures, metamaterials/metasurfaces, PCB realization, Fabry resonators, EBG/FSS-inspired radiators, microwave filters, reconfigurable structures, and wireless propagation.

Ghanshyam Singh received Ph.D. degree in Electronics Engineering from the Indian Institute of Technology, Banaras Hindu University, Varanasi, India, in 2000. Currently, he is working as a full Professor and Director with the Centre for Smart Information and Communication Systems, Department of Electrical and Electronic Engineering Sciences, Auckland Park Kingsway Campus, University of Johannesburg, South Africa. He was associated with Central Electronics Engineering Research Institute, Pilani, and Institute for Plasma Research, Gandhinagar, India, respectively, where he was Research Scientist. He had also worked as an Assistant Professor at Electronics and Communication Engineering Department, Nirma University of Science and Technology, Ahmedabad, India. He was a Visiting Researcher at the Seoul National University, Seoul, South Korea. He also worked as Professor with the Department of Electronics and Communication Engineering, Jaypee University of Information Technology, Wakanaghat, Solan, India. He has more than 24 years teaching and research experience. His research and teaching interests include Millimeter/THz Wave Technologies it's Applications in Communication and Imaging, Next-Generation Communication Systems (5G/6G)/ Cognitive Radio/NOMA, Resource Allocation and Management, Interference Management, Applications of 5G/6G in Sustainable Smart City, Industry 4.0/5.0, Healthcare 4.0, Intelligent Transport System, Energy Management (IoE) and Digital Farming. He is an author/co-author of more than 370 scientific papers of the peer reviewed Journal and International Conferences and 10 published books by Springer, Wiley and CRC Press. He has supervised many Ph.D. and M. Tech. theses. He has worked as editor/associate editor and reviewer for several reputed Journals and Conferences.

Contributors

Ahmad Saif
Department of Electronics and Communication Engineering
Integral University Lucknow
Uttar Pradesh, India

Babu B Anil
Department of ECE
GMR Institute of Technology
Rajam, AP, India

Bhattacharyya Somak
Department of Electronics Engineering
IIT (BHU)
Varanasi, Uttar Pradesh

C Mahapriya
School of Electronics Engineering
Vellore Institute of Technology
Vellore, India

Chaitanya T Krishna
ECE Department
Bapatla Engineering College Bapatla
Andhra Pradesh

Choubey Arvind
Department of Electronics & Communication Engineering
Indian Institute of Information Technology
Bhagalpur, India

Gehlot Anita
Department of Computer Science and Engineering
Uttaranchal University
Dehradun, India

Karuna Medikonda
CSE Department
Bapatla Engineering College Bapatla
Andhra Pradesh, India

Khan Mohd Javed
Department of Electronics and Communication Engineering
Integral University Lucknow
UP, India

Kumar Arjun
Department of Electronics and Communication Engineering
National Institute of Technology
Jamshedpur, Jharkhand, India

Kumar Prashant
Department of Physics
School of Chemical Engineering and Physical Sciences
Lovely Professional University
Jalandhar, Punjab, India

Kumar Praveen
Department of Electronics & Communication Engineering
National Institute of Technology
Jamshedpur, India

Kumar Ranjeet
Department of Electronics & Communication Engineering
National Institute of Technology
Jamshedpur, India

Kumar P Ravi
Department of ECE
GMR Institute of Technology
Rajam, AP, India

Kumar T Anil
Department of ECE
GMR Institute of Technology
Rajam, AP, India

Madhav BTP
Antennas and Liquid Crystals Research Centre
Koneru Lakshmaiah Education Foundation
Andhra Pradesh, India

Mahto Santosh Kumar
Department of Electronics & Communication Engineering
Indian Institute of Information Technology
Ranchi, India

Malik Praveen Kumar
Department of Electronics
School of Electronics and Electrical Engineering
Lovely Professional University
Phagwara, Punjab, India

Mishra Naveen
School of Electronics Engineering
Vellore Institute of Technology
Vellore, Tamil Nadu, India

Painam Surendra Kumar
ECE Department
Bapatla Engineering College
Bapatla
Andhra Pradesh, India

Pal Pravesh
Department of Electronics & Communication Engineering
National Institute of Technology
Jamshedpur, India

Pandurangan Durai
School of Electronics Engineering
Vellore Institute of Technology
Vellore, Tamil Nadu, India

Samantaray Diptiranjan
Department of ECE
Faculty of Technology
University of Delhi
New Delhi, India

Sangeetha S.
College of Engineering
Guindy, Anna University, India

Sarma Subhra Sankha
School of Electronics Engineering
Vellore Institute of Technology
Vellore, Tamil Nadu, India

Saxena Gaurav
Department of Electronics and Communication Engineering
Ajay Kumar Garg Engineering College
Ghaziabad, Uttar Pradesh, India

Sengar Suverna
Department of Electronics
School of Electronics and Electrical Engineering
Lovely Professional University
Phagwara, Punjab, India

Shafik Wasswa
School of Digital Science
University Brunei Darussalam
Gadong, Brunei;
Dig Connectivity Research Laboratory (DCRLab)
Kampala, Uganda

Singh Ajit Kumar
Department of Electronics & Communication Engineering
Indian Institute of Information Technology
Ranchi, India

Singh Akhilendra Pratap
School of Electronics Engineering
Maharishi University of Information Technology
Lucknow, Uttar Pradersh, India

Singh Indrasen
School of Electronics Engineering
Vellore Institute of Technology
Vellore, India

Singh Rajnish Kumar
Department of Electronics and Communication Engineering
Indian Institute of Information and Technology
Nagpur, Maharashtra, India

Singh Vineet
Department of Computer Science and Engineering
Indian Institute of Technology Roorkee
Greater Noida, India

Sinha Rashmi
Department of Electronics & Communication Engineering
National Institute of Technology
Jamshedpur, India

Srivastava Harshit
Department of Electronics and Communication Engineering
Galgotias College of Engineering College
Greater Noida, Uttar Pradesh, India

Sood Deepak
Department of Electronics
University Institute of Engineering and Technology
Kurukshetra University
Haryana, India

S. Kanthamani
Department of ECE
Thiagarajar College of Engineering
Madurai, Tamil Nadu, India

Thomas Shweta B
School of Electronics Engineering
Vellore Institute of Technology
Vellore, India

Tripathi Prabhakar
Institute for Plasma Research Gandhinagar
Gujarat, India

Vishwakarma Ishan
Department of Electronics and Communication Engineering
Galgotias College of Engineering College
Greater Noida, Uttar Pradesh, India

Yadav Dipanshu
Department of Electronics and Communication Engineering
Galgotias College of Engineering College
Greater Noida, Uttar Pradesh, India

1 Design and Analysis of UWB Patch Antenna for 5G Millimeter Wave Applications

Suverna Sengar, Praveen Kumar Malik, Anita Gehlot and Prashant Kumar

1.1 INTRODUCTION

5G stands for fifth generation, and it is a wireless technology used in digital cellular networks. 5G technology frequency spectrum is divided into three major bands: Millimeter wave and its frequency range from 24 GHz to 40 GHz; it has the fastest speed but smaller coverage area. Mid-band has a frequency band from 1 GHz to 6 GHz; it is also called sub-6 GHz band. Low-band (less than 1 GHz) is more extensive but has slower speeds [1–2]. Antennas for 5G will be major, needing to have a continuous radiation pattern, low path loss, and reduced latency due to the explosive growth in the development of 5G technology, and it is a key sign of the explosion of technology that meets basic needs. The performance of the 5G UWB microstrip patch antenna would enhance the wireless communication systems of the next generation. The key advantages of 5G technologies are faster transmission rates, exceptionally high data rates, increased dependability, lower latency, improved spectral efficiency, and the capacity to connect more wireless devices to the networks. One of the most anticipated 5G innovations is the UWB microstrip patch antenna, which increases spectral efficiency by 50 times while boosting capacity compared to a traditional antenna [3].

The microstrip patch antennas can meet the majority of the specified requirements nowadays. They contain a metallic ground, a radiating patch, a feed coupled to the patch, and a dielectric substrate [4]. Even though microstrip patch antennas are small and have a low profile, they have several drawbacks such as low return loss, narrow bandwidth, and poor strength [5]. The current specifications for antenna designs for use in portable systems call for a size reduction while maintaining high performance.

Since the 2002 approval of the commercial use of unlicensed frequencies from 3.1 to 10.6 GHz by the US Federal Communication Commission (FCC) [6], the ultra-wideband (UWB) technology has received a lot of attention in the wireless industry. Among various benefits of UWB communication systems fast data rate, great power efficiency, minimal complexity, low cost, and high precision [7].

DOI: 10.1201/9781032656021-1

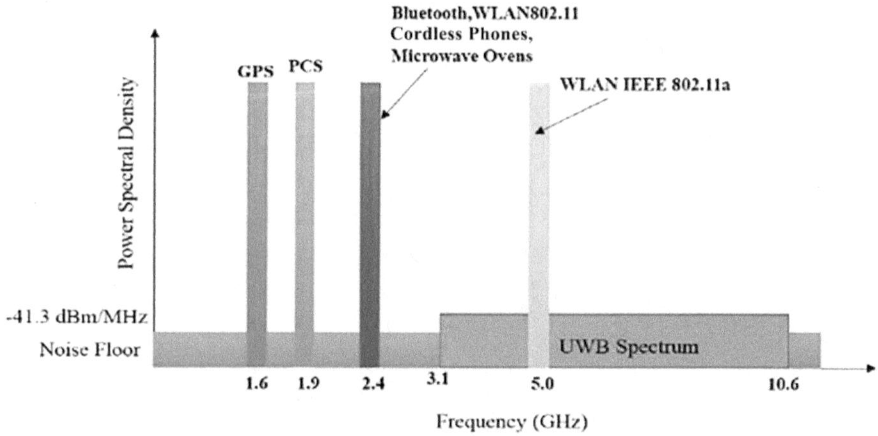

FIGURE 1.1 Comparison power spectral density (PSD) of UWB system with other systems [9].

However, due to signal reflection and diffraction, these systems experience multipath fading effects [8]. The most likely technology to be used for short-range, high-speed indoor data transmission is ultra-wideband (UWB). The power spectrum density of UWB systems is compared to all other radio communication systems in Figure 1.1.

UWB antennas should typically be tiny and affordable while still providing appropriate wideband performance for a variety of industrial applications [10]. Thus, obtaining better bandwidth, efficiency, and low profile within allotted reduced dimensions is one of the key problems of building UWB antennas. By altering the radiating patch's shape, UWB antenna properties can be enhanced. The patch can be round, elliptical, heart-shaped, rectangular, or any other shape.

1.1.1 Motivation for UWB Antenna Design

Many researchers, scientists, and engineers view ultra-wideband (UWB) technology as an intriguing and promising choice for greater speed, high data rate, and short-range indoor wireless communications. The US Federal Communication Commission (FCC) approved the use of the frequency band between 3.1 GHz and 10.6 GHz for UWB radio applications in 2002 [11]. Both academic research and the business world are currently very interested in the development and deployment of UWB systems. The increasing use of UWB systems has led to several improvements in the design of UWB antennas. Among the numerous difficulties facing the installation of a UWB system is the creation of an adequate antenna that can function across the whole UWB spectrum. This is due to the antenna's significant contribution to the performance of the UWB system and the system as a whole. In recent years, a variety of antenna designs, such as the spiral antenna, bi-conical antenna, Vivaldi antenna, log periodic antenna, etc., have been used in UWB systems to achieve broad bandwidth.

A Vivaldi antenna's directional radiation pattern makes it unsuitable for use with either of these applications because mobile/portable devices and indoor wireless communication both need omnidirectional radiation patterns to make it easier and

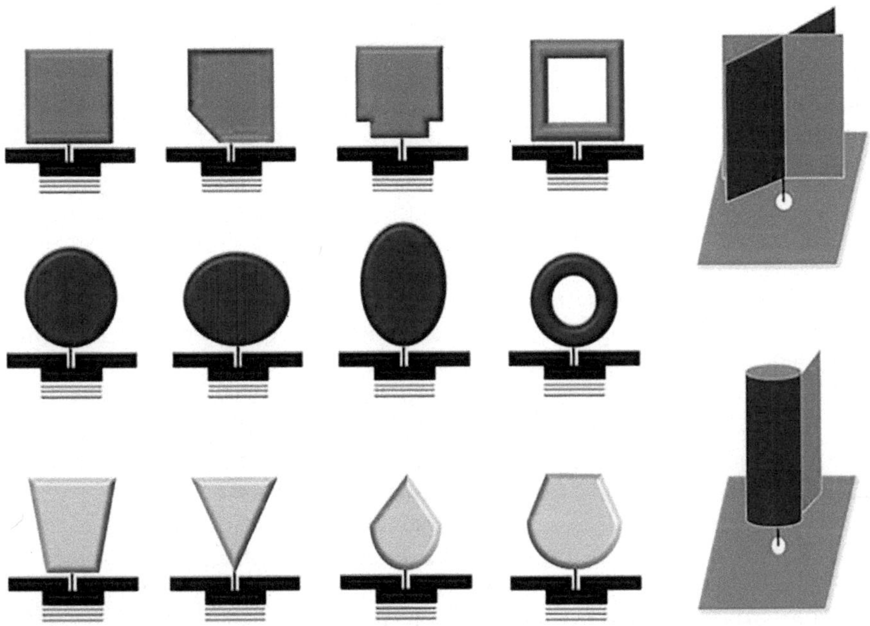

FIGURE 1.2 Different shapes of planar UWB antennas [13].

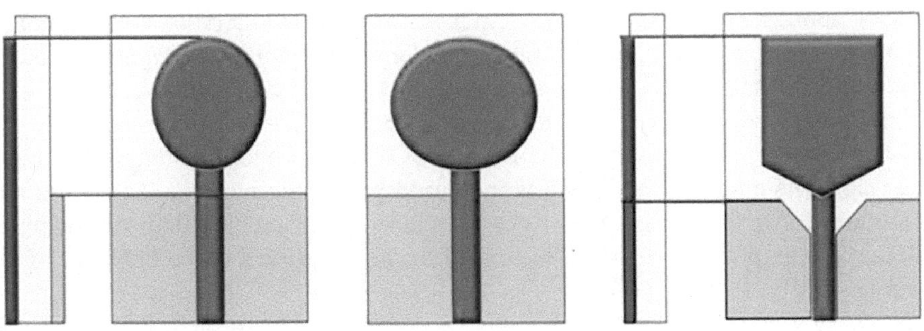

FIGURE 1.3 UWB PCB or printed antenna designs [13].

more efficient for transmitters and receivers to communicate with one another in all directions [12]. The usage of these devices is constrained by the 16 enormous physical characteristics of single- and double-conical antennas' [13] difficult designs. Academics are showing more interest in mobile/portable devices and UWB indoor wireless communications that can address all of these issues. The planar or printed antenna is a different option. Widely shaped printed antennas, including elliptical, circular, rectangular, and trapezoidal ones, have been proposed and are depicted in Figures 1.2 and Figure 1.3, respectively, for use in UWB applications.

Recent years have seen a lot of research into conventional antenna design for ultra-wideband wireless communication. Numerous aspects of antennas are common among them, and it's important to note that each design aims an increased perfectly matched impedance spectrum without losing a radiation pattern that is omnidirectional.

1.2 LITERATURE SURVEY

Additionally, many antennas have large sizes and intricate designs that are inappropriate for portable wireless terminals with limited space. A 28 mm × 29 mm patch was printed with an I-shaped monopole antenna [14]. A 60 mm × 60 mm multiband antenna with an H-shaped slot was displayed [15]. A wideband dual-polarized antenna with strong isolation and minimal cross-polarization is provided here by the author. The two linear polarizations are produced using a reflector ground plane and a pair of orthogonal planar printed dipoles. To increase isolation and cross-polarization, slots are carved out at the corners of the dipoles and brief connections are made between the orthogonal dipoles [16]. The author of this work proposed a printed elliptical patch antenna with a UWB operational bandwidth and a band-notch frequency band at 5 GHz. The antenna performs the band-notching function by being put into a U-shaped slot. This design yields a generally flat gain over the UWB frequency range, an impedance bandwidth from 2.97 to 15.09 GHz, band rejection of 4.90 to 5.86 for −10 dB return loss, and an acceptable radiation pattern [17], with the exception of the band rejection. This article suggested a remarkable antipodal Vivaldi antenna with excellent impedance matching and gain for ultra-wideband (UWB) applications. Two symmetrical sets of rectangular slits are implanted around the tapered slot to increase impedance matching at the low-frequency band, while the slot itself has an elliptical-shaped director placed in it to maximize gain at the high-frequency band [18]. The author proposed a compact dual-band antenna with a special patch inspired by the Fibonacci sequence and a rectangular ground plane slot. The ellipse's bandwidth is increased by connecting circular sectors and an L-shaped strip to the rectangular slot to create a UWB antenna. The ground plane also has a strip in the form of "#" to control the circularly polarized behavior of the antenna [19].

Here, a low-cost UWB elliptic antenna was introduced by the author. The antenna is mounted on FR4 dielectric material, which has a compact size of (21 × 27 × 1.6) mm^3. It has a maximum Gain of 6 dBi and a 16.26 GHz impedance bandwidth [20]. Here, the author has provided an elliptic monopole antenna with a small size of (15 × 25 × 1.6) mm^3. A simple FR4 dielectric material is utilized in the antenna design. The frequency range between 3.1 GHz and 10.6 GHz, where the reflection coefficient is less than −10 dB, is where the antenna operates most effectively. Additionally, the antenna's minimum and maximum gain values in the UWB range are 2.5 dBi and 9 dBi, respectively [21]. In this article, the author presents a small (17 × 14 × 1.6) mm^3 elliptical antenna constructed of FR-4 material and suitable for mobile applications. The suggested antenna has a gain of 6.8 dBi and operates at a bandwidth of 24 GHz [22]. In this article, the author describes a small (18.3 × 23 × 1.6) mm^3 CPW-fed elliptical antenna for ultra-wideband (UWB) that was constructed out of FR-4 material and operational bandwidth of 3.6 GHz to 11.2 GHz. The unique feature of this antenna is its modest size and uniplanar form, which makes fabrication simple and inexpensive [23]. Table 1.1 shows some literature study of UWB antenna respectively.

TABLE 1.1
Some More Literature Survey with Different Shape and Design

Ref. No	Year	Work Done	Findings	Application
[24]	2021	A self-decoupled small dual-band mm-wave MIMO antenna array featuring E-shaped and H-slotted elements was designed. Size of the antenna: $20 \times 24 \times 0.508\ mm^3$. Dielectric material: Rogers-5880	Working Frequency: 27.6–28.6 GHz and 37.4–38.6 GHz Gain = 7.9 dB Isolation = 28 dB ECC < 0.001 Efficiency = 85%	mm-wave for 5G technology
[25]	2021	A 28/38 GHz dual-band inverted -F array antenna system for mobile phones was created by the author (handsets). Size of Antenna: $5.5 \times 04 \times 0.2\ mm^3$ Substrate: FR4, $\varepsilon_r = 4.4$	Working Frequency Range: 27.94–28.83 GHz and 37.97–38.96 GHz Acheived Gains: 16.52 dB at 28.38 GHz, and 15.35 dB at 38.49 GHz Mutual Coupling is < −35dB	Used in both 4G and 5G mobile applications
[26]	2021	Here, an end-fire broadband antenna with a grounded-CPW feed and T-shaped helical inspiration was designed. Size of antenna: $15 \times 25 \times 0.203\ mm^3$ Substrate: Rogers-5880, $\varepsilon_r = 3.38$	Working Frequency Range: 26.25 GHz to 30.14 GHz Gain is 5.83 dB, Mutual Coupling: < −30 dB ECC is < 0.005	Used in pattern diversity applications
[27]	2017	Here, an 8-element linear array and a broadband printed-dipole antenna were created. Substrate Used: Rogers RT/Duroid™ −5880, $\varepsilon_r = 2.2$	Working Frequency Range: 26.5 GHz–38.2 GHz Gain is 4.5–5.8 dBi Mutual coupling is < −20 dB	Used in the 5G wireless cellular network applications

(Continued)

TABLE 1.1 (*Continued*)
Some More Literature Survey with Different Shape and Design

Ref. No	Year	Work Done	Findings	Application
[28]	2021	A 2-port wide-band circularly polarized (CP) small size MIMO antenna array was created in this case. Substrate Used: FR4, $\varepsilon_r = 4.4$	Working Frequency Range: 3.3 GHz–4.2 GHz ECC is < 0.10 dB Isolation is >15 dB TARC: ≤ −15 dB DG: > 9.94 dB Efficiency: 95%	Used in the sub-6 GHz applications

From the aforementioned literature survey some of the important design parameters where we need to do more and better improvements are as following:

The primary goal of this research is to minimize the antenna overall size while maintaining its essential characteristics [11–13, 29]. With the addition of two elliptical slots to the capsule-shape UWB antenna, the total antenna size was reduced, and the radiation patch area was lowered overall by 38% compared to the others and an antenna-radiating structure has been developed for increasing the bandwidth that can be used for 5G wireless applications.

For use in 5G and UWB applications, this article introduces a capsule-shaped microstrip line-fed tiny patch antenna. The major goal of the research is to produce findings that can be applied to more advanced 5G technologies that operate in the 24–28 GHz frequency range. A capsule-shaped patch, a portion of a ground plane, and a line-stripe microstrip feed line make up the suggested design. The ground plane is calibrated to produce the desired results. The simulation findings are provided for the antenna's reflection coefficient is < −10 dB, VSWR < 2, gain, radiation efficiency, surface current distributions, and radiation patterns. In this article, we suggest a UWB antenna that is small in size and has a very wide bandwidth that extends more than 24 GHz. Additionally, it offers improved gain performance that is advantageous for today's 5G wireless applications.

1.3 ANTENNA DESIGN PROCEDURE AND METHODOLOGY

A dielectric substrate is placed between a top layer called a patch and a bottom layer called a ground layer in a microstrip patch antenna (MPA). These MPAs can be made by etching various forms, such as squares, circles, ellipses, E, H, and S, among others, on copper layers of a double-sided PCB. The bottom PCB layer serves as the ground plane and can have a smooth or uneven surface, while the top patch layer serves as both a radiator for broadcasting the antenna and a receptor for receiving it [30–32].

Antenna radiations in MPAs are caused by the fringing effect between the intended patch structure's corners and the ground plane. In order to obtain resonant frequency ranges for specific applications, patch antennas can use a variety of shapes. Recently, elliptical patch antennas are receiving a lot of attention as its patch configuration or

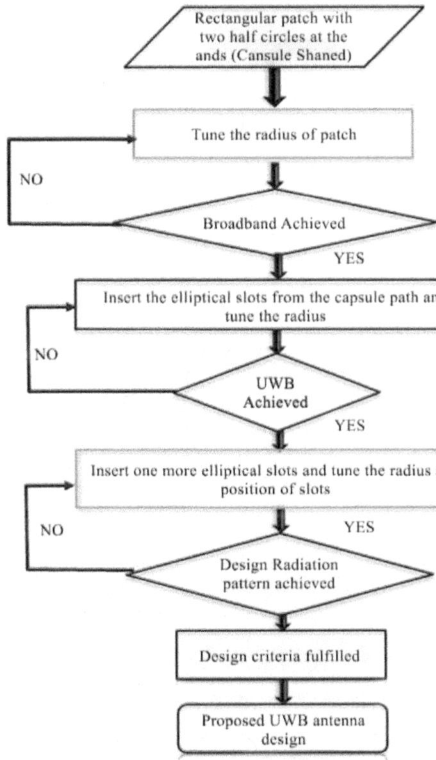

FIGURE 1.4 Flow chart for designing steps of proposed UWB antenna.

geometry, they have circular polarized property. Due to the elliptical patch antenna properties [33], we designed a similar kind of shape called capsule and it is designed with a rectangular patch antenna with two halves of a circle at both ends, which can fulfill all requirements of today's wireless communication systems, broadband and multiple features, etc.

The following conclusion can be drawn from the proposed antenna's design steps:

Step 1: By adjusting the characteristics of the exponentially-shaped, 50 micro strip line, create a capsule-shaped UWB antenna with a realized gain at 24–28 GHz.

Step 2: In order to achieve appropriate impedance matching at 24–28 GHz, embed two sets of elliptical slots/slits from the radiating patch antenna.

Step 3: To enhance the bandwidth and gain at 24–28 GHz, elliptical slits should be inserted into the radiating patch.

Step 4: Obtain the final design that satisfies all parametric requirements.

A flowchart outlining the suggested work is shown in Figure 1.4. Initially, an ultra-wideband antenna is created using a straightforward slit/slots loading approach to function over the band of 24–28 GHz. After that, the structure is simulated to see if the antenna produces the intended outcomes. A parametric analysis is then completed to create the antenna's optimal version.

To get the length (L) and width (W) of the antenna, use the condensed formula from the transmission line model that is given as follows.

The suggested antenna's width (W) is determined using the formula that follows.

$$W = \frac{c}{2f_r} \times \left(\frac{\varepsilon_r + 1}{2} \right)^{-\frac{1}{2}} \tag{1.1}$$

Where, c is the light speed in air (c = 3 × 10^8 meter per second), ε_r is the substrate material permittivity (given as $\varepsilon_r = 2.2$), fr is the resonant frequency, and the equation can be obtained as follows to determine the substrate's effective permittivity [34]:

$$\varepsilon_{reff} = \left(\frac{\varepsilon_r + 1}{2} \right) + \left\{ \frac{\varepsilon_r - 1}{2} \left(1 + \frac{12h}{W} \right)^{-\frac{1}{2}} \right\} \tag{1.2}$$

Where, h is thickness of the substrate. The length (L) can be calculated as:

$$L = \frac{c}{2f_r \sqrt{\varepsilon_{reff}}} - (2 \times \Delta L) \tag{1.3}$$

Elliptical geometry:

$$e = \sqrt{1 - \left(\frac{b}{a} \right)^2} \tag{1.4}$$

In an elliptical patch, the major axis is represented by "a," the minor axis is represented by "b," and the eccentricity is represented by "e."

The equation of Resonance frequency calculation for dual band elliptical slots:

$$f_{11}^{e,0} = \frac{15}{\pi e a_{eff}} \sqrt{\frac{Q_{11}^{e,0}}{\varepsilon_r}} \text{ GHz} \tag{1.5}$$

$$a_{eff} = a\sqrt{\left[1 + \frac{2h}{\pi \varepsilon_r a} \left\{ ln\left(\frac{a}{2h} \right) + (1.41\varepsilon_r + 1.77) + \frac{h}{a}(0.268\varepsilon_r + 1.65) \right\} \right]} \tag{1.5a}$$

$$Q_{11}^{e} = -0.0049e + 3.788e^2 - 0.7228e^3 + 2.2314e^4 \tag{1.5b}$$

$$Q_{11}^{0} = -0.0063e + 3.832e^2 - 1.1351e^3 + 5.2229e^4 \tag{1.5c}$$

Figure 1.5 shows the steps development procedure of the proposed UWB antenna as first starting with basic rectangular patch, but it did not achieve the proper

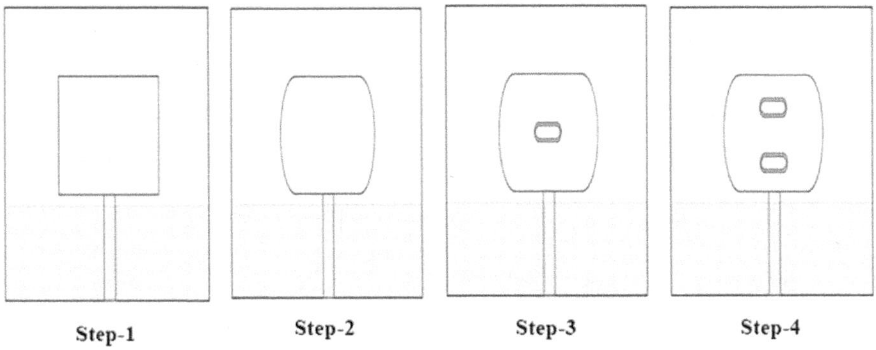

FIGURE 1.5 Steps development procedure of the single cell UWB antenna.

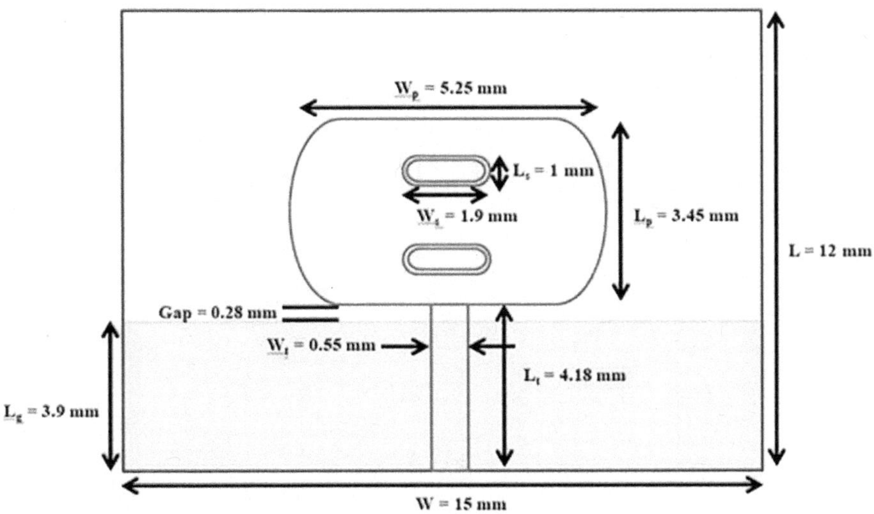

FIGURE 1.6 Unit Cell Antenna design with dimensions.

parameters; then improving the design with added two half circles at the end of rectangle which achieved the parametric results but did not achieve the required bandwidth; then cutting the two elliptical slots from the radiating patch which achieved all the parametric results with good bandwidth.

Figure 1.6 depicts the schematic for the UWB antenna. The initial engraving was done on a partially ground (3.9×15 mm^2) Roger RT Duroid 5880 laminate with the thickness ($t = 0.8$ mm), permittivity ($\varepsilon_r = 2.2$), and tangent loss ($\tan\delta = 0.0009$) Roger RT Duroid 5880 laminate with a capsule patch of radius 1.725 mm coupled with a feed transmission line of 4.18 mm. The suggested antenna has been simulated on Ansoft High-Frequency Structure Simulator (HFSS).

Table 1.2 lists all dimensions pertaining to the antenna design's specifications. It should be mentioned that the proposed antenna has a small dimensions; this

TABLE 1.2
Dimensions of Proposed Capsule-Shaped UWB Antenna

Parameter of Antenna Used	Dimensions in mm Unit
Dielectric Substrate	Roger RT Duroid 5880
Thickness of Substrate (h)	0.8 mm
Dielectric Constant (ε_r)	2.2
Length of Substrate (*Ls*)	12
Width of Substrate (*Ws*)	15
Transmission Line Length (*Lt*)	4.18
Transmission Line Width (*Wt*)	0.55
Ground Length (*Lg*)	3.9
Ground Width (*Wg*)	15
Capsule Patch Length (*Lp*)	3.45
Capsule Patch Width (*Wp*)	5.25
Elliptical Slots (E_1 *and* E_2) Length (*Ls*)	1
Elliptical Slots (E_1 *and* E_2) Width (*Ws*)	1.9

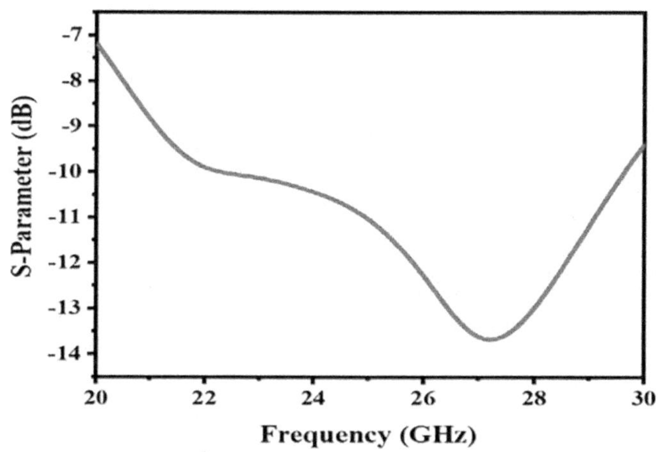

FIGURE 1.7 Simulated reflection coefficient (S_{11}) performance for capsule-shaped UWB antenna.

miniature architecture permits its integration into a range of 5G wireless devices. The proposed antenna has exceptional performance since it can attain the maximum bandwidth possible while maintaining the smallest possible size.

1.4 ANALYSIS OF ANTENNAS AND FINDINGS

The radiation and return loss characteristics of any antenna can be used to evaluate its performance as shown in Figure 1.7. The scattering and reflection characteristics

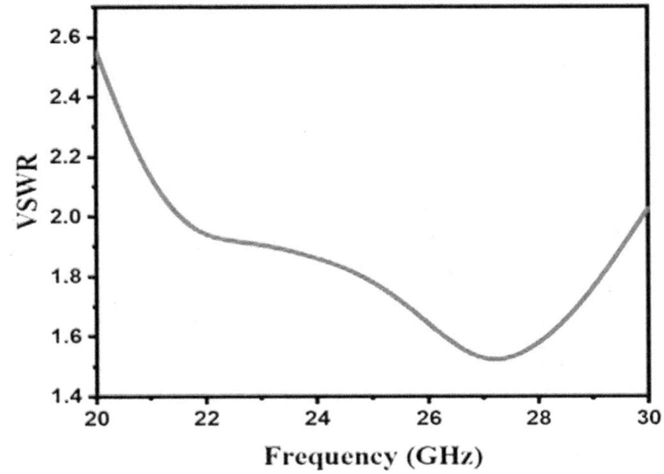

FIGURE 1.8 VSWR vs. Frequency Plot for capsule-shaped UWB antenna.

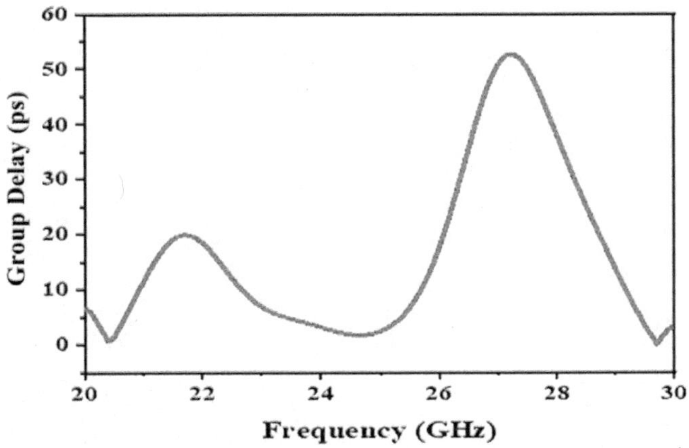

FIGURE 1.9 Group time-delay plot of the suggested UWB antenna.

of an antenna determine its capacity to transmit and receive electricity. The directivity, gain, and power density of an antenna are determined by its radiating characteristics for the frequency bands in which it resonates. Radiation patterns in 2D and 3D can graphically display several properties. The Ansoft HFSS simulator has been used to simulate antenna performance in resonant band of 24 GHz to 28 GHz with a start frequency of 20 GHz and a stop frequency of 30 GHz. As indicated in Table 1.2, the return loss, bandwidth, voltage standing wave ratio (VSWR) shown in Figure 1.8, gain, and radiation pattern are used to evaluate an antenna's performance.

The suggested antenna predicted group time delay, which must be able to transmit an electrical pulse with the least amount of distortion, is in picoseconds (ps), which is extremely little given the frequency range of (24–28) GHz as shown in Figure 1.9 and it's showing extremely good time delay.

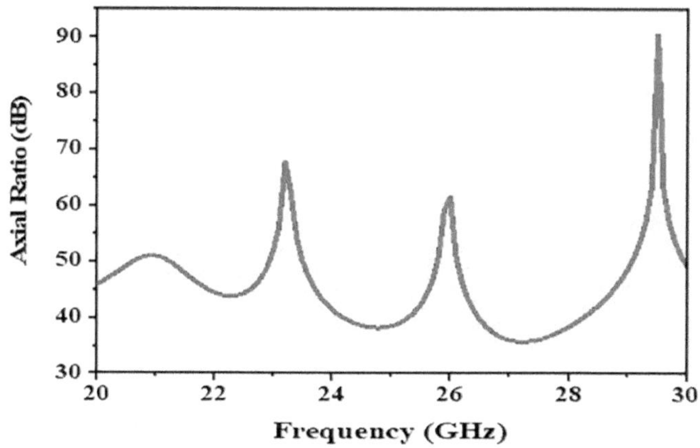

FIGURE 1.10 Axial ratio plot of the proposed UWB antenna.

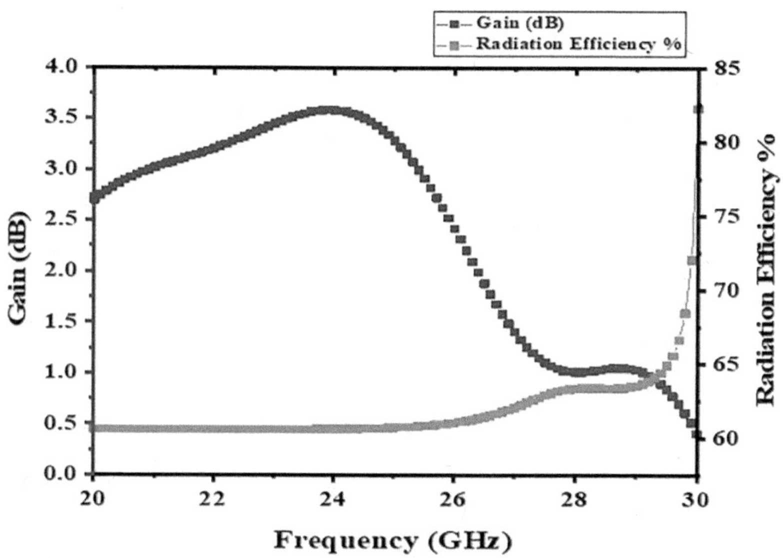

FIGURE 1.11 Radiation efficiency of proposed UWB antenna.

In a circular polarization antenna, the major axis to minor axis ratio determines the axial ratio. An antenna's axial ratio (AR) indicates whether or not the proposed antenna is circular polarization. The circularly polarized antenna has an AR of more than 10 dB. Figure 1.10 shows the suggested antenna axial ratio at operating frequencies.

The ratio of an antenna's output power to the power input into its excitation port is known as the antenna's radiation efficiency of the antenna [35–38]. A perfect lossless antenna has a radiation efficiency of 1. Figure 1.11 shows the radiation efficiency

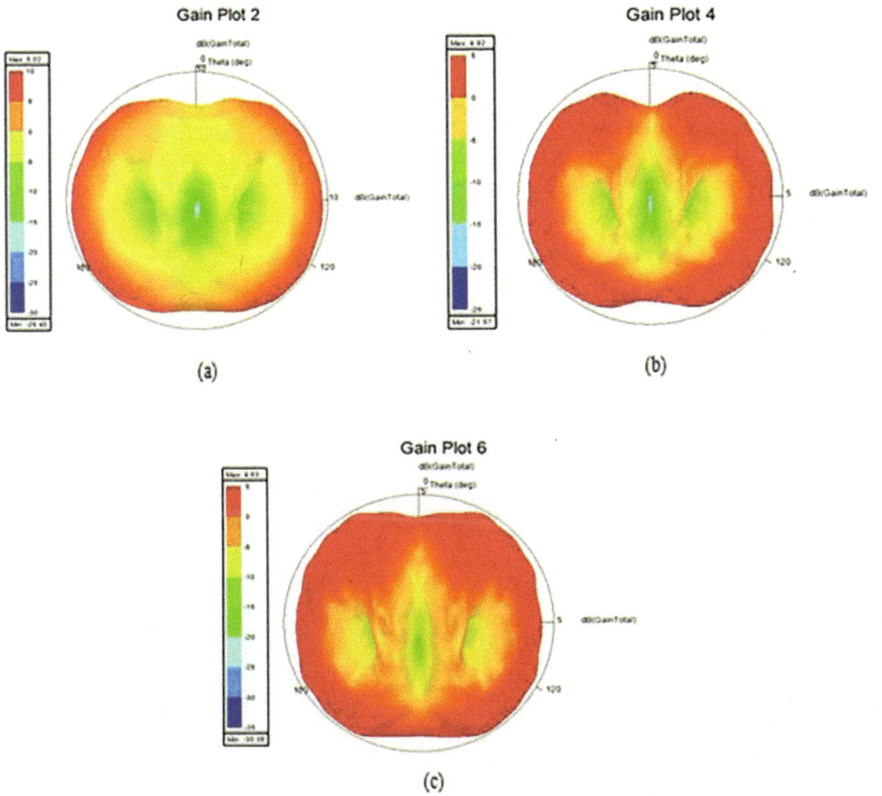

FIGURE 1.12 The radiation pattern plot of the proposed UWB antenna in 3D: (a) 24 GHz, (b) 26 GHz, and (c) 28 GHz.

of the proposed UWB antenna, here the radiation efficiency improved by 60%. Figure 1.12 shows the proposed UWB antenna 3D gain radiation patterns at different frequencies in terms of XZ (E-plane) elevation plane and YZ (H-plane) azimuthal plane which is expected from an ultra-wideband antenna. Here at 24 GHz gain is 5.02dBi, at 26 GHz frequency gain is 5.05dBi, and at 28 GHz frequency gain value is 4.53dBi. So, here at 26 GHz obtained the maximum gain.

Figure 1.13 shows the 2D directivity radiation pattern plot at 27 GHz for UWB antenna, which has a resonant frequency of 27 GHz, clearly shows a figure-eight shape pattern in the XZ-plane (phi 0°) and a pattern that is almost omnidirectional in the YZ-plane (phi 90°), respectively. These findings unequivocally demonstrate the suitability of the UWB bandwidth radiation patterns. The suggested antenna directivity is omnidirectional at Phi = 90° and concentrated between 0° and −180° at Phi = 0° it's a figure of eight. All directions experience the same level of directivity.

Analysis of the electric current distribution density and antenna current distribution can be exhibited by varying different operating frequencies throughout a range of phi and theta values in the HFSS simulator in order to analyze the return loss

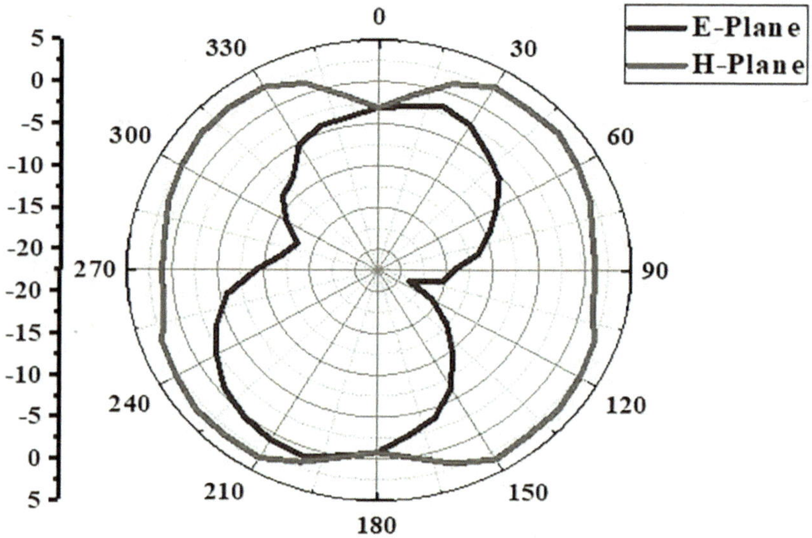

FIGURE 1.13 Simulated 2D directivity plot at 27 GHz for capsule-shaped UWB antenna.

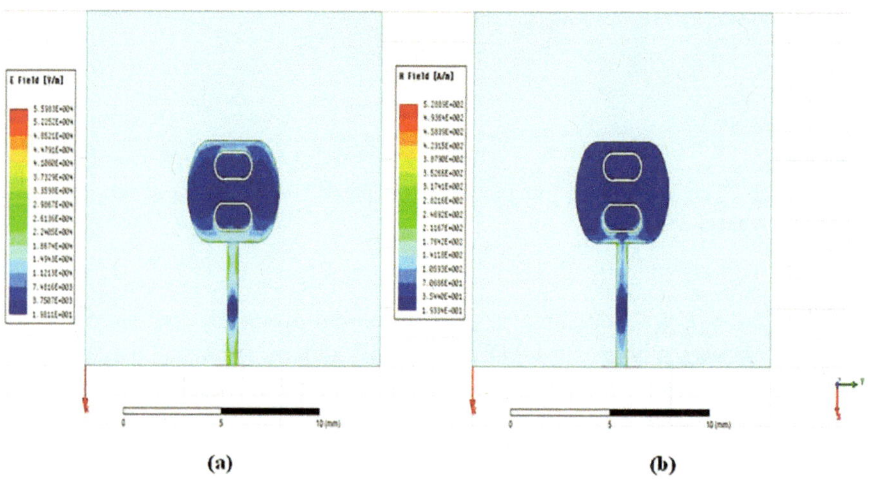

FIGURE 1.14 HFSS simulation E-field and H-field current distribution of proposed antenna at 27 GHz frequency.

behavior of the proposed construction. The graphs of the proposed antenna's electric and magnetic field densities at resonance are depicted in Figure 1.14 and Figure 1.15. The highest current is shown to be concentrated along the feed line boundaries and at the edges of the capsule patch, which gives the proposed antenna its wide bandwidth properties. Additionally, the current is evenly dispersed across the capsule patch surface. The feed line electrical length is increased by the current distribution

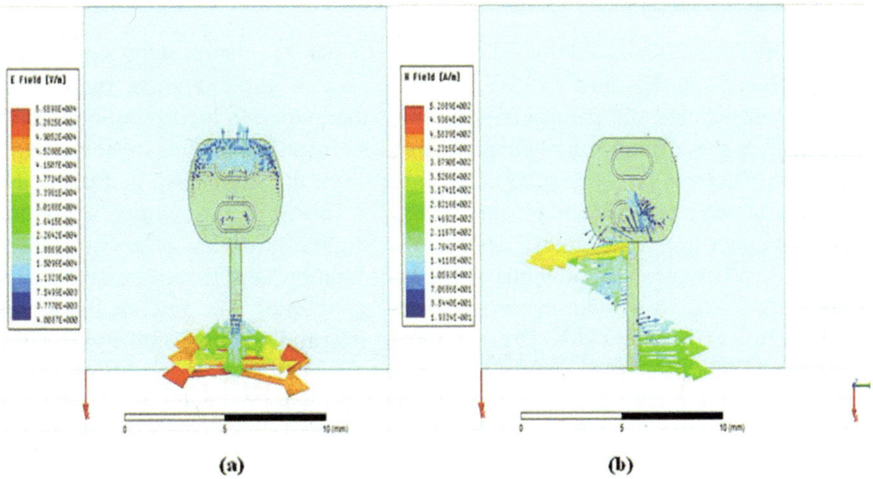

FIGURE 1.15 HFSS simulation E-field and H-field vector current distribution of suggested antenna at 27 GHz frequency.

TABLE 1.3
Comparison Table of UWB Antenna

Reference No.	Frequency (GHz)	No. of Elements	Efficiency (%)	Board Size (L × W) (mm²)	Gain (dBi)
[39]	2.45/3.9/5	1	NA	NA	3.6626 dB
[40]	3.1–10.6	1	94%	40 × 26	3.7
[41]	1.575–1.665, 2.4–2.545, 3.27–3.97 and d 5.17–5.93	1	85.5%	48 × 18	3.55–1.57 GHz 3.93–2.45 GHz 5.02–3.5 GHz 4.86–5.2 GHz
[42]	2.5–8.5	1	NA	82 × 55	3.26 dBi
Proposed Antenna	24–28	1	>60%	12×15 × 0.8	3.5 dB (or 6 dBi)

along the patch's borders, which causes the antenna to resonate with a satisfactory return loss value.

The capsule radiating element's edge has the largest current density, as can be observed. Additionally, the current spread is made constantly throughout the finite ground plane, with an immediate increase in current strength due to electromagnetic coupling effects with the ground at the bottom of the capsule patch.

As compared with all previous antenna designs in Table 1.3 the proposed antenna is compact in size and also fulfills all requirements like gain, VSWR, and return loss, etc. So, here the proposed design is the appropriate candidate for high frequencies and 5G wireless as well as UWB applications.

1.5 CONCLUSION

A simple and novel capsule-shaped micro strip patch antenna with slot loading technique has been designed for 5G applications at millimeter wave frequencies. At higher frequencies, the compactness ensures the antenna's incorporation into the constrained space around microwave electronics. In order to improve the antenna design, two elliptical slots are added, and the current density is also increased. The proposed antenna's optimal dimensions are (12 × 15 × 0.8) mm^3, which shows that it has a compact and very simple design. The suggested design offers advantages over other UWB micro strip antennas like less radiation loss, less dispersion, a new design, and good gain without the usage of any techniques. The antenna has a good VSWR (< 2), a wide bandwidth (10 GHz between 20 and 30 GHz), and a gain that is satisfactory (3.53 dB). Additionally, the radiation patterns in the E- and H-planes are acquired at various frequencies and described in the paper. To obtain high bandwidth at the SHF range, a capsule-shaped UWB antenna with elliptical slots has been proposed. One of the key advantages of the suggested design is its miniaturization. The suggested antenna is a good choice for both 5G and UWB applications at millimeter wave frequencies because of its appropriate geometry and its simple structure.

REFERENCES

[1] Meer Zafarullah Noohani and Kaleem Ullah Magsi, "A Review of 5G Technology: Architecture, Security and Wide Applications", *International Research Journal of Engineering and Technology (IRJET)*, vol. 7, no. 5, May 2020.

[2] Asvin Gohil, Hardik Modi, and Shobhit K. Patel, "5G Technology of Mobile Communication: A Survey", *2013 International Conference on Intelligent Systems and Signal Processing*. https://ieeexplore.ieee.org/document/6526920.

[3] Suverna Sengar and Praveen Kumar Malik, "A Comprehensive Survey of Massive-MIMO Based on 5G Antennas", *The International Journal of RF and Microwave Computer-Aided Engineering*, vol. 32, p. e23496, 2022. https://doi.org/10.1002/mmce.23496.

[4] C. A. Balanis, *Antennas theory: Analysis and design* (3rd ed.). Hoboken: Wiley, 2005.

[5] Y. Kumar and S. Singh, "Performance Analysis of Coaxial Probe Fed Modified Sierpinski—Meander Hybrid Fractal Heptaband Antenna for Future Wireless Communication Networks", *Wireless Personal Communications*, vol. 94, pp. 3251–3263, 2017.

[6] FCC, *Federal Communications Commission Revision of Part 15 of the Commission's Rules Regarding Ultra-Wideband Transmission System from 3.1 to 10.6 GHz'*. Washington, DC: Federal Communications Commission FCC, ET-Docket, pp. 98–153, 2002.

[7] R. J. Fontana, "Recent System Applications of Short-Pulse Ultra-Wideband (UWB) Technology", *IEEE Transactions on Microwave Theory and Techniques*, vol. 52, no. 9, pp. 2087–2104, 2004.

[8] G. J. Foschini and M. J. Gans, "On Limits of Wireless Communications in a Fading Environment When Using Multiple Antennas", *Wireless Personal Communications*, vol. 6, no. 3, pp. 311–335, 1998.

[9] FCC, *First Report and Order in the Matter of Revision of Part 15 of the Commission's Rules Regarding Ultrawide Band Transmission Systems*. Washington, DC: Federal Communications Commission FCC, No. 190211614895, 2002.

[10] H. M. Jafari, M. J. Deen, S. Hranilovic, and N. K. Nikolova, "A Study of Ultrawideband Antennas for Near-Field Imaging", *IEEE Transactions on Antennas and Propagation*, vol. 55, pp. 1184–1188, 2007.

[11] John Volakis, Chi-Chih Chen, and Kyohei Fujimoto, *Small Antennas: Miniaturization Techniques & Applications*. New York: McGraw Hill Professional, 2009.

[12] C.A. Balanis, *Antenna Theory: Analysis and Design* (2nd ed.). New York: John Willey & Sons Inc., 1997.

[13] M. Ayyappan and Pragati Patel, "On Design of a Triple Elliptical Super Wideband Antenna for 5G Applications", *IEEE Access*, vol. 10. Received 18 May 2022, accepted 13 June 2022, date of publication 22 June 2022, date of current version 25 July 2022. https://doi.org/10.1109/ACCESS.2022.3185241.

[14] Y. Jee and Y.-M. Seo, "Triple-Band CPW-Fed Compact Monopole Antennas for GSM/PCS/DCS/WCDMA Applications", *Electronics Letters*, vol. 45, no. 9, pp. 446–448, 2009.

[15] T.-H. Chang and J.-F. Kiang, "Compact Multi-Band H-Shaped Slot Antenna", *IEEE Transactions on Antennas and Propagation*, vol. 61, no. 8, pp. 4345–4349, 2013.

[16] Shi-Gang Zhou, Peng-Khiang Tan, and Tan-Huat Chio, "Low-Profile, Wideband Dual-Polarized Antenna with High Isolation and Low Cross Polarization", *IEEE Antennas and Wireless Propagation Letters*, vol. 11, pp. 1032–1035, 2012.

[17] Abdallah A. Alshehri and A. R. Sebak, "A Band-Rejected Printed Elliptical Antenna for UWB Wireless Communications", *The International Union of Radio Science*, 2008.

[18] Xinrong Shi, Yunfei Cao, Yougen Hu, Xudong Luo, Hua Yang, and Liang Hua Ye, "A High-Gain Antipodal Vivaldi Antenna with Director and Metamaterial at 1–28 GHz", *IEEE Antennas and Wireless Propagation Letters*, vol. 20, no. 12, December 2021.

[19] Ananda Kumar Behera, et al., "A Circularly Polarized Fibonacci Series-Inspired-Elliptical Patch Antenna for UWB Radiation", *2021 IEEE 2nd International Conference on Applied Electromagnetics, Signal Processing, & Communication (AESPC)*. IEEE, 2021. https://www.researchgate.net/publication/358623597_A_Circularly_Polarized_Fibonacci_Series-Inspired-Elliptical_Patch_Antenna_for_UWB_Radiation.

[20] M. S. Karoui, N. Ghariani, M. Lahiani, and H. Ghariani, "A Compact UWB Elliptic Antenna for Indoor Localization System", *IEICE Electronics Express*, vol. 16, p. 20190425, 2019.

[21] P. Hedayati, "Ultra Wideband Elliptic Monopole Antenna", *Transactions on Electromagnetic Spectrum*, vol. 1, no. 1, pp. 13–17, 2022. https://doi.org/10.5281/zenodo.6644967.

[22] Ali Khalid Jassim and Raad H. Thaher, "Calculate the Optimum Slot Area of Elliptical Microstrip Antenna for Mobile Application", *Indonesian Journal of Electrical Engineering and Computer Science*, vol. 16, no. 3, pp. 1364–1370, December 2019, ISSN: 2502-4752. https://doi.org/10.11591/ijeecs.v16.i3.pp1364-1370.

[23] Garima Srivastava and Manju Khari, "An Elliptical CPW Fed UWB Slot Antenna", *Wireless Personal Communications*, vol. 119, pp. 2253–2263, 2021. https://doi.org/10.1007/s11277-021-08329-y.

[24] Kiran Raheel, et al., "E-Shaped H-Slotted Dual Band mm Wave Antenna for 5G Technology", *Electronics*, vol. 10, no. 9, p. 1019, 2021.

[25] Irtiza Abbas Awan, et al., "Single Patch Fractal-Shaped Antenna with Small Footprint Area and High Radiation Properties for Wide Operation Over 5G Region", *2021 46th International Conference on Infrared, Millimeter and Terahertz Waves (IRMMW-THz)*. IEEE, 2021. https://www.researchgate.net/publication/355459450_Single_Patch_Fractal-Shaped_Antenna_with_Small_footprint_Area_and_High_Radiation_Properties_for_Wide_Operation_Over_5G_Region?_tp=eyJjb250ZXh0Ijp7ImZpcnN0UGFnZSI6InB1YmxpY2F0aW9uIiwicGFnZSI6InB1YmxpY2F0aW.

[26] Hijab Zahra, et al., "A 28 GHz Broadband Helical Inspired End-Fire Antenna and Its MIMO Configuration for 5G Pattern Diversity Applications", *Electronics*, vol. 10, no. 4, p. 405, 2021.

[27] Son Xuat Ta, Hosung Choo, and Ikmo Park, "Broadband Printed-Dipole Antenna and Its Arrays for 5G Applications", *IEEE Antennas and Wireless Propagation Letters*, vol. 16, 2183–2186, 2017.

[28] Ajay Kumar Dwivedi, et al., "Two Port Circularly Polarized MIMO Antenna Design and Investigation for 5G Communication Systems", *Wireless Personal Communications*, vol. 120, no. 3, pp. 2085–2099, 2021.

[29] P. Malik, J. Lu, B. T. P. Madhav, G. Kalkhambkar, and S. Amit, eds., *Smart Antennas: Latest Trends in Design and Application*. Cham: Springer. ISBN 978-3-030-76636-8, https://doi.org/10.1007/978-3-030-76636-8.

[30] R. Roges and Praveen Kumar Malik, "Planar and Printed Antennas for Internet of Things-Enabled Environment: Opportunities and Challenges", *The International Journal of Communication Systems*, vol. 34, no. 15, p. e4940, September 2021. (IF: 2.047) ISSN: 10991131. https://doi.org/10.1002/dac.4940.

[31] Abdul Rahim and Praveen Kumar Malik, "Analysis and Design of Fractal Antenna for Efficient Communication Network in Vehicular Model", *Sustainable Computing: Informatics and Systems, Elsevier*, vol. 31, p. 100586, 2021, ISSN 2210-5379. https://doi.org/10.1016/j.suscom.2021.100586.

[32] N. Shaik and Praveen Kumar Malik, "A Comprehensive Survey 5G Wireless Communication Systems: Open Issues, Research Challenges, Channel Estimation, Multi Carrier Modulation and 5G Applications", *Multimedia Tools and Applications*, vol. 80, 2021. https://doi.org/10.1007/s11042-021-11128-z.

[33] Praveen Kumar Malik, D. S. Wadhwa, and J. S. Khinda, "A Survey of Device to Device and Cooperative Communication for the Future Cellular Networks", *International Journal of Wireless Information Networks, Springer*, vol. 27, pp. 411–432, 2020. https://doi.org/10.1007/s10776-020-00482-8.

[34] P. Tiwari and P. K. Malik, "Wide Band Micro-Strip Antenna Design for Higher "X" Band", *International Journal of e-Collaboration (IJeC)*, vol. 17, no. 4, pp. 60–74, October 2021. ISSN: 1548-3673. http://doi.org/10.4018/IJeC.2021100105.

[35] Amandeep Kaur and Praveen Kumar Malik, "Multiband Elliptical Patch Fractal and Defected Ground Structures Microstrip Patch Antenna for Wireless Applications", *Progress in Electromagnetics Research B*, vol. 91, pp. 157–173, 2021. ISSN: 1937-6472. https://doi.org/10.2528/PIERB20102704.

[36] Nilofer Shaik and Praveen Kumar Malik, "A Retrospection of Channel Estimation Techniques for 5G Wireless Communications: Opportunities and Challenges", *International Journal of Advanced Science and Technology*, vol. 29, no. 5, pp. 8469–8479, June 2020. ISSN: 2005-4238.

[37] Praveen Kumar Malik and Madam Singh, "Multiple Bandwidth Design of Micro strip Antenna for Future Wireless Communication", *International Journal of Recent Technology and Engineering*, vol. 8, no. 2, pp. 5135–5138, July 2019. ISSN: 2277–3878. https://doi.org/10.35940/ijrte.B2871.078219.

[38] Suverna Senger and Praveen Kumar Malik, "Design of a Novel Capsule-Shaped Compact UWB Antenna for 5G Wireless Applications", *Journal of Nano- and Electronic Physics*, vol. 13, no. 4, pp. 04031-1–04031-5, 2023.

[39] P. Tamil Manoj Prasanth and G. Velmathi, "Design and Development of Cylindrical Dielectric Resonator (CDR) MIMO Antenna for Multiband Applications", *Proceedings of 2016 IEEE International Conference of Wireless Communication Signal Processing Networking, WiSPNET 2016*, pp. 670–673, 2016. https://doi.org/10.1109/WiSPNET.2016.7566217.

[40] J. P. Kumar and G. Karunakar, "Ultra Wideband MIMO Notched Antenna for WLAN and Mobile Applications", *International Journal of Pure and Applied Mathematics*, vol. 118, no. 9, pp. 929–934, 2018.

[41] Y. F. Cao, S. W. Cheung, and T. I. Yuk, "A Multiband Slot Antenna for GPS/WiMAX/WLAN Systems", *IEEE Transactions on Antennas and Propagation*, vol. 63, no. 3, pp. 952–958, March 2015.

[42] T. Tewary, S. Maity, S. Mukherjee, A. Roy, P. P. Sarkar, and S. Bhunia, "Design of High Gain Broadband Microstrip Patch Antenna for UWB/X/Ku Band Applications", *AEU International Journal of Electronics and Communications*, vol. 139, pp. 1–13, 2021.

2 Material Characterization of Graphene-Based Circular Patch Terahertz (THz) Antenna for Biomedical Applications

Ranjeet Kumar, Rashmi Sinha, Arvind Choubey, Santosh Kumar Mahto, Pravesh Pal, Praveen Kumar and Ajit Kumar Singh

2.1 INTRODUCTION

In the 19th century, terahertz (THz) waves was not an independent spectrum; it came between electromagnetic and infrared spectrum as pictured in Figure 2.1 [1]. But with the advent of THz sources, the practical applications were made possible during the 20th century [2]. It was then that terahertz technology was evolved, wherein terahertz wave 0.3–10 THz shifted from a no man's land to a new era of revolutionary development in wireless communication. It has brought a renaissance in a true sense where lists of constraints of microwave technology could be resolved [3]. Ultra-high date rate, spatial diversity, high resolution, compactness, non-ionizing nature of signal, and spectral efficiency, are a few aspects which have attracted industries and researchers to invest in this technology. At present, a large number of practical applications are migrating from analog to digital or computers to cable-free devices. This has led to a deluge of users and heavy data tariffing. According to statistics, data rate of communication systems will reach Gbps to Tbps in the next decade [4–1]. In [5], the terahertz technology provides Gbps data rate, and it is predicted that Tbps will be the emerging phase of wireless communication systems and it can also be obtained by polarization multiplexing. The excellent characteristics of terahertz wave are: (i) High Resolution Spectrum—The non-destructive detection of dangerous things such as explosives, pistols, chemical, virus etc. provides a great solution [6]. (ii) Low Damage—The energy of a single photon is one part per million lower than X-ray. Therefore, it is used in biomedical fields for scanning the body without harming the body cells [7]. (iii) Visualization—Due to the short wavelength of terahertz wave, it can penetrate non-polar and non-metallic objects, so it can scan the materials and give high-definition images [8]. Therefore, this application is widely utilized

 DOI: 10.1201/9781032656021-2

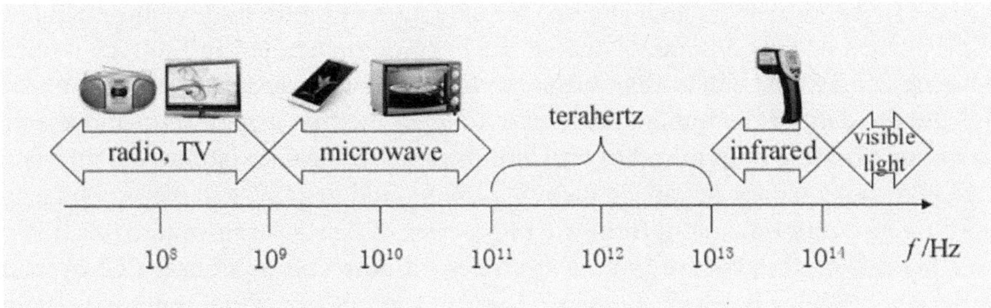

FIGURE 2.1 The location of the terahertz (THz) wave.

in crowed places such as airports, railways stations, and high-security zones. (iv) Wide Bandwidth—Due to the highest frequency band, it can be used as a signal carrier which transfers the information rate to a new level of Gbps or Tbps. In [9], a graphene-based square shape THz antenna with $\frac{\lambda}{4}$ wave transformer is built on Sio_2 substrate of 1.8 μm for WLAN application, which has bandwidth of 12.3% ranging from 12.2 to 13.8 THz. This THz antenna shows significant improvement of radiation characteristic by the help of graphene as radiator, but does not cover the lower range of frequency. In [10], a graphene-based hexagonal-shaped antenna is designed on quartz substrate. U-shaped and elliptical slots are etched out from radiating patch, which provides significant radiation performances and have an impedance bandwidth ranging from 0.68–1.63 THz. In [11] authors have designed and analyzed a graphene-based circular patch antenna with different dielectric materials, which is operated in the frequency range of 6.8–7.2 THz, and provides outstanding performance with a gain of 16.7 dB with polyimide substrate is achieved. A double layer dielectric material dual band antenna for surveillance [12], an elliptical ring fractal antenna of bandwidth 9 THz from 0.3 to 9.3 THz [13], a square-shaped patch with U-shaped slot with RT/Duriod 6010 substrate [14], a rectangular microstrip array antenna on liquid crystalline polymer dielectric material [16], a tree-shaped graphene-based MIMO antenna [16], a Fractal loaded planar MIMO antenna [17], and a proximity coupled MIMO antenna [18] have been demonstrated in the literature highlighting the most recent developments in the area of THz antennas.

In the field of medical sciences, many research teams and industries are dedicatedly working to develop and explore antennas for medicine and healthcare units. Nowadays, wearable sensors are measuring several biological signals as well as activities. So, wearable devices permit biological signals to regularly monitor day-to-day activity. For example, a patient suffering from a lung problem, chronic diseases, heart issue etc. will be regularly and easily monitored through wearable devices and the updated report is sent to their respective physician with the help of a centralized monitoring system such as WBAN. This will overcome the problem of frequent clinical visits and doctors can provide health care remotely. Biosensors and wearable antennas are integrated onto the human body to monitor for electromyography (EMG) [19], electroencephalography (EEG) [20–22], and electrocardiogram

(ECG) [23]. A microstrip patch antenna for detection of biotin [24], vitamin K2 [25], and properties of GaAS semiconductor has been demonstrated in [26]. To monitor the biological signals with a wearable antenna, a two-stage system is needed to communicate the data to the remote healthcare unit. In the first step, a short-range communication protocol is followed to transmit the data to the user interaction interface, which is responsible for advanced data processing and display, and the second stage is long-range communication, from where processed data are transmitted to a distant server placed in a healthcare unit. The processed data can be transmitted over the internet and cellular network. Figure 2.2 shows the outline of the remote medicine and healthcare unit [27].

2.2 GRAPHENE-BASED PATCH ANTENNA DESIGN AND ANALYSIS

Based on previous literature survey, a microstrip transmission line fed THz patch antenna consists of (stack form) ground plane (copper), thin dielectric substrate, and radiating patch (graphene). Transmission line is matched to radiating patch through $\frac{\lambda}{4}$ wave transformer of 50 Ω characteristic impedance, so that maximum input power is transferred to patch (load). Antenna efficiency relies on dielectric substrate and patch materials. The operating frequency band depends on electrical path of a radiating patch antenna [28–29]. The actual radius (*a*) of a circular patch is given by Equation (2.1) [29]. The fringing makes the circular patch look electrically large, so its effective radius (*ae*) is calculated using Equation (2.2) [29].

$$a = \frac{F}{\sqrt{1 + \frac{2h}{(\pi \varepsilon_r F)\left[ln\left(\frac{\pi F}{2h} + 1.7726\right)\right]}}} \tag{2.1}$$

$$a_e = a\sqrt{1 + \frac{2h}{\pi \varepsilon_r a}\left[ln\left(\frac{\pi a}{2h} + 1.7726\right)\right]} \tag{2.2}$$

Where, $F = \frac{8.791 \times 10^9}{f_r \sqrt{\varepsilon_r}}$. Here, h, fr, ε_r are the height of substrate, resonant frequency, dielectric constant of material, respectively. The Computer Simulation Technology (CST) microwave studio software tool is used to simulate the SISO antenna. The performance parameters such as S-parameters, VSWR, 2D and 3D radiation patterns, and diversity performances are analyzed in depth.

2.3 SINGLE INPUT SINGLE OUTPUT (SISO) ANTENNA LAYOUT AND ANALYSIS

A graphene-based single circular shape SISO antenna is captured in Figure 2.3. The circular patch of radius 5.04 μm is carved on substrate of extremely compact

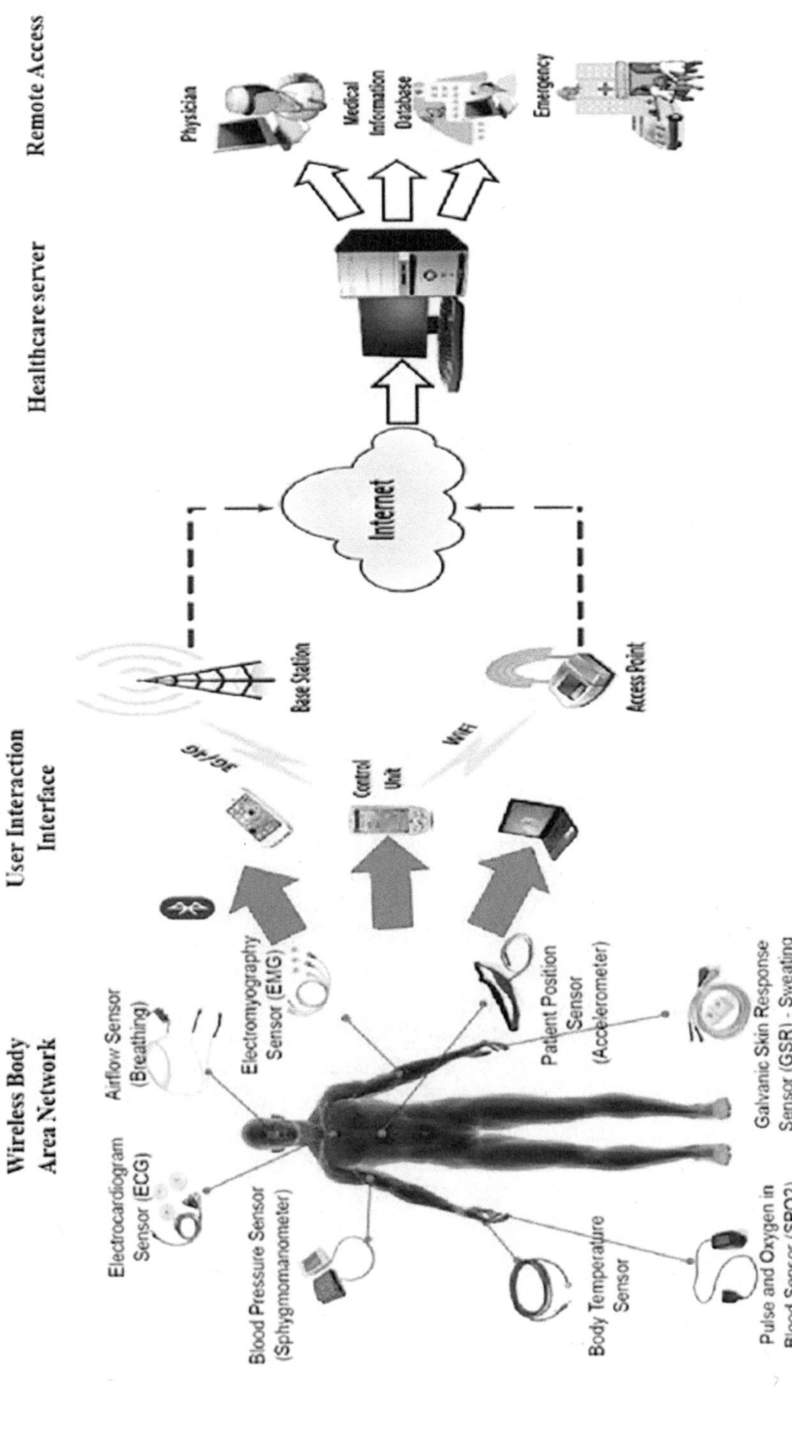

FIGURE 2.2 Outline of the remote medicine and healthcare unit.

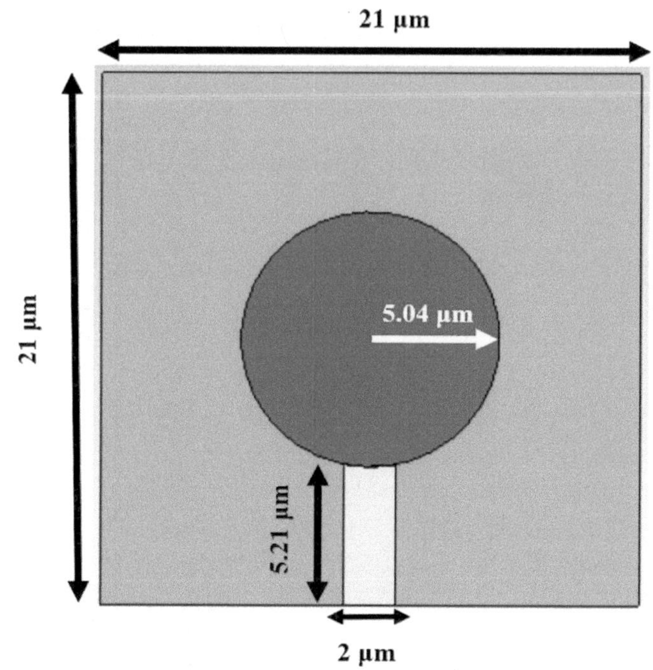

FIGURE 2.3 Graphene-based SISO antenna.

dimension 21×21×2 μm³. The excitation to radiating patch is given by microstrip feedline on the top of dielectric materials. The opposite side of the substrate has a conducting ground plane of the same dimension as that of the substrate. The single circular patch is made of graphene placed at the center of the substrate. The thickness of the graphene patch and ground plane is 0.0485 μm. Graphene is preferred primarily because its resonating frequency can be tuned by altering and optimized parameters such as temperature, chemical potential, dimension and relaxation time [28]. Graphene parameters are: Chemical Potential—2eV, Relaxation Time -10–^{12} sec, Temperature —300K, Thickness —10 nm. Different types of substrates are considered for analysis purpose and their return loss characteristic is plotted in Figure 2.4. Taconic RF-35TCA and silicon dioxide substrate has maximum and minimum return loss value, respectively. The polyamide has moderate return loss among all considered materials. The complete specification of different dielectric materials, and its respective resonate frequency are listed in Table 2.1.

2.4 RESULT AND DISCUSSION

The far field 2D and 3D radiation pattern with different substrates such as Polyimide, Taconic RF-35TCA, Arlon AD 350A, Quartz, Rogers RO3035, and silicon dioxide are portrayed in Figure 2.5. Radiation pattern delineates the relative field strength diverse in different direction. The main lobe pattern at Phi = 0°, and Phi = 90° is more than 4 dBi and at Theta = 90° it is about –3.5 dBi. Angular width (3 dB) at Phi = 0°,

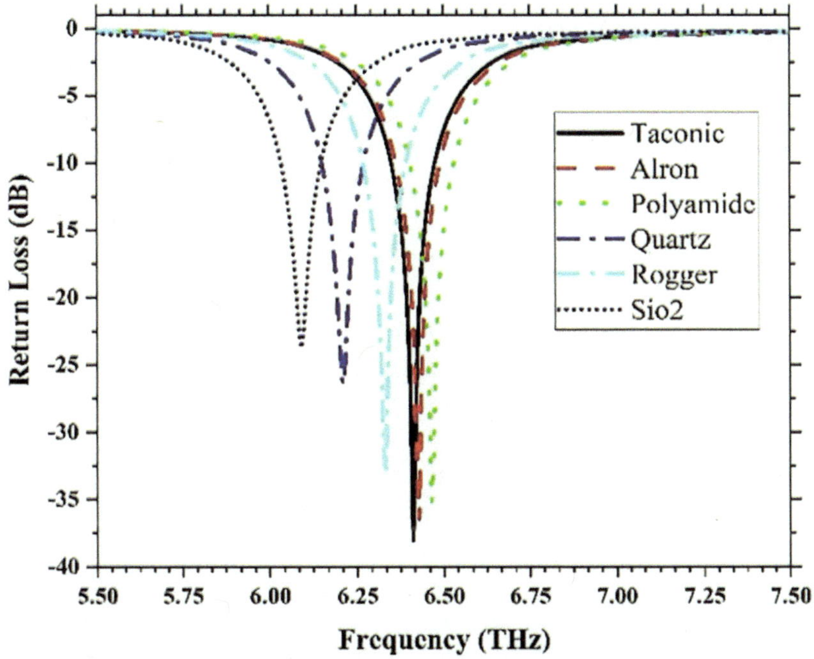

FIGURE 2.4 Return loss of graphene antenna with different substrate materials.

TABLE 2.1
Different Substrate Specification and Its Notch Frequency Coordinate

Material: Polyimide	Material: Taconic RF-35TCA
Epsilon: 3.5	Epsilon: 3.5
Electric tangent loss: 0.0027	Electric tangent loss: 0.0017
Rho: 1400 [kg/m^3]	Rho: 000 [kg/m^3]
Thermal cond.: 0.2 [W/K/m]	Thermal cond.: 0.83 [W/K/m]
Heat capacity: 1 [kJ/K/kg]	Heat capacity: 0 [kJ/K/kg]
Diffusivity: 1.42857e-07 [m^2/s]	Diffusivity: 0 [m^2/s]
Young's modulus: 2.5 [kN/mm^2]	Young's modulus: 0 [kN/mm^2]
Poisson's ratio: 0.4	Poisson's ratio: 0
Thermal expansion: 25 [1e-6/K]	Thermal expansion: 25 [1e-6/K]
Notch frequency: 6.46, –35.64 dB	**Notch frequency: 6.41, –38.06 dB**
Material: Quartz (Fused)	Material: Rogers RO3035
Epsilon: 3.75	Epsilon: 3.6
Electric tangent loss: 0.0004	Electric tangent loss: 0.0015
Rho: 2200 [kg/m^3]	Rho: 1400 [kg/m^3]
Thermal cond.: 5 [W/K/m]	Thermal cond.: 0.66 [W/K/m]
Heat capacity: 0.7 [kJ/K/kg]	Heat capacity: 0 [kJ/K/kg]
Diffusivity: 3.24675e-06 [m^2/s]	Diffusivity: 0 [m^2/s]
Young's modulus: 75 [kN/mm^2]	Young's modulus: 0 [kN/mm^2]
Poisson's ratio: 0.17	Poisson's ratio: 04

(Continued)

TABLE 2.1 ***(Continued)***
Different Substrate Specification and Its Notch Frequency Coordinate

Thermal expan: 0.5 [1e-6/K]	Thermal expansion: 0 [1e-6/K]
Notch frequency: 6.2, –26.2 dB	**Notch frequency: 6.33, –32.86 dB**
Material: Arlon AD 350A	Material: silicon dioxide
Epsilon: 3.5	Epsilon: 3.9
Electric tangent loss: 0.003	Rho: 0 [kg/m^3]
Rho: 1400 [kg/m^3]	Thermal cond.: 0 [W/K/m]
Thermal cond.: 0.235 [W/K/m]	Heat capacity: 0 [kJ/K/kg]
Heat capacity: 0 [kJ/K/kg]	Diffusivity: 0 [m^2/s]
Diffusivity: 0 [m^2/s]	Young's modulus: 0 [kN/mm^2]
Young's modulus: 0 [kN/mm^2]	Poisson's ratio: 0
Poisson's ratio: 0	Thermal expansion: 0 [1e-6/K]
Thermal expansion: 25 [1e-6/K]	**Notch frequency: 6.09, –23.74 dB**
Notch frequency: 6.42, –36.64 dB	

Phi = 90°, and Theta = 90° are 98°, 90°, and about 245° respectively. Observation from Figure 2.5–2.10 reveals that more radiation is directed away from the antenna due to conducting ground plane and side and back lobe radiations are very less. This radiation characteristic ensures that its hazardous effect on the human body is minimal. Therefore, the proposed antenna can be deployed in a wearable antenna for implemented health monitoring.

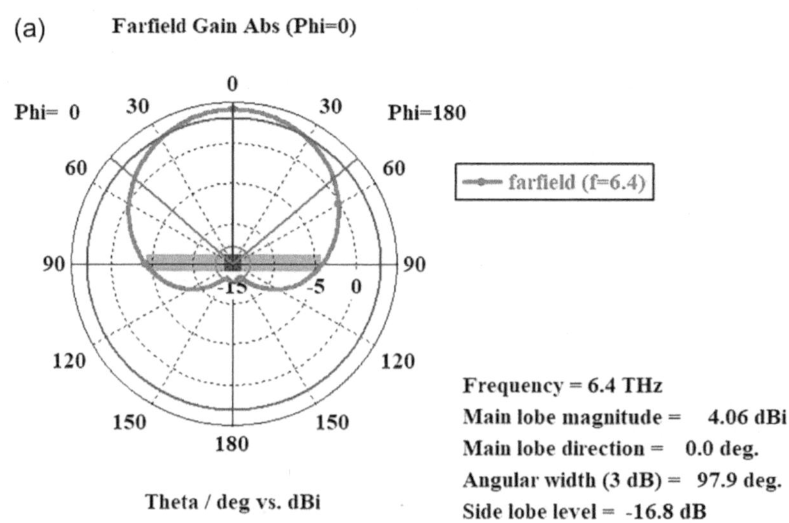

FIGURE 2.5 Polyamide.

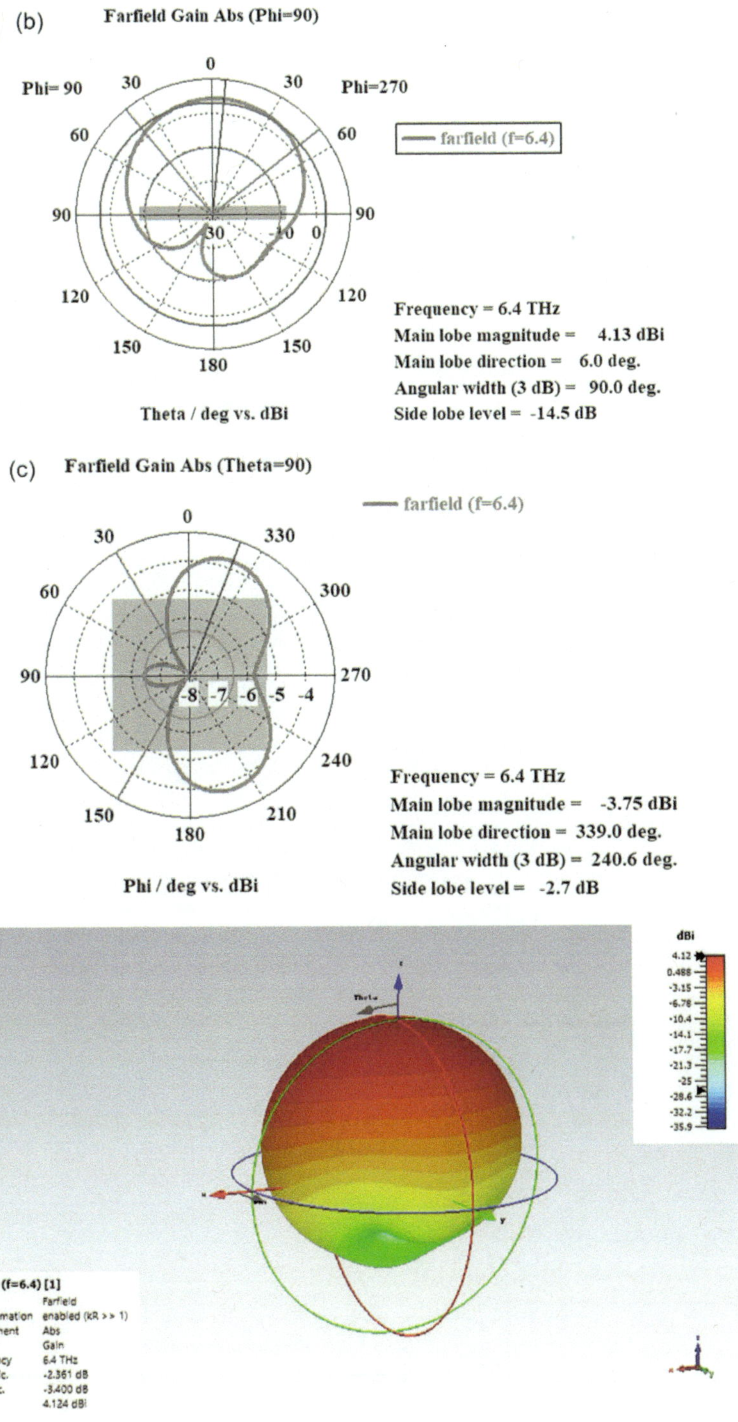

FIGURE 2.5 (continued)

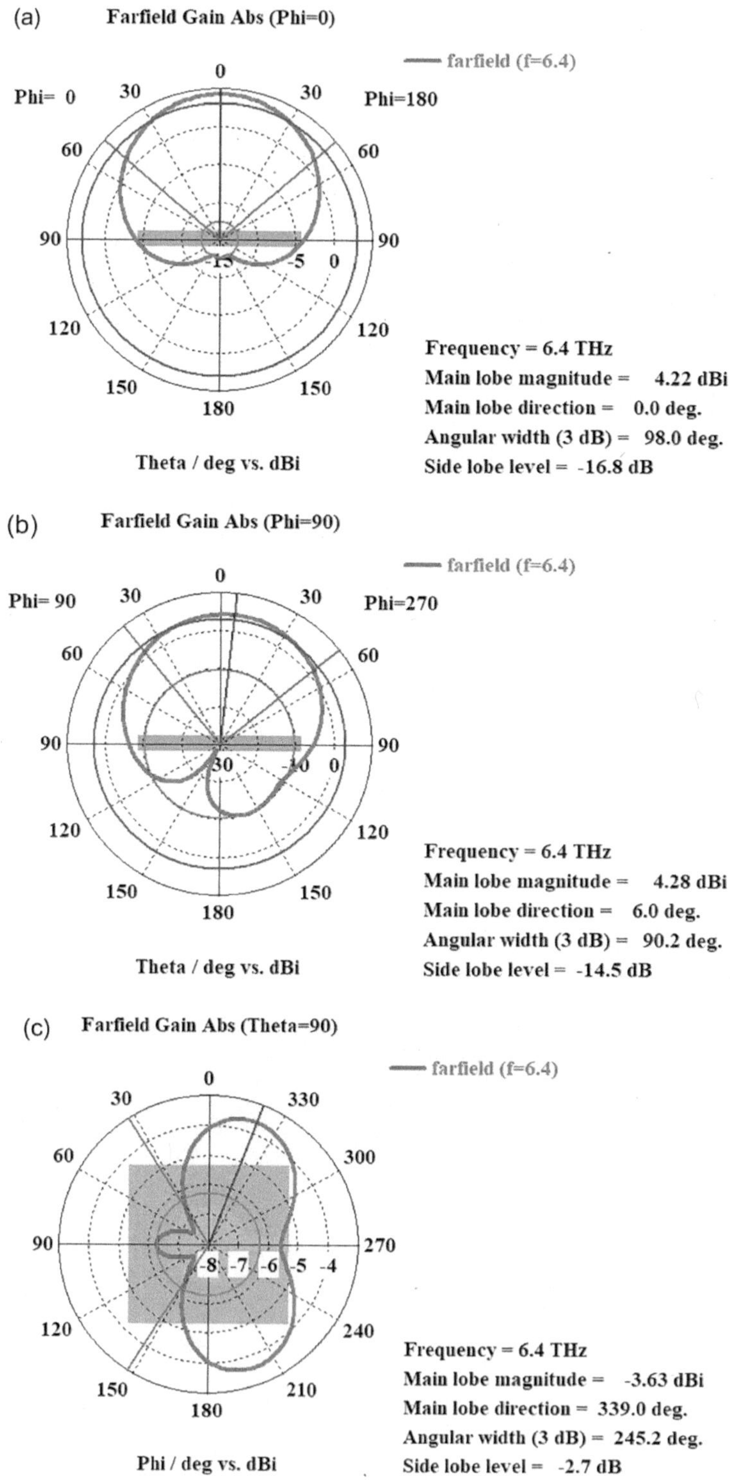

FIGURE 2.6 Taconic RF-35TCA.

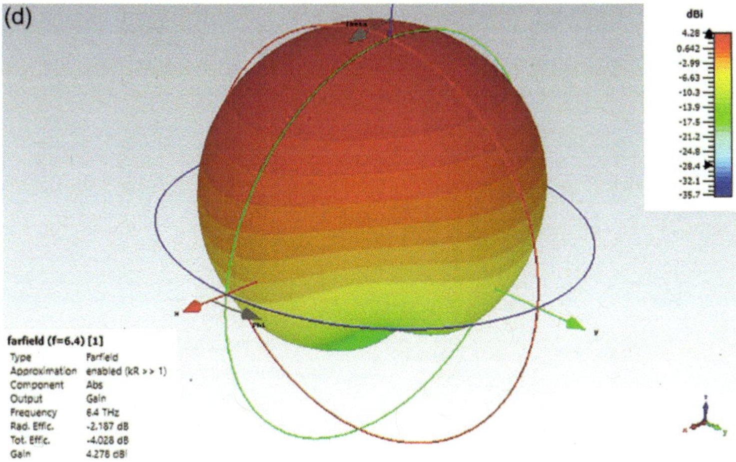

FIGURE 2.6 (Continued)

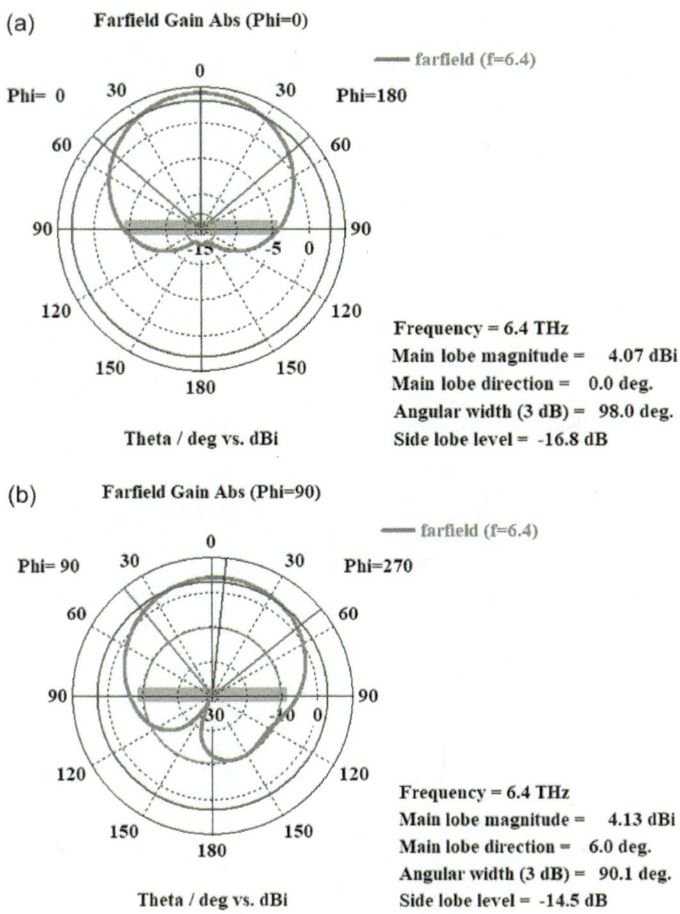

FIGURE 2.7 Arlon AD 350A.

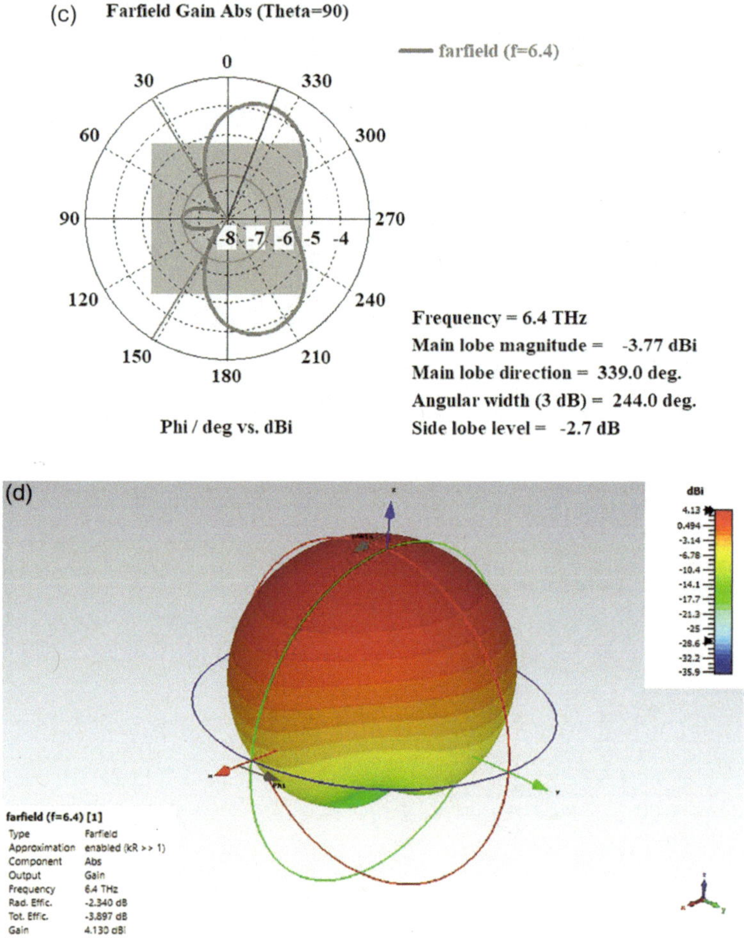

FIGURE 2.7 (Continued)

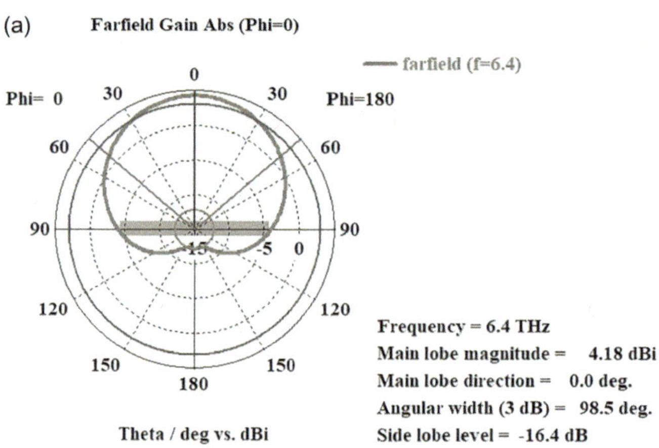

FIGURE 2.8 Quartz.

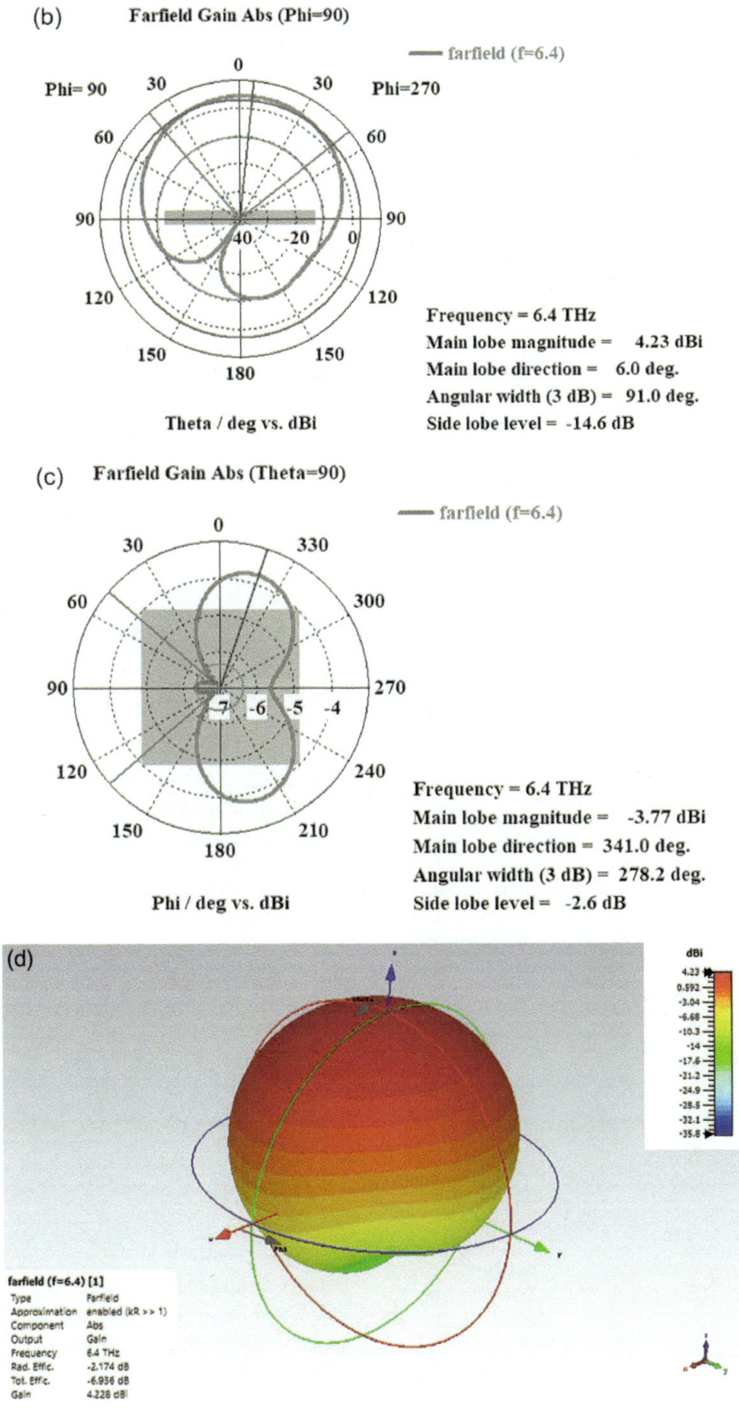

FIGURE 2.8 (Continued)

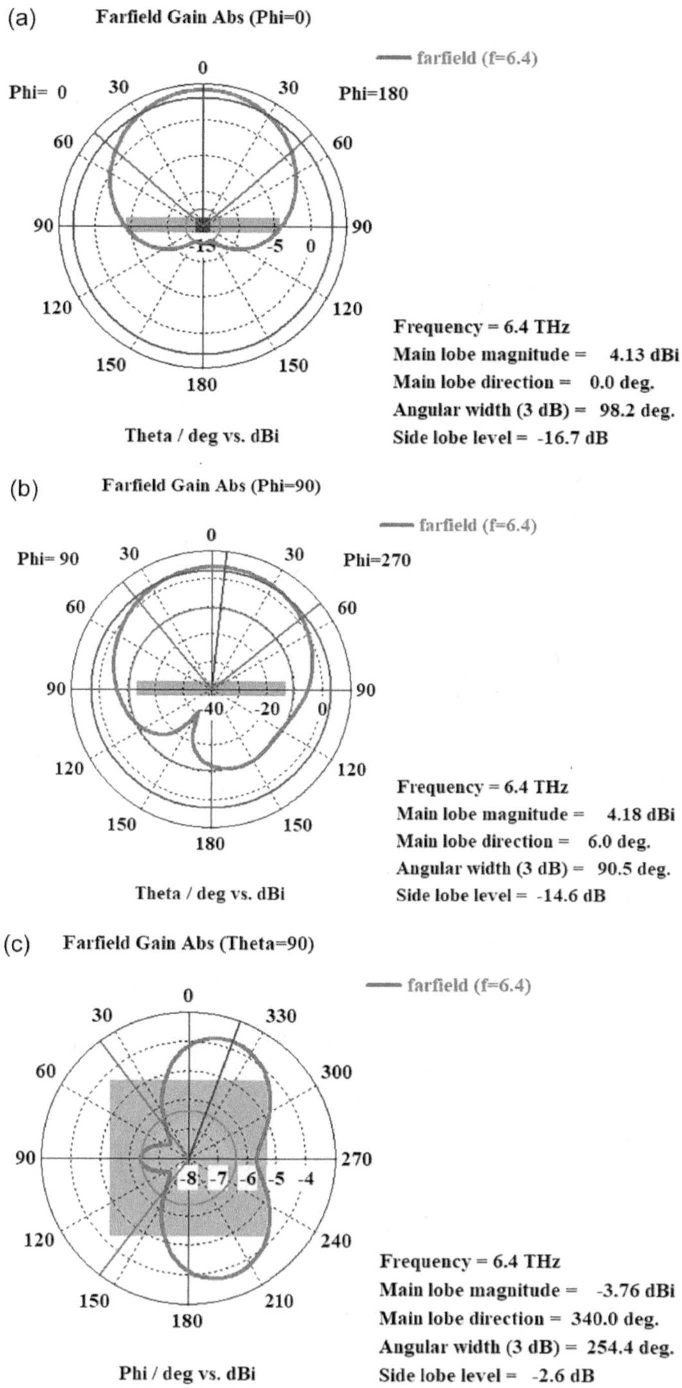

FIGURE 2.9 Rogers RO3035.

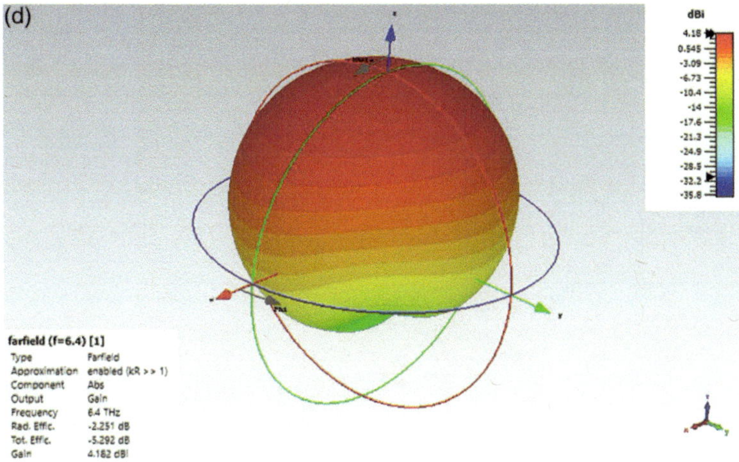

FIGURE 2.9 (Continued)

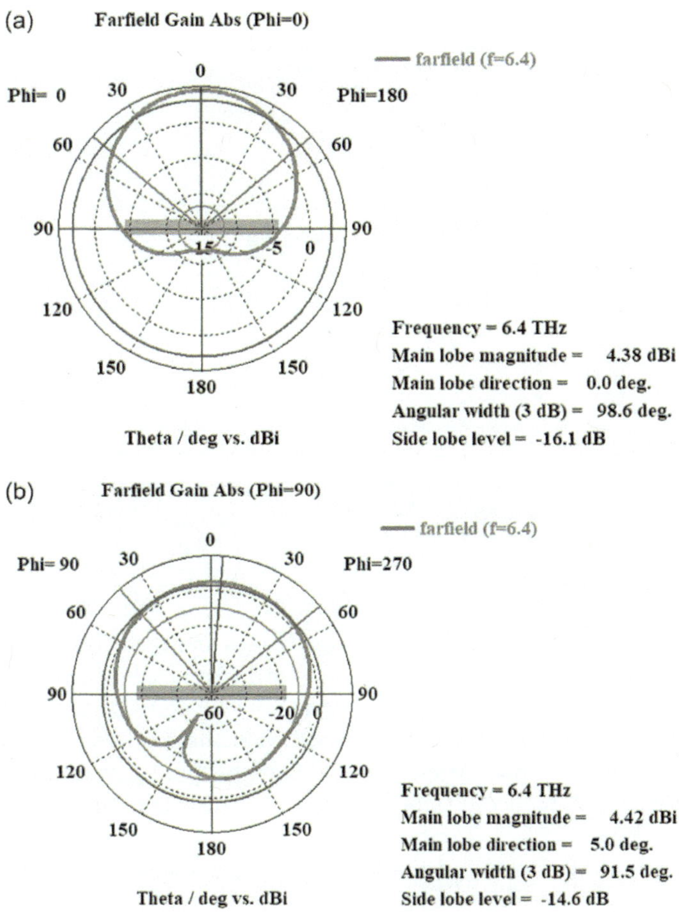

FIGURE 2.10 Silicon Dioxide.

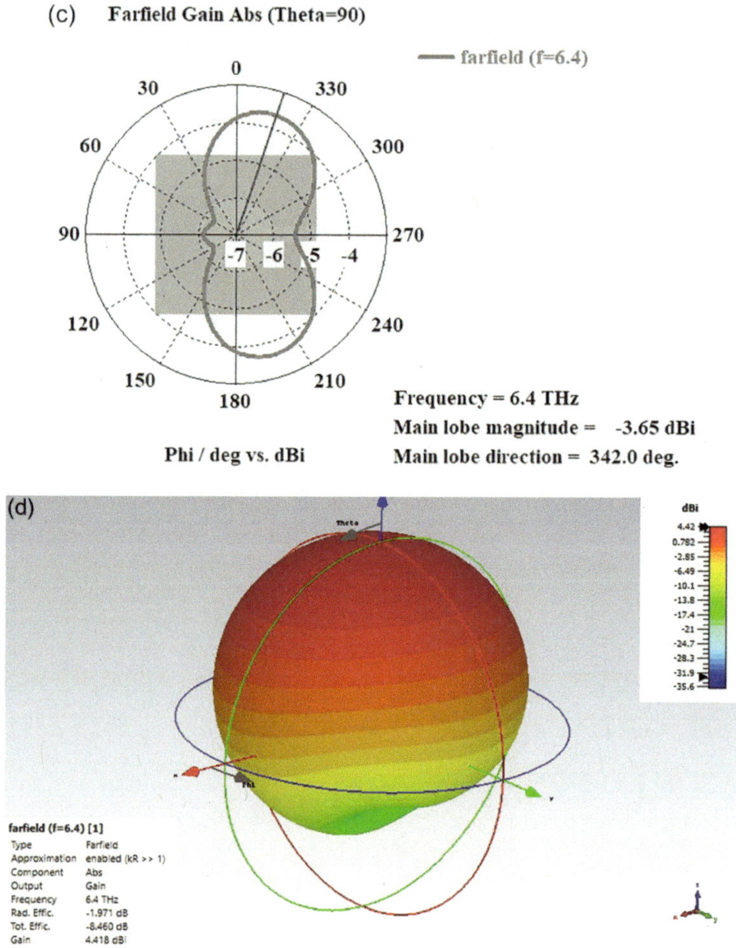

FIGURE 2.10 (Continued)

2.5 CONCLUSION

A graphene-based circular patch terahertz antenna is analyzed. The performance of SISO antenna of compact size 21x 21 x 2 μm3 with different substrates like Polyimide, Taconic RF-35TCA, Arlon AD 350A, Quartz, Rogers RO3035, and Silicon dioxide is presented. Performance of the proposed antenna is investigated thoroughly by adopting analyzing S-parameters, and radiation patterns. The antenna has favorable radiation pattern, moderate gain and bandwidth, excellent impedance matching which makes it perfect for wearable devices, which requires short-range wireless communication in the vicinity of the human body for non-stop health monitoring. This idea is a step towards improving the lifestyle of mankind by providing to them through the proposed antenna embedded in wearable devices. Besides monitoring the physiological parameters, the THz antenna proposed in this chapter can be used for vitamin k2, and biotin level detection. The THz radiation emitted by antennas can detect skin cancer and other skin diseases through imaging.

REFERENCES

[1] He Y, Chen Y, Zhang L, Wong SW, Chen ZN. An overview of terahertz antennas. *China Communications*. 2020 Jul 24;17(7):124–165.

[2] Grischkowsky DG, et al. Far-infrared time-domain spectroscopy with terahertz beams of dielectrics and semiconductors. *Journal of the Optical Society of America B-Optical Physics*. 1990;7:2006–2015.

[3] Raimundo X, et al. Channel characterization at THz frequencies for high data rate indoor communications. *Proceedings of the 2018 12th European Conference on Antennas and Propagation (EuCAP)*. 2018:1–2.

[4] Chen Z, et al. A survey on terahertz communications. *China Communications*. 2019;16(2):1–35.

[5] Nagatsuma T. Advances in terahertz communications accelerated by photonics technologies. *Proceedings of the 2019 24th Opto Electronics and Communications Conference (OECC) and 2019 International Conference on Photonics in Switching and Computing (PSC)*. 2019:1–3.

[6] Petrov NV, Kulya MS, Tsypkin AN, Bespalov VG, Gorodetsky A, Application of terahertz pulse time-domain holography for phase imaging. *IEEE Transactions on Terahertz Science and Technology*. 2016;6(3):464–472.

[7] Tabata H. Application of terahertz wave technology in the biomedical field. *IEEE Transactions on Terahertz Science and Technology*. 2015;5(6):1146–1153.

[8] Grade J, et al. Electronic terahertz antennas and probes for spectroscopic detection and diagnostics. *Proceedings of the IEEE*. 2007;95(8):1583–1591.

[9] Bala R, Marwaha A. Characterization of graphene for performance enhancement of patch antenna in THz region. *Optik*. 2016 Feb 1;127(4):2089–2093.

[10] Shamim SM, Das S, Hossain MA, Madhav BT. Investigations on graphene-based ultra-wideband (UWB) microstrip patch antennas for terahertz (THz) applications. *Plasmonics*. 2021 Oct;16(5):1623–1631.

[11] Azam S, Khan MA, Shaem TA, Khan AZ. Graphene based circular patch terahertz antenna using novel substrate materials. In *2017 6th International Conference on Informatics, Electronics and Vision & 2017 7th International Symposium in Computational Medical and Health Technology (ICIEV-ISCMHT)*. 2017 Sep 1 (pp. 1–6). IEEE.

[12] Jha KR, Singh G. Dual-band rectangular microstrip patch antenna at terahertz frequency for surveillance system. *Journal of Computational Electronics*. 2010 Mar; 9:31–41.

[13] Singhal S. Elliptical ring terahertz fractal antenna. *Optik*. 2019 Oct 1; 194:163129.

[14] Rubani Q, Gupta SH, Pani S, Kumar A. Design and analysis of a terahertz antenna for wireless body area networks. *Optik*. 2019 Feb 1; 179:684–690.

[15] Rabbani MS, Ghafouri-Shiraz H. Liquid crystalline polymer substrate-based THz microstrip antenna arrays for medical applications. *IEEE Antennas and Wireless Propagation Letters*. 2017 Jan 9; 16:1533–1536.

[16] Vasu Babu K, Das S, Varshney G, Sree GN, Madhav BT. A micro-scaled graphene-based tree-shaped wideband printed MIMO antenna for terahertz applications. *Journal of Computational Electronics*. 2022 Feb;21(1):289–303.

[17] Das S, Mitra D, Chaudhuri SR. Fractal loaded planar super wide band four element MIMO antenna for THz applications. *Nano Communication Networks*. 2021 Dec 1; 30:100374.

[18] Varshney G, Gotra S, Pandey VS, Yaduvanshi RS. Proximity-coupled two-port multi-input-multi-output graphene antenna with pattern diversity for THz applications. *Nano Communication Networks*. 2019 Sep 1; 21:100246.
[19] Linz T, Gourmelon L, Langereis G. Contactless EMG sensors embroidered onto textile. In *4th International Workshop on Wearable and Implantable Body Sensor Networks (BSN 2007) March 26–28, 2007 RWTH Aachen University, Germany*. 2007 (pp. 29–34). Springer.
[20] Löfhede J, Seoane F. Thordstein: Soft textile electrodes for EEG monitoring. In *10th IEEE International Conference on Information Technology and Applications in Biomedicine (ITAB)*. 2010 (pp. 1–4). IEEE.
[21] Löfhede J, Seoane F, Thordstein M. Textile electrodes for EEG recording—a pilot study. *Sensors*. 2012 Dec 7;12(12):16907–16919.
[22] Chi YM, Cauwenberghs G. Wireless non-contact EEG/ECG electrodes for body sensor networks. In *2010 International Conference on Body Sensor Networks*. 2010 Jun 7 (pp. 297–301). IEEE.
[23] Cheng J, Amft O, Lukowicz P. Active capacitive sensing: Exploring a new wearable sensing modality for activity recognition. In *International Conference on Pervasive Computing*. 2010 May 17 (pp. 319–336). Springer.
[24] Bhatoa R, Sidhu E. Novel terahertz microstrip patch antenna design for detection of biotin applications. In *2017 International Conference on Big Data Analytics and Computational Intelligence (ICBDAC)*. 2017 Mar 23 (pp. 289–292). IEEE.
[25] Kaur G, Mehta V, Sidhu E. Rectangular terahertz microstrip patch antenna design for vitamin K2 detection applications. In *2017 1st International Conference on Electronics, Materials Engineering and Nano-Technology (IEMENTech)*. 2017 Apr 28 (pp. 1–3). IEEE.
[26] Brar GS, Singh V, Sidhu E. Stacked decagon shaped THz Microstrip Patch antenna design for detection of GaAs semi-conductor properties. In *2016 International Conference on Automatic Control and Dynamic Optimization Techniques (ICACDOT)*. 2016 Sep 9 (pp. 771–774). IEEE.
[27] Roudjane M, Messaddeq Y. Innovative wearable sensors based on hybrid materials for real-time breath monitoring. In *Wireless Sensor Networks-Design, Deployment and Applications*. 2020 Sep 23. IntechOpen.
[28] Pozar DM. Microstrip antennas. *Proceedings of the IEEE*. 1992 Jan;80(1):79–91.
[29] Balanis CA. *Antenna Theory: Analysis and Design*. 2016 Feb 1. John Wiley & Sons.

3 Various Mutual Coupling Reduction Techniques for 5G *MIMO Antenna*

Durai Pandurangan, and Naveen Mishra

3.1 INTRODUCTION

Wireless communication in the twenty-first century is witnessing the amazing advancement of technology. High-definition television (HDTV), multimedia applications, modern mobile phone devices, and internet services demand high data rate, large capacity, wide range of coverage, low power consumption, large bandwidth, good quality of services, radio link reliability, low latency etc. These have produced tremendous advancements in telecommunication, from first generation (1G) analog communication (advanced mobile phone system) to fifth generation (5G) digital communication with compact wireless devices [1–2].

For both indoor and outdoor applications, as well as for uninterrupted line-of-sight (LOS) and non-line-of-sight (NLOS) connections, modern technology in 5G demands gigabits per second (Gbps) data speeds. Multi-path fading is the main factor affecting 5G wireless connectivity. An electromagnetic signal may travel through several pathways, each with different angles, delays, and/or frequency variations, before reaching its intended recipient due to electromagnetic waves dispersion in the surrounding environment. As a result, the received signal power fluctuates in space, frequency, and time through the random combination of the impinging multi-path components. This random variation in received EM signal level, known as fading, may significantly impair the reliability and data rate (quality) of wireless communication [3–4].

Designing 5G wireless communication systems with high data rates and high reliability is also quite difficult due to the limitations imposed by high frequency bandwidth and restricted power. The single input and single output (SISO) antenna system needs an extremely broad bandwidth and exorbitant power consumption in order to meet these criteria. The SISO antenna system is shown in Figure 3.1. It consists of a single antenna at the transmitter side and a single antenna at the receiver side in order to establish the datalink [5–6].

A very wide frequency spectrum is needed for the gigabits per second (Gbps) data throughput with SISO, and the technology is even incompatible with non-line-of-sight (NLOS) networks. Also, in order for SISO to function, realistic receivers must

DOI: 10.1201/9781032656021-3

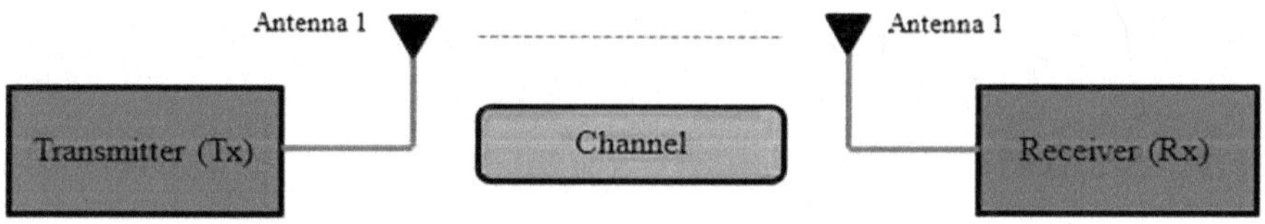

FIGURE 3.1 SISO antenna.

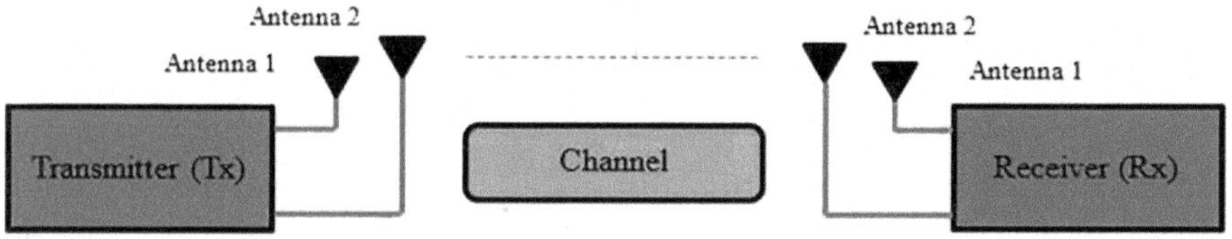

FIGURE 3.2 MIMO antenna.

have a very high signal to noise ratio (SNR). In SISO, it is crucial to attain an average signal to interference with noise ratio (SINR) of at least 10 dB, even when spectral efficiencies are less than 4 bits/s/Hz. Higher gain SISO also causes dispersion and is unable to support NLOS because of the power limitation. Reduced fading margin and range limitations result from expanding SISO bandwidth to reach Gbps data rate. Consequently, extensive coverage demands high power [1, 7–9]. MIMO is an innovative technique for WLAN, WiMAX, LTE, 5G, and future generations that solve all of the aforementioned issues.

3.2 WHAT IS MIMO?

Multiple input multiple output (MIMO) is a wireless technique that combines several transmitter and receiver antenna, fed separately using a micro strip line, and coaxial feeding technique to send more data concurrently. As seen in Figure 3.2, MIMO is a wireless technique that combines several transmitter and receiver antenna fed separately using a microstrip line, and coaxial feeding technique to send more data concurrently. Present 4G and 5G cellphone devices are installed with MIMO antenna systems to support the high data rate.

MIMO was introduced in the early '90s as an alternate solution in 3G to solve the data rate limitation faced by single input and single output (SISO) systems that support time diversity, frequency diversity, polarization diversity, and spatial diversity [10–11]. In addition, MIMO supports spatial multiplexing. In spatial multiplexing, the input data are divided into multiple packets and each packet is transmitted through different transmit antenna (example antenna 1,2,3 . . .) using the same frequency channel. Spatial multiplexing is a highly powerful technique in boosting channel capacity even at high signal-to-noise interference ratio (SNIR). Transmitting and receiving data streams through multiple antennas at different time, different frequency, and independent polarization (linear and circular) reduces signal-to-noise interference ratio (SNIR), reduces fading effect, increases data rate, channel capacity, and spectrum efficiency [12, 13].

Advantages

The benefits of MIMO are as follows:

- MIMO supports spatial multiplexing. As a result, higher data rate can be achieved.
- Signal fading due to multi-path propagation is less due to multiple transmission and reception environment.
- MIMO provides improved channel capacity and spectral efficiency with increased quality of service.
- There is low BER (Bit Error Rate) and high SNR due to redundant received data symbols by multiple antennas.

Disadvantages

- Multiple antennas cause mutual coupling effect between MIMO antenna elements.
- MIMO based systems cost more compared to SISO due to increased antenna elements and advanced digital signal software requirements.

3.3 MIMO APPLICATIONS

For satellite, base station (BS), car, pedestrian, and radar applications, MIMO antenna technology provides the necessary data rates. MIMO antenna systems are used in smart and modern technologies such as local area networks (WLANs), global system for mobile communication (GSM), applications for the global positioning system (GPS), 4G-long-term evolution (LTE), 5G, wireless interoperability for microwave access (WiMAX) bands, and future generations (6G) [14].

3.4 MUTUAL COUPLING IN MIMO

MIMO is considered to be an important component of current wireless technologies for the performance enhancement, i.e., gain, efficiency, data rate, and capacity, etc. However, the small and lightweight devices place a minimum and maximum boundary on the space limitations for the wireless products. So, the close placement of radiating elements in MIMO results in compactness but causes electromagnetic wave interaction between the elements. Such wave interaction is combinedly known as a "mutual coupling", which seriously alters the MIMO antenna's radiation characteristics [15].

The mutual coupling is caused by three factors, coupling due to the spatial electromagnetic fields, the signal leaking along the substrate via surface waves, and the coupling between the feed lines due to conduction current on the common metallic ground plane.

Mutual coupling is also described as the EM energy absorbed by a neighbor antenna when other antennas of the MIMO are operating. It alters the input impedance, increases side lobe radiation, reflection coefficients, transmission coefficients, and radiation pattern of each array element. For communication system, mutual coupling can also increase signal interference, reduce channel capacity, and decrease radiation efficiency. The empirical model [16] for the mutual coupling is given by the equations (3.1 & 3.2),

$$S_{mn} = exp\left(-\frac{2d_{mn}}{\lambda}\right)(\alpha + j\pi), m \neq n \tag{3.1}$$

$$S_{mm} = 1 - \frac{1}{N}\sum\nolimits_{m}\sum\nolimits_{n,m} S_{mn} \tag{3.2}$$

where, S_{mn} is the mutual coupling coefficient and d_{mn} is the distance between the m_{th} and n_{th} antenna elements, N is the number of antennas in the array, S_{mm} is the reflection coefficient and α is the parameter controls the coupling level.

Mutual coupling in practice depends not only on the MIMO array configuration but also depends on the excitations of other array elements. Equation (3.3) is typically used to estimate the S-parameter in decibels between the m_{th} and n_{th} elements. Lower the S_{mn} value shows better the isolation and vice versa [17–19].

$$S_{mn}(dB) = 20\log_{10}(|S_{mn}|). \tag{3.3}$$

3.5 MUTUAL COUPLING MECHANISM

Following is the detailed explanation of the mutual coupling in transmitting and receiving side.

3.5.1 Mutual Coupling in the Transmitting Side

As seen in Figure 3.3a, let us examine the mutual coupling effect of two elements MIMO antenna *m* and *n*, during transmission. Suppose that element *n* has a time-varying excitation signal source connected to it. The energy **A** generated by source partially radiates into space as **B** and partially as **C** toward the *mth* element. A portion of the energy that the *mth* antenna receives is reflected into space as **D**, while the remaining energy moves in the direction of the generator as **E**. The *nth* element will absorb a portion of the rescattered energy **D** as **F**. This process of mutual EM energy interaction will keep happening repeatedly. Yet, because the rescattered energy decreases exponentially after each iteration, it is typically sufficient to take into account the first few iterations.

The vector sum of the radiated and rescattered EM fields forms the total far-field of MIMO antenna. From this it is evident that the mutual coupling changes the antenna radiation pattern. Similarly, an enhancement in the standing wave results from the **E** waves being added vectorially to the reflected and incident waves of the *mth* element. Hence, the input impedance of the *mth* element is altered. Therefore, mutual coupling changes the mutual impedance and antenna's self-impedance. In MIMO antennas, random phase excitation of port may produce significant effect on the mutual coupling and impedance matching [15].

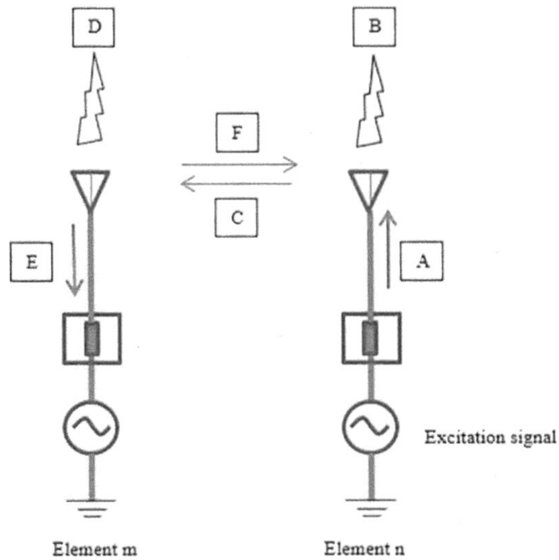

FIGURE 3.3a Transmitting side.

3.5.2 Mutual Coupling in the Receiving Side

Assume that, as illustrated in Figure 3.3b, plane electromagnetic wave **A** impedes upon the MIMO and reaches the *mth* element. First a current is induced in the *mth* element. The incident energy is partially transmitted into the *mth* receiver as **B**, while the remaining portion is redirected into space as energy **C**. A portion of the rescattered wave is sent towards the *nth* element as **D,** where it sums vectorially with wave **E**, the incident plane. The wave that each MIMO element receives is therefore the vector sum of the waves that it receives directly from the incident wave and the linked waves that it receives from other elements as a result of scattering. This superposition of incident and rescattered wave in the receiver antenna is called mutual coupling. In order to minimize the rescattered energy, i.e., maximize the received energy, the load impedance of the antenna element should be selected in a way that the rescattered wave **C** is minimized by the reflected wave **E**. By stimulating the antenna element with a matched termination load in the receiving mode, one may examine the characteristics of the antenna under test (AUT) [15].

3.6 PERFORMANCE METRICS OF MIMO ANTENNA

Due to its ability to leverage multiple channels for data transmission and reception, MIMO antenna have drawn a lot of attention in the context of modern wireless communication systems, resulting in increased output performances and range. It has been stated that the maximum isolation between MIMO system components is essential, in order for the MIMO antenna's elements to function independently and send and receive signals simultaneously without affecting the antenna's characteristics.

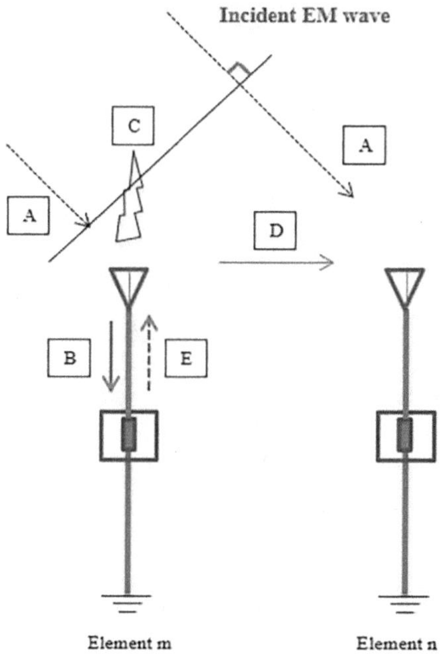

FIGURE 3.3b Receiving side.

However, the coupling effect is a major concern in MIMO design because it mitigates the MIMO antenna performances. So, before learning coupling reduction techniques, it is crucial to learn about MIMO's performance metrics.

Reflection coefficient, envelope correlation coefficient (ECC), total active reflection coefficient (TARC), far-field gain, diversity gain (DG), mean effective gain (MEG), channel capacity, and spectrum efficiency are the important metrics of MIMO antenna [20–21].

3.6.1 Reflection Coefficient (S_{11}) and VSWR

Reflection coefficient or S_{11} is defined as the amount of power that is returned to port 1 for a given power input at port 1. It is calculated using the equations (3.4 & 3.5),

$$S_{11} = \frac{Reflected\ Power}{Incident\ power}, \tag{3.4}$$

$$S_{11}\left(\text{in dB}\right) = 10\log_{10}\left(S_{11}\right)\text{dB}, \tag{3.5}$$

Voltage standing wave ratio (VSWR) in terms of S_{11} can be obtained by equation (3.6),

$$\text{VSWR} = \frac{1+|S_{11}|}{1-|S_{11}|}. \tag{3.6}$$

3.6.2 Envelope Correlation Coefficient

The envelope correlation coefficient is the important parameter based on diversity that shows how independent array elements are in terms of radiation pattern. For example, if one antenna uses horizontal polarization (HP), and the nearby antenna is completely vertically polarized (VP), then the two antennas would have a zero correlation ideally. It can be calculated in two ways, (i) using radiation patterns and (ii) S-parameters. In MIMO systems, the ECC determines the degree to which multiple radiating elements are independent of each other in their radiation pattern; hence, the ECC value determined using the far-field radiation pattern is highly recommended. The accepted standard value is ECC < 0.5 (practically).

The mathematical equation for calculating ECC using the far-field pattern of MIMO is given by equation (3.7),

$$e = \frac{\left|\iint_{4\pi} d\Omega F_1(\theta,\phi)^* . F_2(\theta,\phi)\right|^2}{\iint_{4\pi} d\Omega \left|F_1(\theta,\phi)\right|^2 . \iint_{4\pi} d\Omega \left|F_2(\theta,\phi)\right|^2} \tag{3.7}$$

where $F_1(\theta,\phi)$ is the i_{th} antenna radiation fields, θ is the elevation angle and ϕ is the azimuth angle of the spherical co-ordinate system.

The mathematical equation for calculating ECC using the scattering parameters of MIMO is given by equation (3.8),

$$\left|\rho(i,j,N)\right| = \frac{\left|\sum_{n=1}^{N} S_{i,n}^* S_{n,j}\right|}{\sqrt{\left|\prod_{k=i,j}\left[1-\sum_{n=1}^{N} S_{i,n}^* S_{n,k}\right]\right|}} \tag{3.8}$$

where i and j are the antenna elements and S is the scattering parameter. For the 2-element MIMO, equation (3.9) can be rewritten as,

$$\rho_{12} = \frac{\left|S_{11}S_{12} + S_{21}S_{22}\right|^2}{\left(1-\left|S_{11}\right|^2\left|S_{21}\right|^2\right)\left(1-\left|S_{22}\right|^2\left|S_{12}\right|^2\right)}, \tag{3.9}$$

where S_{11}, S_{22} are reflection coefficients at port 1 & 2 and S_{21}, S_{12} are the forward and reverse transmission coefficients.

3.6.3 Diversity Gain

Diversity gain indicates the increase in signal to noise ratio due to diversity schemes used. For example, spatial diversity, time diversity, and spatial multiplexing. The *DG* indicates MIMO antenna's quality and reliability. The maximum value of *DG* for the MIMO antenna is 10dB. When diversity gain is high, mutual coupling is low and vice versa. Equation (3.10) is used to calculate the *DG* based on the ECC value,

$$DG = 10 \times \sqrt{1 - \left|\rho_e\right|^2} \tag{3.10}$$

3.6.4 Total Active Reflection Coefficient

Adjacent antenna elements in multiport antenna systems influence one another and, when they function simultaneously, affect the efficiency and operating frequency. Therefore, S_{11}- parameter by itself won't be sufficient to forecast MIMO's real response. To account for this impact, a new parameter known as the total active reflection coefficient (TARC) has been added. In the instance of a four-port MIMO with a specified phase angle between neighboring ports, it determines the real behavior of the scattering parameters such as S12 (between ports 1 and 2), S13 (between ports 1 and 3), and S14 (between ports 1 and 4).

The general expression for the TARC is given in the equation (3.11),

$$\Gamma_a^t = \frac{\sqrt{\left|S_{ji} + S_{jj} * e^{j\theta}\right|^2 + \left|S_{ii} + S_{ij} * e^{j\theta}\right|^2}}{\sqrt{N}} \tag{3.11}$$

where θ is the phase angle between the antenna.

3.6.5 Mean Effective Gain

The directive gain is insufficient to produce accurate findings under NLOS circumstances. Mean effective gain (MEG) is therefore employed. The performance parameter, known as the MEG, is the ratio of the MIMO antenna element's mean incident power to mean received power.

Equations (3.12 & 3.13) may be used to determine the average received signal intensity of each antenna using the MEG,

$$MEG_i = \oint \left[\frac{XPR.G_{\theta i}(\Omega) + G_{\theta i}(\Omega).P_{\phi}(\Omega)}{1 + XPR} \right] d\Omega \tag{3.12}$$

$$= 2\pi_0\pi_0 \frac{XPR}{1+XPR} C_{\theta}(\theta,\phi)P_{\theta}(\theta,\phi) + \frac{1}{1+XPR} G_{\phi}(\theta,\phi)P_{\phi}(\theta,\phi))\sin\theta d\theta d\phi \tag{3.13}$$

where *XPR* is the cross-polarization power ratio, $P\theta$ and $P\phi$ are the angular density functions of incoming power.

3.6.6 Spectral Efficiency

Spectral efficiency, often referred to as bandwidth efficiency, is the highest quantity of data bits that can be sent to a given number of users in a given amount of time without degrading service quality. Spectral efficiency is higher with MIMO. 4.08 bits/second/Hz for 4G-LTE and 16.32 bits/second/Hz for 4 × 4 MIMO technology are the respective spectral efficiencies in SISO. Comparably, with 4G-LTE-A (Long Term Evolution—Advanced) technology, the spectral efficiency is 3.75 bits/second/Hz for SISO and 30.0 bits/second/Hz for 8 × 8 MIMO. Equation (3.14), which is used to compute spectral efficiency (bits/second/Hz),

$$\eta = \frac{R}{B}, \tag{3.14}$$

where η is spectral efficiency, R is data rate, and B is the bandwidth of channel.

3.6.7 Channel Capacity

The maximum rate at which information can be transmitted through a radio channel is called channel capacity. The channel capacity is another factor that defines the MIMO's performance. Let us consider, MIMO is having Nt transmitting antennas and Nr receiving antennas then the formula used to calculate channel capacity is given in equation (3.15),

$$C = log_2\left({}_{Nr} \frac{P}{NNt} HH^* \right), \tag{3.15}$$

where P is the signal power, I is the identity matrix, N is the noise power, H is the channel matrix and its transpose is H^*.

Equation (3.16) shows the channel matrix for a two-port MIMO,

$$H = \begin{bmatrix} h_{11} & h_{12} \\ h_{21} & h_{22} \end{bmatrix} \tag{3.16}$$

where, h_{11}—channel between transmitter 1 and receiver 1 h_{12}—channel between transmitter 1 and receiver 2

h_{21}—channel between transmitter 2 and receiver 1 h_{22}—channel between transmitter 2 and receiver 2.

3.7 MUTUAL COUPLING REDUCTION TECHNIQUES

MIMO antennas are needed by modern wireless communication devices for high data rate and high-capacity applications. Gain, efficiency, directivity, and correlation need high attention during the antenna design for obtaining good quality of services. However, antenna designers face challenges due to mutual coupling phenomena. These issues are addressed using the reduction techniques shown in Figure 3.4a [20].

3.7.1 Diversity Techniques

In this section, a brief discussion of the various diversity techniques utilized in MIMO antenna designs is provided.

3.7.1.1 Space Diversity

Space diversity is the first mutual coupling reduction technique used in the MIMO antenna. The primary source of coupling is the close packing of array elements. To achieve minimal mutual coupling or de-correlation, this technique relies heavily on variations in space. This holds true for all different kinds of MIMO antenna

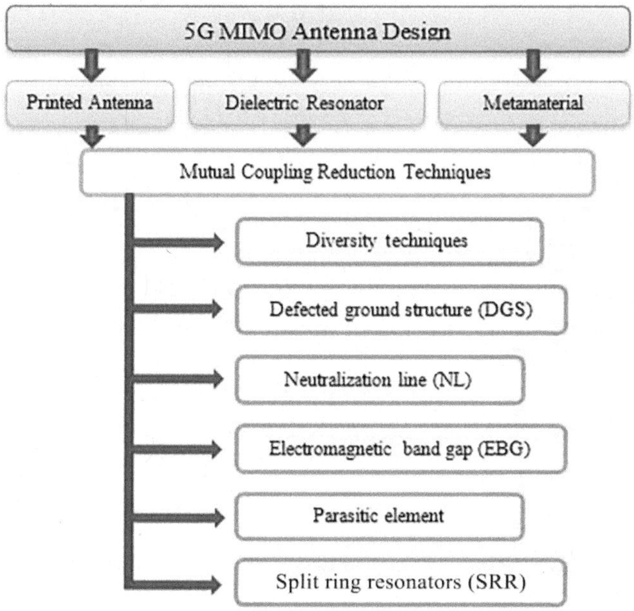

FIGURE 3.4a Mutual coupling reduction techniques.

configurations, including patch, dipole, conventional, monopole, and planar inverted frequency antennas (PIFA) [22]. A sufficient distance between the radiating elements guarantees that the design operates correctly and generates the required performance characteristics. It also limits the coupling current, responsible for mutual coupling. The radiator's placement and orientation—such as anti-face (180°), face-to-face, diagonal placement, etc.—are crucial for improving the MIMO antenna's isolation and performance metrics [23]. However, increasing distance between MIMO elements increases the overall size of the antenna that makes space diversity techniques unsuitable for compact device applications.

3.7.1.2 Polarization Diversity

Space diversity is suitable for applications where space is not a constraint. For example, base station. However, serious space limitation is required for present wireless and mobile entities. Another technique, so-called polarization diversity, has attracted more attention on the wireless devices design. In the polarization diversity each and every element in the MIMO uses unique polarization like vertical/horizontal linear polarization and left hand/right hand circular polarization. As a result, it increases isolation. Low correlation between MIMO radiators and better link utilization are the results of the joined effect of polarization and pattern diversity [24–25].

3.7.1.3 Pattern Diversity

Two or more nearby antennas operating with different radiation pattern is called pattern diversity. Pattern diversity ensures the quality of services and links reliability in

FIGURE 3.4b Basic DGS shapes.

MIMO environments by creating different far-field radiation patterns. The radiation patterns of far-fields depend on the position and orientation of radiating elements. The symmetry/asymmetry in radiation patterns may lead to the enhanced isolation and return-loss. Bandwidth enhancement and reconfigurability is also achieved due to pattern diversity [26].

3.7.2 Defected Ground Structure

The term "defected ground structure" (DGS) refers to one or more number of slots that are deliberately etched onto a printed circuit board's ground plane in order to prevent wave propagation over a range of frequencies. In general, the term "defect" in this technology refers to the introduction of a "slot" into the ground plane, which approximates an infinitely perfect conducting current sink. The DGSs may be roughly categorized into two groups based on their configurations: single unit cell (DGS) and periodic uniform or non-uniform arrangement of multiple unit cells called periodic defected ground structure (PDGS). Using DGS geometries, the resonance's tunability characteristic can be achieved. The DGS was developed in 2000 and has since been included in nearly all large printed MIMO antennas to improve performance [27].

DGS is implemented by etching off a simple slot in the ground plane that regulates electromagnetic wave excitation and propagation via the substrate layer results from disruption of the shielded current distribution in the ground plane, which is dependent on the size and shape of the defect. For better results, the defect's shape might be altered from its original, basic shape to a more complex one.

The literature has described a variety of DGS structural forms, including rectangular, hairpin DGS, circular, hexagonal DGS, cross-shaped DGS, square, dumbbell, spiral, L-shaped, triangle, T-shape, concentric ring, U-shaped, and V-shape as shown in Figure 3.4b. Periodic forms of these structures are also employed [28].

An S-shaped periodic defect ground structure device is placed between the radiating elements of the antenna ground plane to reduce mutual coupling. The introduction of PDGS significantly reduces the EM field and current induced between the patch elements. Without a PDGS device, a large surface current is formed in the array elements, resulting in higher EM wave coupling. The coupling between two patch antennas is a function of the relative alignment of two adjacent beam elements. However, the current induced on the etched PDGS surface is greatly reduced because it is confined to a limited substrate area [29]. Figure 3.4c shows the periodic DGS

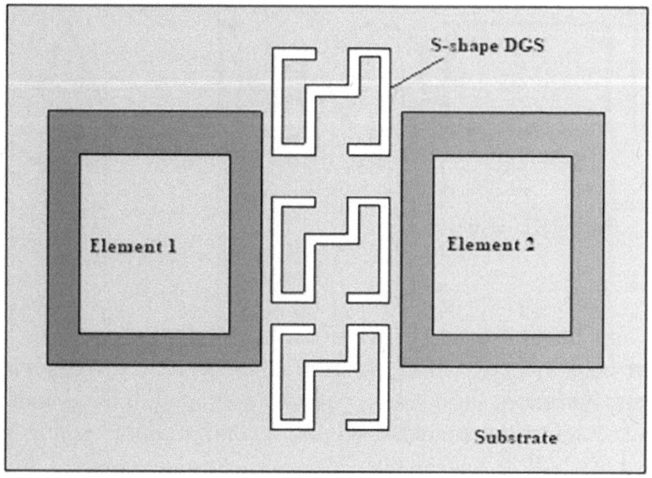

FIGURE 3.4c MIMO antenna with S-shaped PDGS (Image courtesy: [29]).

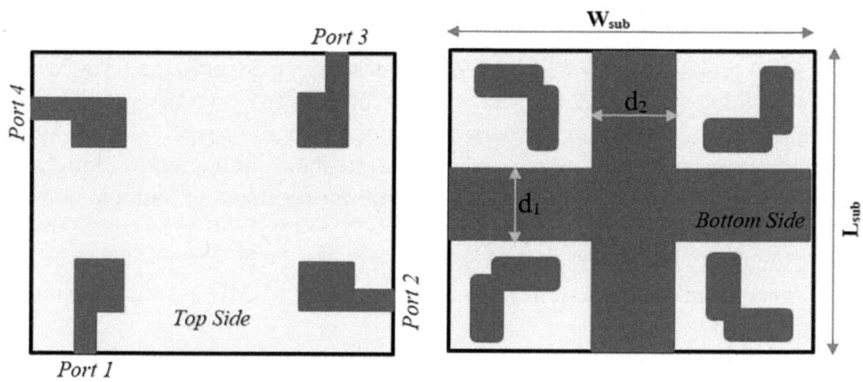

FIGURE 3.4d 4-Element MIMO antenna radiating patch (Top View and Bottom View) (DGS) (Image courtesy: [30]).

based MIMO antenna and scattering parameters of defected ground structure on the mutual coupling reduction. Isolation of -55dB is obtained using S-shaped DGS.

A 4-port multiple input multiple output, circularly-polarized (CP) wideband antenna is shown in Figure 3.4d. In this antenna four L-shaped patch and four orthogonally placed slotted ground plane called "defected ground structure" is used. Through polarization diversity, the orthogonally positioned curved rectangular L-shaped etched structure in the bottom ground plane serves to reduce the mutual coupling between MIMO components by producing a circularly polarized waveform in the broadside direction. High isolation (>33 dB), ECC is less than 0.001, diversity gain is (9.99 dB), mean effective gain is 7 dB and channel capacity loss (CCL) of 0.15 bits/s/Hz is achieved [30].

TABLE 3.1
Performance Comparison of DGS Based MIMO Antenna

Ref.	Antenna Dimension (mm³)	Frequency (GHz)	DGS Type	Isolation Level (dB)	ECC/DG	Gain/ Radiation Efficiency
[29]	100×72×3.81	2.57	S-shaped PDGS	>–55dB	NA/NA	3.75dBi/93%
[30]	20×20×1.6	3.81–7.48 GHz	Orthogonally placed L-shaped DGS	>33dB	<0.001/9.99dB	NA/NA
[31]	50×160×0.8	1.0	Open Ended DGS slots	>–20dB	<0.075/9.97dB	2dBi/80%
[32]	60.2×60.2×1.6	2.45 GHz	Square ring DGS	≥–25dB	≤0.1/9.94dB	2.1dBi/81%
[33]	22×26×0.8	11.8 GHz	Open ended slot and trident shaped strip	≥20dB	≤0.03/9.99dB	3.6 – 6dBi/85%

3.7.3 Neutralization Line Approach

Neutralization line (NL) is another simple and straightforward coupling reduction technique used in the 5G MIMO antennas. A neutralization line is a thin solid metallic line connected between the feed line to reflect the coupling current induced due to wave interaction radiated from another antenna. This additional line in the antenna increases isolation by reducing the coupling current. NL is usually placed between feed lines, between radiators, between ground plane and radiators. Sometimes more than one neutralization line is also used to improve the isolation. The position and structure of the antenna determines the size and orientation of the neutralization line. However, searching a path for neutralization is complicated [34].

The geometry of the two-monopole antenna system constructed on a single-layered FR4 substrate for MIMO applications is shown in Figure 3.4e. On the top opposing sides of the substrate, the two monopoles are printed. Monopoles 1 and 2 are connected to ports 1 and 2 (at points A & B) and short-circuited to the system ground plane via an L-shaped strip, at points C and D, respectively, as shown in Figure 3.7e. Between points E and F, a thin printed neutralization line made of coated copper is etched to connect the monopoles. The neutralizing line is there to improve isolation by reducing the current induced due to close positioning of MIMO elements. Isolation level varies with respect to the NL location connected between points E and F [35].

Figure 3.4f shows a planner antenna utilizing a neutralization line between radiating elements. This MIMO antenna consists of four circular monopole patch feeds by four microstrip lines respectively. Two antennas are placed on the upper side of the substrate, and two antennas are placed on the bottom side of the substrate. The four radiating patches are perpendicular to one another. A partial ground plane with

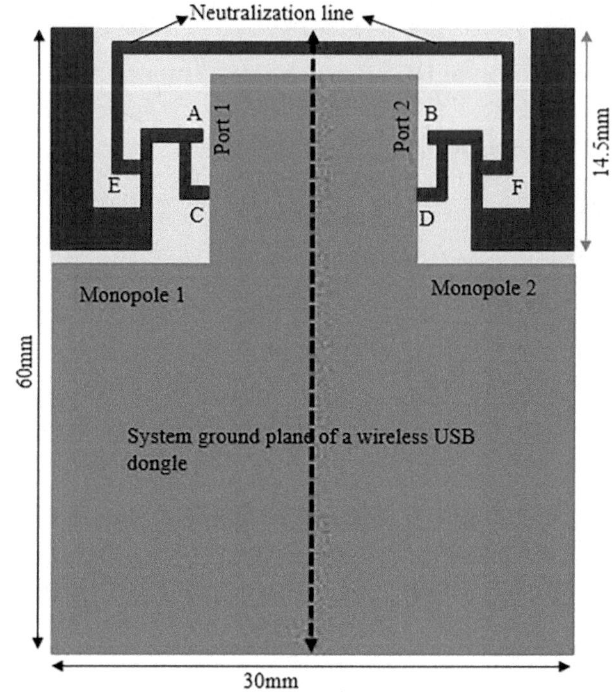

FIGURE 3.4e Geometry of the two monopole MIMO antennas connected by a neutralization line [35].

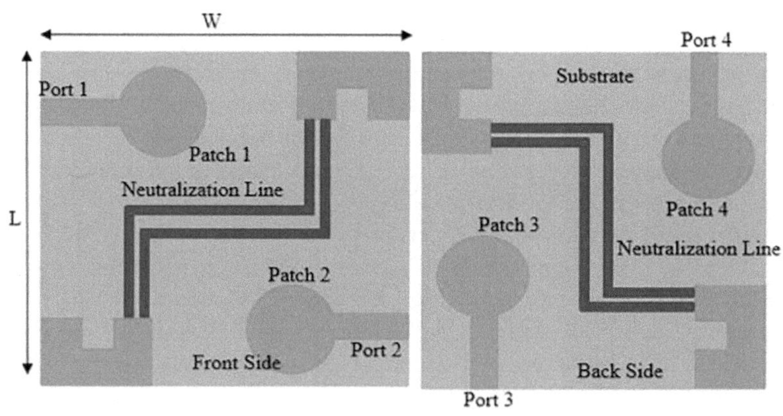

FIGURE 3.4f Double neutralization lines based 4-port MIMO antenna (Image courtesy: [36]).

small rectangular slots is constructed in order to increase bandwidth. Additionally, to improve isolation between antenna parts, the partial ground plane is connected using the microstrip multimode resonator (MMS) principle. The neutralization line used in this MIMO acts as a band reject filter to minimize surface current thereby reducing the mutual coupling effect [36].

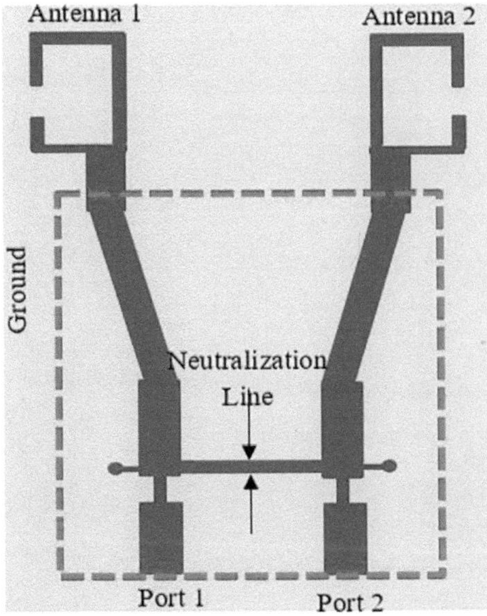

FIGURE 3.4g C-shaped radiator MIMO antenna using neutralization line (Image courtesy: [37]).

In Figure 3.4g, a MIMO antenna is designed by implementing the neutralization lines in between two C-shaped antennas. Two closely spaced antennas with phase-shifting components are part of this design. Additionally, for lossless networks, admittance matrix transformation is employed. Mutual admittance is a complex number, and the lossless admittance matrix needs to be entirely imaginary for complete impedance matching. The diagonal elements of the admittance matrix should be zero for improved port isolation. To convert mutual admittance to a pure imaginary number, a two-section stepped impedance transformer is utilized as a phase-shifting network. A neutralizing line that is linked in parallel to the C-shaped radiator antenna also serves as a phase-shifting element. This thin, straight neutralization line can generate an imaginary admittance number by producing an opposite coupling. Therefore, the neutralization line vanishes out the unwanted coupling at first port when port 2 is excited and vice versa [37].

Figure 3.4h depicts a pair of symmetrical tri-band multiple input multiple output (MIMO) antennas with strong isolation that are supplied by microstrip lines through ports 1 and 2. There are two NL in this design. The first U-shaped NL is linked to the antenna's microstrip feeder line, while the second U-shaped NL is inverted and connected to two radiation elements. U-shaped and inverted U-shaped NLs act as the stop band filter, and they can decouple surface currents in the medium and high frequency bands [38].

3.7.4 Electromagnetic Band Gap Structure (EBG)

The abundance of periodic structures in nature has drawn the attention of several researchers. Amazing features and exciting new phenomena arise when they interact with electromagnetic radiation. Behaviors such as frequency pass bands, band gaps, and stop bands in particular might be recognized. The "electromagnetic band gap

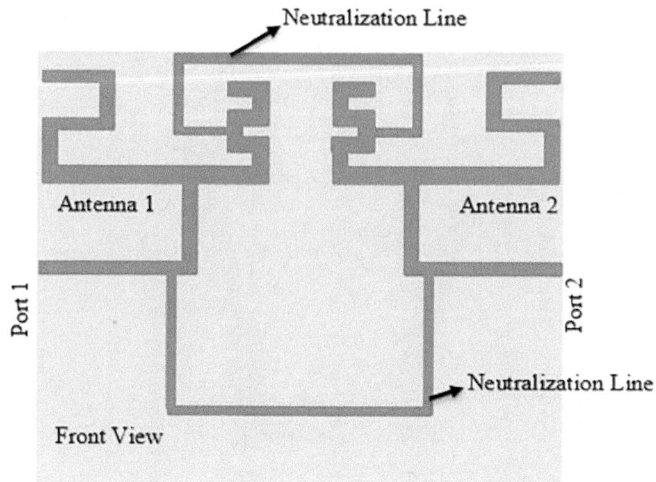

FIGURE 3.4h U-shaped neutralization line MIMO antenna (Image courtesy: [38]).

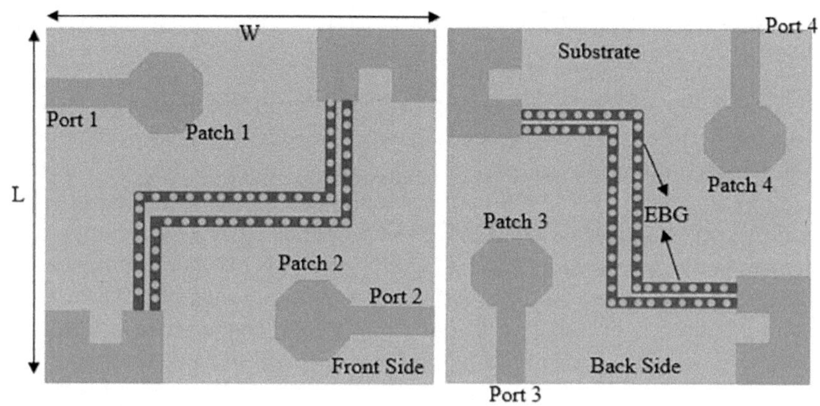

FIGURE 3.4i Geometry of the polygon-shaped MIMO antenna front and back side (Image courtesy: [41]).

(EBG)" arrangement is one example of a periodic structure that has these properties. It offers good isolation between radiating elements in MIMO and displays band pass/stop characteristics. A2D/3D metal structure or a combination of metal and dielectric components makes up the EBG structure. Photonic band gap (PBG) structures and frequency selective surfaces (FSS) are subclasses of electromagnetic band gaps. EBG structures are usually placed on the upper side of the substrate material particularly between the radiating elements [39–40].

Figure 3.4i illustrates a new, compact, four-port, double-side multiple input multiple output antenna with an electromagnetic band gap (EBG) structure designed to improve isolation for Ultra-Wide Band (UWB) applications. The EBG structure and four polygon-shaped radiating elements form this MIMO design. The substrate

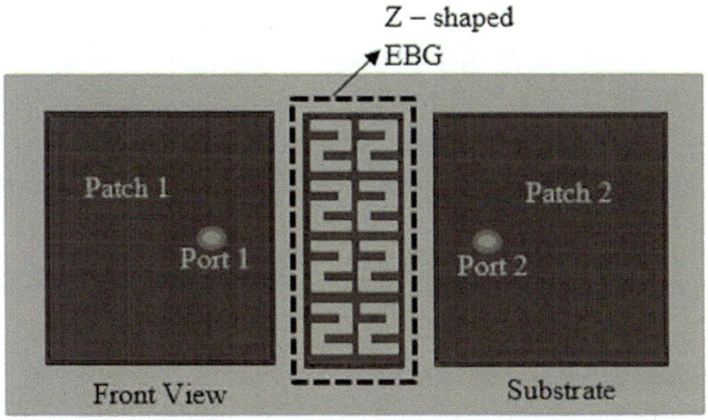

FIGURE 3.4j Z-shaped EBG for isolation enhancement (Image courtesy: [42]).

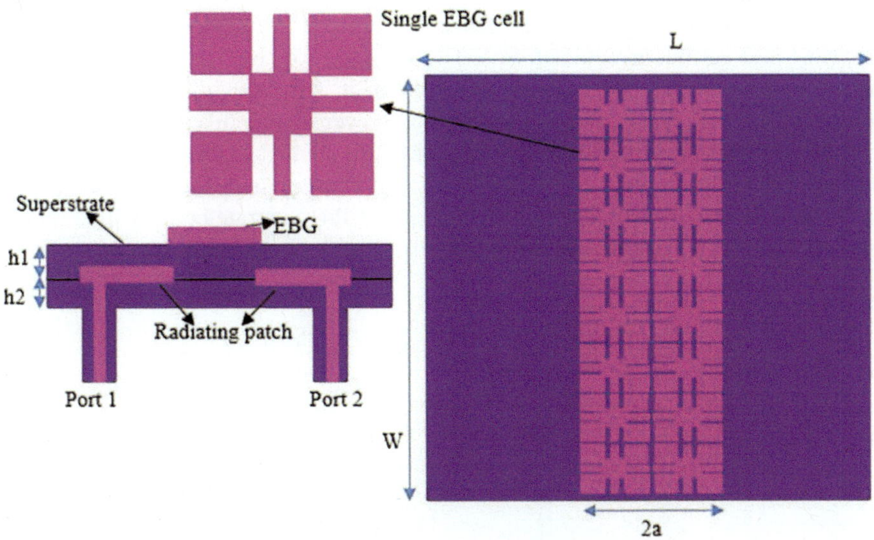

FIGURE 3.4k Uniplanar compact-EBG MIMO antenna (Image courtesy: [43]).

is FR4, with two radiators on each side and a partial ground plane connected to the other two components attached on the other side. The EBG decoupling structures are employed on the front and back surfaces of the substrate in order to reduce the correlation coupling between the radiators [41].

A small planner antenna featuring the special capability of wide-angle scanning is shown in Figure 3.4j. In order to minimize mutual coupling, 4×2 EBG cell elements are positioned in between microstrip square patches. Using co-axial probes, the square radiating patches are fed.

Metallic components with etched Z shapes are arranged in a periodic pattern on the EBG. Additionally, the periodic boundary conditions and suitable phase shifts in this structure lead to improved isolation [42].

A MIMO configuration of uniplanar compact electromagnetic band gap (UC-EBG) structure to lessen mutual coupling between the radiating components is shown in Figure 3.4k. The UC-EBG is printed on the top of the antenna layer using another substrate called "superstrate" made up of similar antenna dielectric substrate material Roger RT/Duroid 6010 with dielectric constant of 10.2. Because the patch antenna fringing fields are not lowered by the placement of the EBG structure on the top layer, the resonant frequency remains unchanged [43]. Surface waves within a specific frequency range can be reduced by the UC-EBG superstrate. Thus, when the UC-EBG superstrate is employed, the radiation pattern improves and the mutual coupling practically vanishes.

3.7.5 Parasitic Element Method

An antenna's parasitic element, also known as a passive radiator, is any conductive component that isn't electrically coupled to the radiator. Usually, this is a metal rod, tube, or line. It is often positioned in the space between the MIMO antennas. It takes a framework that can range from simple to complex and that can regulate the mutual coupling effect. Common shapes include dumbbells, squares, rectangles, circles, meander line, ring, and various irregular forms. The parasitic elements operate as resonators and link electromagnetically with the driven element. They also influence the radiation pattern and resonant frequency of the antenna, preventing/modifying the propagation of surface current and radio waves to other MIMO elements. The location, size, and shape of a parasitic element ensure the diversion of certain undesired current [44].

The antenna is printed on a 95 mm × 60 mm FR4 substrate board with a relative permittivity of 4.4 and a thickness of 0.8 mm. The antenna has two symmetric slot

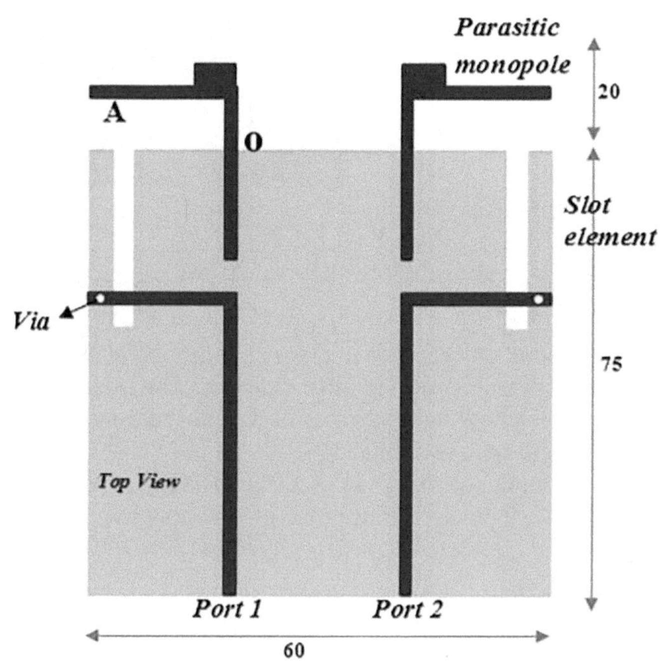

FIGURE 3.4l Geometry of the dual-slot-element monopole antenna (Image courtesy: [45]).

elements, as seen in Figure 3.4l, and a 50-Ω microstrip line with one via-grounded end feeds each slot element. The ground plane, which measures 75 mm × 60 mm, is printed on the substrate's bottom layer. The addition of two parasitic monopoles reduces mutual coupling. To make the design simpler, the parasitic monopoles' metal strips were made to 1.5 mm in width, similar to the feeding lines. One rectangular component, referred to as a stub, is inserted at the corner of each parasitic monopole in order to modify matching. The entire length of each parasitic monopole from point "O" to "A" is approximately 26 mm, and the length of each slot element is 25.3 mm, both implying a λ/4 mode [45].

Assuming that the conductor diameter of the two dipoles is tiny and their electrical length is equal to half-wavelength (λ/2); two dipoles are positioned closely together and parallel to each other. Port A is connected to the current source J, while Port B of the dipole B is matched with a load of 50 ohms. Assume that dipole A receives a sinusoidal excitation signal and current through both the legs of dipole flows in the same direction. The strong coupling current (*I coupled-direct*) on dipole B is induced around 180 degrees out of phase with the excited current on dipole A (*I Excited*) because dipole B is located extremely near to dipole A, as seen in Figure 3.4m. The stimulated current and the coupled current are exactly proportionate. Arranging the two dipoles to be orthogonal to one another or separating them by a significant distance is the simplest and most straightforward method of reducing the coupling current. However, such arrangements require a large area [46].

To eliminate out the original coupling, a parasitic element C is placed between the two dipoles. The element C gets excited by the current (*I Excited*) and creates another coupled current with phase and magnitude differences and thereby reduces the dipole B coupling current as shown in Figure 3.7m.

Two symmetric antenna components supplied by ports 1 and 2 are shown in Figure 3.4n [47]. There is a λ/2 gap between the antenna components. Every antenna element is made up of a radiator in the shape of a staircase that has a bottom shorted strip. It is clear from the image that poor isolation results from increased mutual coupling between radiators because of the decreased port-to-port distance (without

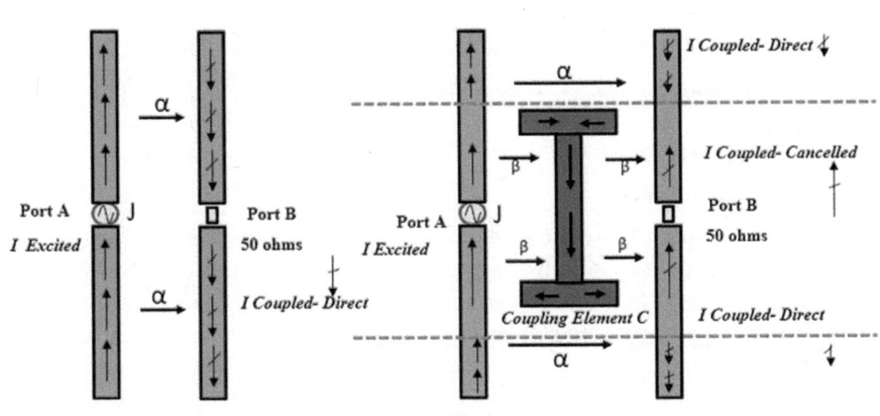

FIGURE 3.4m Two dipole antennas with coupling element (Image courtesy: [46]).

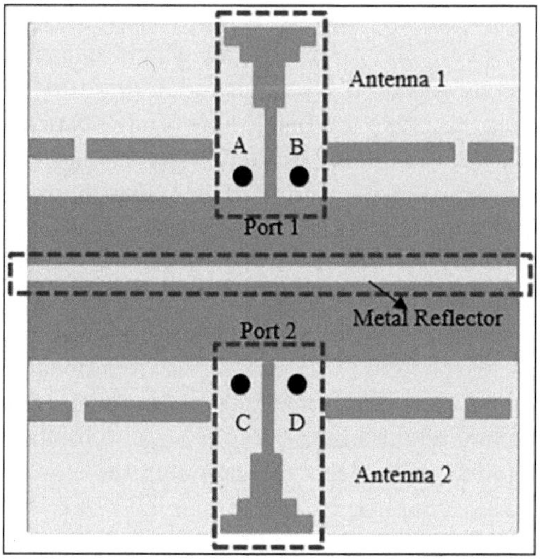

FIGURE 3.4n Staircase-shaped MIMO and metal reflector for mutual coupling reduction (Image courtesy: [47]).

parasitic element). Hence, to reduce mutual coupling a metallic strip 0.4 mm thick is placed at a distance of 0.10–0.25λ from the feeding point. This metal strip acts as EM wave reflector between both radiating elements. The radiating elements and metal strip are printed on the top side substrate material. Isolation of -20dB is achieved using metal reflector technique.

In a broadband design, surface currents were created via a parasitic structure between the MIMO components, canceling out the existing coupling channel and reducing ohmic losses in the process [48]. By positioning a horizontal parasitic element at the antenna element's bottom side, the mutual coupling was decreased. By acting as a passive resonator, this component lessens the coupling between the antenna parts. The design's benefits were omnidirectional radiation patterns, 15dB isolation, and a peak gain of 5.75 dBi [49]. The performance of the system can be enhanced by adding a meandering line cut in between two slot antennas. It contributes to the array antenna's increased efficiency and bandwidth. The omnidirectional radiation patterns yield a 15dB isolation and an ECC of less than 0.5 [50].

3.7.6 SRR/CSRR Approach

Split ring resonator (SRR) is a resonant approach based metamaterial structure (MTM). There are two approaches used to categorize metamaterial structures named as resonant approach and transmission line approach. Metamaterial can be broadly defined as artificially designed homogeneous structure which is designed to achieve special properties like negative refractive index [50–53]. Split ring resonator is made up of two concentric metallic rings in a square or circular shape, etched on the ground plane or patch and a small gap at opposite ends of the rings. The rings act as inductance (L) and the splits are the structural inhomogeneities act as capacitance (C) produces radiation at particular frequency called resonant frequency (f_r). It is given by the equation (3.17)

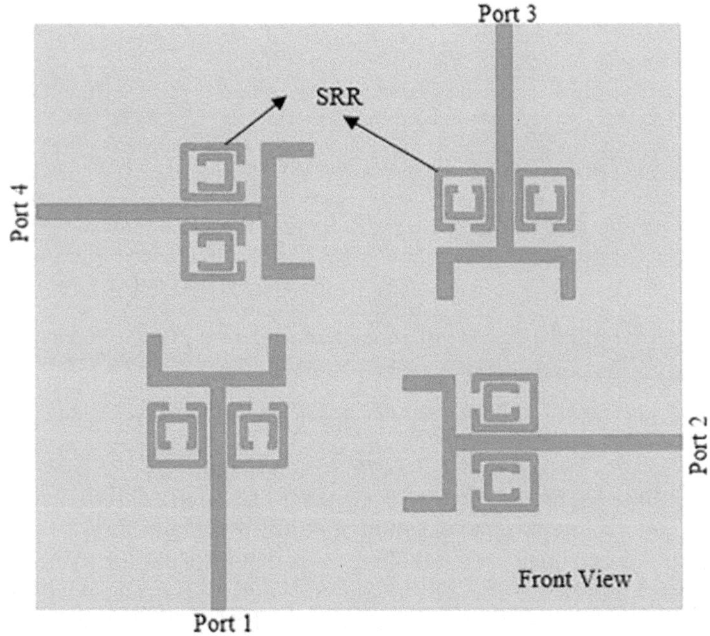

FIGURE 3.4o 4-Port MIMO antenna using square SRR (Image courtesy: [55]).

$$f_r^1 = \frac{1}{2\pi\sqrt{LC}} \tag{3.17}$$

The gap between the rings offers high capacitance. Because of the inverse relationship between the capacitance and the resonant frequency, resonance happens at frequencies far higher than its dimension. Quality rises because the resonant wavelength is significantly higher than the SRR's dimension. In the case of metamaterials (SRR), the magnetic and electric dipoles created in the rings as a result of electromagnetic field interactions provide unique negative material characteristics such as electric permittivity, magnetic permeability, and refractive index [54].

The electromagnetic wave propagation through SRR exhibits novel phenomena as a result of these negative material properties. Because of its unique properties, SRR (metamaterial) may be used to increase isolation in MIMO systems and manufacture smaller antennas. The most often used split ring resonators are the square split ring resonator, circular split ring resonator, and complementary split ring resonator (CSRR).

Figure 3.4o depicts a dual-band SRR-based antenna geometry in a MIMO configuration that radiates at 28 GHz and 36 GHz. A basic topology forms the basis of the single antenna's design. A significant problem in the design of MIMO antennas is the reduction of mutual coupling between the antenna components, which is achieved by placing the antenna elements orthogonally to one another. It is composed of two square split ring resonators (SSRR) symmetrically positioned on either side of the

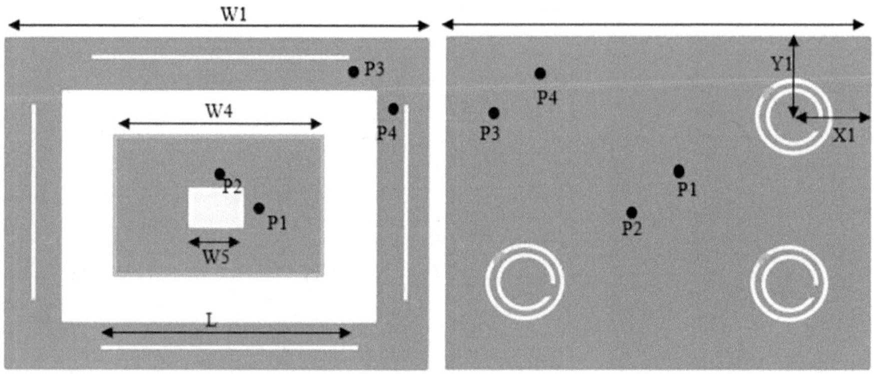

FIGURE 3.4p Quad port MIMO antenna using CSRR (Image courtesy: [56]).

feed line and four U-shaped monopole elements. Envelope correlation coefficient (ECC) 0.008, total gain 4.81, and isolation of -70dB is achieved [55].

Figure 3.4p shows a quad port MIMO antenna that resonates in several transverse magnetic modes and has a circular square ring radiating patch. Additionally, the ground plane has three CSRR resonant structures etched in it. At 2.45 GHz, which is the resonance frequency, these CSRRs exhibit a sub-wavelength value. Since there is less energy stored in the metallic portions, the concentric square ring's objective is to increase bandwidth. The inner ring is enclosed by the second square, which also reduces the overall size of the construction. In order to excite higher order modes orthogonally and obtain a greater degree of isolation between them, the feed location is chosen accordingly. Also, surface wave propagation is high for higher modes with non-zero substrate thickness. In order to enhance this performance, reactive loading in antennas is implemented by etching a slot on each square ring. The slots have a length of $\lambda/2$ and a resonance frequency of 2.45 GHz. Additionally, the reciprocal coupling reduction between higher order modes created in rings is facilitated by this slot. Around the antenna's corners, at a 45-degree angle, are three CSRRs etched into the ground plane.

Higher coupling current is produced by CSRRs, which alters the standing wave distribution. This also demonstrates the resonant characteristic in fields with vertical polarization. As a result, the antenna's two P2 and P3 ports handle horizontally polarized electromagnetic waves, while the other two P1 and P4 ports handle vertically polarized signals. Because of this design, there is less wave interaction between them, which leads to good isolation [56].

Figure 3.4q illustrates an efficient method of lowering the mutual coupling between millimeter-wave dielectric resonator antennas (DRAs) by use of a metamaterial polarization-rotator (MPR) wall. A split ring resonator (SRR) on one side, a coupling strip in the middle, and another SRR on the other side make up the periodic structure of the metamaterial unit cell that makes up the MPR wall [57]. The SRR structure is frequently employed in the design of metamaterials in order to achieve negative effective permeability. Inductance is produced by the ring, while capacitance is produced by the space between the rings. The basic mode, known as the hybrid electromagnetic mode or $HEM_{11\delta}$, is stimulated by means of a slot located in

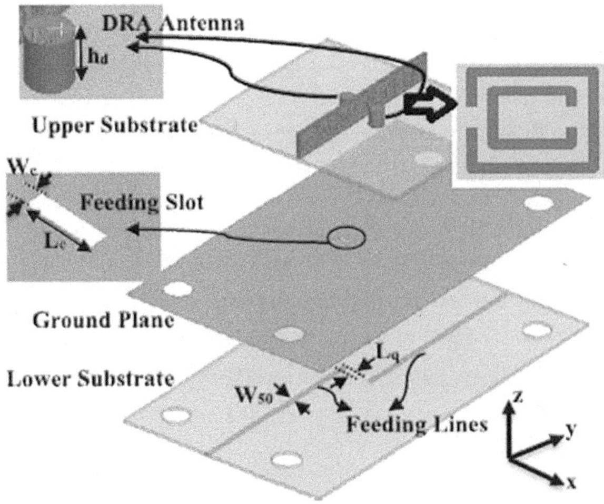

FIGURE 3.4q Metamaterial polarization rotator-based DRA-MIMO antenna (Image courtesy: [57]).

TABLE 3.2
Comparison of Mutual Coupling Reduction Techniques

Techniques	Advantages	Disadvantages
Diversity techniques	• Simple • Improves isolation • Link reliability • Complementary pattern • Better efficiency	• Consumes large space • Polarization mismatch
Defected ground structure	• Offers compact antenna size • Results in broadband • Good spatial diversity	• Low gain • High back lobe radiation
Neutralization line	• Easy to design • Connected between feeding strips • Reduces coupling current • Good impedance matching	• Placement of neutralization line is challenging
Parasitic element	• Produces less ohmic losses	• Highly sensitive to the position of parasitic elements
Split ring resonator	• Significant antenna size reduction • Channel capacity improvement	• Suitable only for narrow band applications

the center of the dielectric resonator in the DRA antenna. When the E-field rotates 90 degrees, the coupling between the two antennas becomes orthogonal, which results in a reduction in mutual coupling.

Table 3.2 summarizes the advantages and disadvantages of different coupling reduction techniques discussed in this chapter.

3.8 CONCLUSION

A MIMO antenna system is the essential part in modern day wireless communication like 4G and 5G. The need for a multiple antenna system is unavoidable, and when MIMO is added, the design becomes challenging due to the mutual coupling effect. In this chapter, the fundamental concepts of MIMO antenna including all the performance metrics that characterize its behaviors are discussed. Several mutual coupling reduction techniques have also been discussed to help the students, researchers, and antenna designers to be familiar with it. The second part of the chapter is focused on the MIMO antenna design with different isolation techniques like DGS, NL, EBG, parasitic element, metamaterial, and diversity techniques to control the effect of mutual coupling. Finally, the merits and demerits of various coupling reduction techniques are discussed.

REFERENCES

[1] J. Paulraj, D. A. Gore, R. U. Nabar, and H. Bolcskei, "An Overview of MIMO Communications-A Key to Gibabit Wireless," *Proceedings of the IEEE*, vol. 92, no. 2, pp. 198–218, February 2004.

[2] Rakhesh Singh Kshetrimayum, *Fundamentals of MIMO Wireless Communications*. Cambridge University Press, 2017.

[3] D. Gesbert, M. Kountouris, R. W. Heath, Jr., C.-B. Chae, and T. Sälzer, "Shifting the MIMO Paradigm: From Single User to Multiuser Communications," *IEEE Signal Processing Magazine*, vol. 24, no. 5, pp. 36–46, October 2007.

[4] E. Biglieri, R. Calderbank, A. Constantinides, A. Goldsmith, A. Paulraj, and H. Poor, "Introduction," in *MIMO Wireless Communications* (pp. 1–23). Cambridge University Press, 2007. doi:10.1017/CBO9780511618420.003.

[5] M. S. Sharawi, "A Dual-Band Dual-Element Compact MIMO Antenna System for Mobile 4G Terminals," *Microwave and Optical Technology Letters*, vol. 55, pp. 325–329, 2013.

[6] J. R. James and P. S. Hall, *Handbook of Microstrip Antennas* (vol. 2, pp. 1079–1151). Short Run Press Ltd, 1989.

[7] S. Dey and N. C. Karmakar, "Design of Novel Super Wide Band Antennas Close to the Small Antenna Limitation Theory," *IEEE MTT-S International Microwave Symposium (IMS2014)*, pp. 1–4, June 2014.

[8] N. C. Karmakar, "Investigations into a Cavity Backed Circular Patch Antenna," *IEEE Transactions on Antennas and Propagation*, vol. 50, no. 12, pp. 1706–1715, December 2002.

[9] N. C. Karmakar, P. Hendro, and L. S. Firmansyah, "Shorting Strap Tunable Single Feed Dual-Band PIFA," *IEEE Microwave and Wireless Component Letters*, vol. 13, no. 1, pp. 13–15, January 2003.

[10] L. Zheng and D. N. C. Tse, "Diversity and Multiplexing: A Fundamental Tradeoff in Multiple-Antenna Channels," *IEEE Transactions on Information Theory*, vol. 49, no. 5, pp. 1073–1096, May 2003.

[11] A. Lozano and N. Jindal, "Transmit Diversity vs. Spatial Multiplexing in Modern MIMO Systems," (PDF). *IEEE Transactions on Wireless Communications*, vol. 9, no. 1, pp. 186–197, 2010.

[12] J. Tao and Q. Feng, "Compact Ultra-Wideband MIMO Antenna with Half-Slot Structure," *IEEE Antennas and Wireless Propagation Letters*, vol. 16, pp. 792–795, 2016.

[13] A. F. Molisch, *Wireless Communications* (2nd ed.). Wiley Publication Ltd, 2010.

[14] C. A. Balanis, *Advanced Engineering Electromagnetics* (2nd ed.). John Wiley and Sons Publication, 2012.

[15] X. Chen, S. Zhang, and Q. Li, "A Review of Mutual Coupling in MIMO Systems," *IEEE Access*, vol. 6, pp. 24706_24719, 2018.

[16] J. L. Allen and B. L. Diamond, *Mutual Coupling in Array Antennas* (Tech. Rep. TR-424). MIT Lincoln Lab, October 1966.

[17] H. T. Hui, M. E. Bialkowski, and H. S. Lui, "Mutual Coupling in Antenna Arrays," *International Journal of Antennas and Propagation*, vol. 2010, Art. no. 715138, February 2010.

[18] X. Chen, "Antenna Correlation and Its Impact on Multi-Antenna System," *Progress in Electromagnetics Research B*, vol. 62, pp. 241_253, January 2015.

[19] P. Sharma, R. N. Tiwari, P. Singh, P. Kumar, and B. K. Kanaujia, "MIMO Antennas: Design Approaches, Techniques and Applications," *Sensors (Basel)*, vol. 22, no. 20, p. 7813, 2022 October 14. doi:10.3390/s2220781.

[20] I. Nadeem and D.-Y. Choi, "Study on Mutual Coupling Reduction Technique for MIMO Antennas," *IEEE Access*, vol. 7, pp. 563–586, 2019. doi:10.1109/ACCESS.2018.2885558.

[21] Sanjay Chouhan, Debendra Panda, Manish Gupta, and Sarthak Singhal, "Multiport MIMO Antennas with Mutual Coupling Reduction Techniques for Modern Wireless Transreceive Operations: A Review," *International Journal of RF and Microwave Computer-Aided Engineering*, vol. 28, p. e21189, 2017. doi:10.1002/mmce.21189.

[22] R. G. Vaughan and J. B. Anderson, "Antenna Diversity in Mobile Communications," *IEEE Transactions on Vehicular Technology*, vol. 36, no. 4, pp. 149–172, 1987.

[23] M. P. Karaboikis, V. C. Papamichael, G. F. Tsachtsiris, C. F. Soras, and V. T. Makios, "Integrating Compact Printed Antennas onto Small Diversity/MIMO Terminals," *IEEE Transactions on Antennas and Propagation*, vol. 56, no. 7, pp. 2067–2078, July 2008.

[24] W. C. Y. Lee and Y. U. S. Yeh, "Polarization Diversity System for Mobile Radio," *IEEE Transactions on Communications*, vol. 20, no. 5, pp. 912–923, 1972.

[25] L. Malviya, R. K. Panigrahi, and M. V. Kartikeyan, "A 2 × 2 Dualband MIMO Antenna with Polarization Diversity for Wireless Applications," *Progress in Electromagnetics Research C*, vol. 61, pp. 91–103, 2016.

[26] L. Malviya, R. K. Panigrahi, and M. V. Kartikeyan, "Pattern Diversity-Based MIMO Antenna for Low Mutual Coupling," in *IEEE Applied Electro-Magnetic Conference (AEMC), Guwahati, India* (pp. 96–97). IEEE, 2015.

[27] F. Yang and Y. Rahmat Samii, *Electromagnetic Band Gap Structures in Antenna Engineering*. Cambridge University Press, 2009.

[28] Ashwini Arya, Kartikeyan Machavaram, and A. Patnaik, "Defected Ground Structure in the Perspective of Microstrip Antennas: A Review," *Frequenz*, vol. 64, pp. 79–84, 2010. doi:10.1515/FREQ.2010.64.5-6.79.

[29] K. Wei, J. Li, L. Wang, Z. Xing, and R. Xu, "S-Shaped Periodic Defected Ground Structures to Reduce Microstrip Antenna Array Mutual Coupling," *Electronics Letters*, vol. 52, no. 15, pp. 1288–1290, 2016.

[30] Basma M. Yousef, Allam M. Ameen, Arpan Desai, Heng-Tung Hsu, Vigneswaran Dhasarathan, Ahmed A. Ibrahim, "Defected Ground Structure-Based Wideband Circularly Polarized 4-Port MIMO Antenna for Future Wi-Fi 6E Applications," *AEU International Journal of Electronics and Communications*, vol. 170, p. 154815, 2023, ISSN:1434-8411. doi:10.1016/j.aeue.2023.154815.

[31] Y.-S. Chen and C.-P. Chang, "Design of a Four-Element Multiple-Input_Multiple-Output Antenna for Compact Long-Term Evolution Small-Cell Base Stations," *IET Microwaves, Antennas & Propagation*, vol. 10, no. 4, pp. 385–392, 2016.

[32] R. Anitha, V. P. Sarin, P. Mohanan, and K. Vasudevan, "Enhanced Isolation with Defected Ground Structure in MIMO Antenna," *Electronics Letters*, vol. 50, no. 24, pp. 1784–1786, 2014.

[33] C. R. Jetti and V. R. Nandanavanam, "Trident-Shape Strip Loaded Dual Band-Notched UWB MIMO Antenna for Portable Device Applications," *AEU International Journal of Electronics and Communications*, vol. 83, pp. 11–21, January 2018.

[34] L. Malviya, R. Panigrahi, and M. Kartikeyan, "MIMO Antennas with Diversity and Mutual Coupling Reduction Techniques: A Review," *International Journal of Microwave and Wireless Technologies*, vol. 9, no. 8, pp. 1763–1780, 2017. doi:10.1017/S1759078717000538.

[35] S. Wen Su, C. Tse Lee, and Fa Shian Chang, "Printed MIMO Antenna System Using Neutralization Line Technique for Wireless USB Dongle Applications," *IEEE Transactions on Antennas and Propagation*, vol. 60, no. 2, pp. 456–463, February 2012.

[36] W. A. E. Ali and A. A. Ibrahim, "A Compact Double-Sided MIMO Antenna with an Improved Isolation for UWB Applications," *AEU International Journal of Electronics and Communications*, vol. 82, pp. 7–13, December 2017.

[37] Y. Ou, X. Cai, and K. Qian, "Two-Element Compact Antennas Decoupled with a Simple Neutralization Line," *Progress in Electromagnetics Research*, vol. 65, pp. 63–68, 2017.

[38] R. Liu, X. An, H. Zheng, M. Wang, Z. Gao, and E. Li, "Neutralization Line Decoupling Tri-Band Multiple-Input Multiple-Output Antenna Design," *IEEE Access*, vol. 8, pp. 27018–27026, 2020. doi:10.1109/ACCESS.2020.2971038.

[39] Yang Fan and Y. Rahmat-Samii, "Reflection Phase Characterizations of the EBG Ground Plane for Low Profile Wire Antenna Applications," *IEEE Transactions on Antennas and Propagation*, vol. 51, no. 10, pp. 2691–703, 2003.

[40] Palanisamy Prabhu and Subramani Malarvizhi, "Compact Multiple-Input Multiple-Output Antenna with Low Correlation for Ultra-Wide-Band Applications," *IET Microwaves, Antennas & Propagation*, vol. 9, no. 8, pp. 822–829, 2015.

[41] Palanisamy Prabhu and Subramani Malarvizhi, "Novel Double-Side EBG Based Mutual Coupling Reduction for Compact Quad Port UWB MIMO Antenna," *AEU—International Journal of Electronics and Communications*, vol. 109, pp. 146–156, 2019, ISSN:1434-8411.

[42] B. Mohamadzade and M. Afsahi, "Mutual Coupling Reduction and Gain Enhancement in Patch Array Antenna Using a Planar Compact Electromagnetic Bandgap Structure," *IET Microwaves, Antennas & Propagation*, vol. 11, no. 12, pp. 1719–1725, 2017.

[43] H. S. Farahani, M. Veysi, M. Kamyab, and A. Tadjalli, "Mutual Coupling Reduction in Patch Antenna Arrays Using a UC-EBG Superstrate," *IEEE Antennas and Wireless Propagation Letters*, vol. 9, pp. 57–59, 2010, doi:10.1109/LAWP.2010.2042565.

[44] K. S. Min, D. J. Kim, and Y. M. Moon, "Improved MIMO Antenna by Mutual Coupling Suppression Between Elements," *European Conference on Wireless Technologies*, vol. 2005, pp. 125–128, 2005.

[45] Z. Li, Z. Du, M. Takahashi, K. Saito, and K. Ito, "Reducing Mutual Coupling of MIMO Antennas with Parasitic Elements for Mobile Terminals," *IEEE Transactions on Antennas and Propagation*, vol. 60, no. 2, pp. 473–481, 2012.

[46] A. C. K. Mak, C. R. Rowell, and R. D. Murch, "Isolation Enhancement Between Two Closely Packed Antennas," *IEEE Transactions on Antennas and Propagation*, vol. 56, no. 11, pp. 3411–3419, 2008.

[47] T. K. Roshna, U. Deepak, V. R. Sajitha, K. Vasudevan, and P. Mohanan, "A Compact UWB MIMO Antenna with Reflector to Enhance Isolation," *IEEE Transactions on Antennas and Propagation*, vol. 63, no. 4, pp. 1873–1877, April 2015.

[48] R. Addaci, A. Diallo, C. Luxey, P. Le Thuc, and R. Staraj, "Dual-Band WLAN Diversity Antenna System with High Port to Port Isolation," *IEEE Antennas and Wireless Propagation Letters*, vol. 11, pp. 244–247, 2012.

[49] M. S. Khan, A. D. Capobianco, M. F. Shafique, B. Ijaz, A. Naqvi, and B. D. Braaten, "Isolation Enhancement of a Wideband MIMO Antenna Using Floating Parasitic Elements," *Microwave and Optical Technology Letters*, vol. 57, pp. 1677–1682, 2015.

[50] M. Ayatollahi, Q. Rao, and D. Wang. "A Compact High Isolation and Wide Bandwidth Antenna Array for Long Term Evolution Wireless Devices," *IEEE Transactions on Antennas and Propagation*, vol. 60, pp. 4960–4963, 2012.

[50] N. Mishra and R. K. Chaudhary, "A Compact CPW-Fed CRR Loaded Four Element Metamaterial Array Antenna for Wireless Applications," *Progress in Electromagnetics Research*, vol. 159, pp. 15–26, 2017.

[51] N. Mishra, A. Gupta and R. K. Chaudhary, "A Compact CPW-Fed Wideband Metamaterial Antenna Using Ω-Shaped Interdigital Capacitor for Mobile Applications," *Microwave and Optical Technology Letters*, vol. 57, pp. 2558–2562, 2015.

[52] N. Mishra and R. K. Chaudhary, "A Miniaturized ZOR Antenna with Enhanced Bandwidth for WiMAX Applications," *Microwave and Optical Technology Letters (MOTL), Wiley Journals*, vol. 58, no. 1, pp. 71–75, 2016.

[53] N. Mishra and R. K. Chaudhary "A Miniaturized Directive High Gain Metamaterial Antenna using ELC Ground for Wi-MAX Application," *International Journal of Electronics Letters*, vol. 7, no. 1, pp. 68–76, 2019.

[54] A. K. Singh, M. P. Abegaonkar, and S. K. Koul, *Metamaterials for Antenna Applications* (1st ed.). CRC Press, 2021. doi:10.1201/9781003045885.

[55] Utkarsha Verma, Maharana Pratap Singh, and Saptarshi Ghosh, "Dual Band Millimeter Wave MIMO Antenna Loaded with SRRs for 5G Applications," in *Wireless Antenna and Microwave Symposium (WAMS)* (pp. 1–3). IEEE, 2022. doi:10.1109/WAMS54719.2022.9847846.

[56] A. Ramachandran, S. Valiyaveettil Pushpakaran, M. Pezholil, and V. Kesavath, "A Four-Port MIMO Antenna Using Concentric Square-Ring Patches Loaded with CSRR for High Isolation," *IEEE Antennas and Wireless Propagation Letters*, vol. 15, pp. 1196–1199, 2016. doi:10.1109/LAWP.2015.2499322.

[57] M. Farahani, J. Pourahmadazar, M. Akbari, M. Nedil, A. R. Sebak, and T. A. Denidni, "Mutual Coupling Reduction in Millimeter-Wave MIMO Antenna Array Using a Metamaterial Polarization-Rotator Wall," *IEEE Antennas and Wireless Propagation Letters*, vol. 16, pp. 2324–2327, 2017. doi:10.1109/LAWP.2017.2717404.

4 Design and Simulation Study of Bi-Frequency and Dual-Band Magnetically Insulated Line Oscillator (MILO)

Arjun Kumar, Prabhakar Tripathi, Vineet Singh and Rajnish Kumar Singh

4.1 INTRODUCTION

High-power microwave devices that are capable of generating dual frequencies have been considered as a new research topic in recent years [Benford *et al.* (2007)]. Potential applications of such devices are in plasma heating, large-range energy transmission, linearly charged particle accelerators, and electronic warfare systems [Tang *et al.* (2012)]. Various HPM sources developed in recent years to generate dual frequency are mainly Magnetically Insulated Line Oscillator (MILO), Transit Time Oscillator (TTO) and Cherenkov Oscillator, and Relativistic Backward Wave Oscillator (RBWO) [Yang *et al.* (1999), He *et al.* (2011) and Ju *et al.* (2014)]. Among these devices, the researchers are more attracted towars MILO due to its self-magnetic insulation, light weight, and compact size features [Lemke *et al.* (1997)].

The dual-frequency generation via the HPM source mainly uses two types of design methodology. In the first type of design, two different HPM device concepts are used in a single HPM device. Yang *et al.* first proposed a dual-frequency Cherenkov Oscillator considering RBWO and the relativistic Orotron concept in a single device [Yang *et al.* (1999)]. The proposed device had used a T-type structure, and the output microwave was taken out perpendicular to each other. Zhang *et al.* reported experimentally using the MILO and virtual cathode oscillator (VCO) concept in a single device that the device generates three frequencies (i.e., 2.1 GHz, 3.8 GHz, and 4.1 GHz) with total efficiency of ~13.44% [Zhang *et al.* (2015)]. Xiao *et al.* reported simulation work by using the concept of coaxial TTO and MILO in a single device which generates RF power in three frequencies with an overall efficiency of 25.4% [Xiao *et al.* (2009)]. The use of different device concepts in a single device makes the design more complex and RF extraction becomes very difficult.

 DOI: 10.1201/9781032656021-4

The second type of design method uses two different interaction structures which are separated either axially or azimuthally. Chen *et al.* reported experimentally an azimuthally partitioned bi-frequency MILO which generates 620 MW of RF power in L-band at two frequencies (i.e., 1.27 GHz and 1.49 GHz) with an overall efficiency of 4.3% [Chen *et al.* (2009)]. Azimuthal partitioning of the interaction structure produces asymmetric modes inside the device which causes a drop in overall device efficiency. A dual-frequency coaxial RBWO designed by Tang *et al.* generates two frequencies as 10.06 GHz and 10.46 GHz with overall efficiency of 16.1% [Tang *et al.* (2012)]. Jin-Chuan *et al.* have used two different interaction structures (i.e., which are separated by a baffle cavity) in a single MILO device and reported RF power generation in *C*-band and *X*-band with an efficiency of 11.8% [Jin *et al.* (2009)]. A dual-band MILO generating frequencies in two bands (i.e., L-band (1.72 GHz) and *Ku*-band (14.6 GHz)) has been designed by J-C Ju *et al.* which provided overall efficiency of the device as 12.8% [Ju *et al.* (2014)]. But such literature lacks proper interpretation of device design methodology and proper analysis to support device beam-wave interaction mechanisms. To bridge this literature gap, the dual-band MILO device has been designed with a proper explanation of device design methodology and beam-wave interaction analysis in this chapter. The MILO device can generate dual-frequency (i.e., in *S*-band (2 GHz) and *Ku*-band (14 GHz)) using two separate interaction structures. The generation of dual band microwave through a single device shows high pf^2 factors where f stands for frequency and p for microwave power [Ju *et al.* (2014)].

In this chapter, the device design parameters and oscillation conditions for bi-frequency MILO have been derived in section 4.2. Section 4.4 explains the simulation of the different bi-frequency MILO which include cold simulation and PIC simulation of *S*-band bi-frequency MILO along with *S/Ku* dual-band MILO. Finally, in the last section, conclusion of the chapter is presented.

4.2 DESIGN PROCEDURE OF BI-FREQUENCY MILO

The design of a bi-frequency or dual-band MILO starts with a selection of bands in which the RF should be generated. The cathode radius (*rc*) is generally considered as the design normalization parameter. The first design parameter, disc inner radius (i.e., *rd*), is selected here for calculation. The ratio of disc inner radius and cathode radius (i.e., *rd/rc*) is calculated using the input anode current (*Ia*), and the expression is given as [Lemke *et al.* (1997)]:

$$I_a = \frac{I_\alpha}{\ln\left(r_d / r_c\right)} \gamma_a \ln\left(\gamma_a + \sqrt{\gamma_a^2 - 1}\right) \tag{4.1}$$

here, $I_\alpha = 8.5$ kA and $\gamma_a = 1 + V_0 / 511$. The interaction cavity for two different frequencies desired in the same band or a different band is designed generally for π-mode of operation and the depth of the resonant cavity is considered as $\lambda/4$. Thus the outer anode radius can be designed as [Ju *et al.* (2009)]:

$$r_a = r_d + \lambda / 4 \tag{4.2}$$

here, λ is the operating wavelength. The choke radius (*rch*) is selecting such that the cut-off frequency of the choke section becomes ~80% of the operating frequency for the different band of the device [Kumar *et al.* (2019)]. Similarly, the extractor radius (*rE*) should be selected such that it helps in matching the impedance between the interaction section and the output section of the device [Lemke *et al.* (1997)]. For the given input voltage (*V0*) and current (*Ia*), the drift velocity (*ve*) of electron between the anode and cathode structure is calculated as [Dwivedi and Jain (2013), and Cousin (2005)]:

$$v_e = \frac{2\pi\varepsilon_0 c^2}{\ln(r_a / r_c)} \frac{V_0}{I_a} \tag{4.3}$$

here, c is the speed of light. The efficient beam-wave interaction occurs by selecting accurate disc loading periodicity (*L*). The interaction starts between RF wave and electron beam when the drift velocity of electron and the phase velocity (i.e., *vp*) of the desired mode becomes equal. The phase velocity can be calculated as [Dwivedi and Jain (2013), and Cousin (2005)]:

$$v_p = \omega / \beta_0 = 2Lf \tag{4.4}$$

here, $\beta_0$is the phase constant. The self-magnetic insulation inside the MILO is the crucial factor of the designing and it develops due to the DC current generated inside the device. The required magnetic field generates when the DC current becomes greater than or equal to the critical current (i.e., *Icr*) which is calculated as [Lemke *et al.* (1997), and Cousin (2005)]:

$$I_{cr} = \frac{4\pi m_0 c^3 \varepsilon_0 \sqrt{\gamma_a^2 - 1}}{e \ln(r_a / r_c)} \tag{4.5}$$

here, *m0* and *e* are the mass and charge of an electron, respectively, and ε_0 is the permittivity. This critical current (i.e., *Icr*) generates the cut-off magnetic field (i.e., *Bc*) which is calculated as [Cousin (2005)]:

$$B_c = \frac{\mu_0 I_{cr}}{2\pi r_c \ln(r_a / r_c)} \tag{4.6}$$

After the calculation of the magnetic field, it becomes important to verify the region of operation of the device. The region of operation is defined between the Hull cut-off voltage (i.e., *VH*) and threshold voltage (i.e., *VBH*) which can be found by these expressions [Cousin (2005)]:

$$V_H = \frac{m_0 c^2}{e} \left(\left[1 + \left(\frac{e B_\theta r_d}{m_0 c} \right)^2 \right]^{1/2} - 1 \right) \tag{4.7}$$

and,

$$V_{BH} = B_{\theta} r_d v_{\phi} - \frac{m_0 c^2}{e}\left[1-\left(1-\frac{v_{\phi}^2}{c^2}\right)^{1/2}\right] \tag{4.8}$$

The design of the segregation cavity is considered one of the most important parts of any dual-frequency device. The single band bi-frequency MILO does not require any segregation cavity as the two frequencies generated inside such MILO device are in the same band and their frequency difference is not very high. The depth and length of the segregation cavity are selected according to the next higher-order mode (i.e., TM_{02} mode) of the main operating mode (i.e., TM_{01} mode) of the device. The segregation cavity oscillates toward the higher-order mode (i.e., TM_{02} mode) and cut-off the main operating mode (i.e., TM_{01} mode) in such a way that it reflected back in the forward direction (i.e., in the direction of propagating wave). The cavity radius (*rw*) of the segregation cavity is calculated as [Davis and Gibson (2006)]:

$$r_w = \frac{\chi_{np} c}{2\pi f_r} \tag{4.9}$$

here, *fr* is the resonating frequency and χ_{np} is the modal root of the nth order Bessel–Neumann combination defined as [Lemke (1989)]:

$$J_n(k_c r_c)Y_n(k_c r_w) - J_n(k_c r_c)Y_n(k_c r_w) = 0 \tag{4.10}$$

Finally, using the above described expressions one can calculate the design parameters for desired frequencies either in single band or dual-band.

4.3 ANALYSIS

In the dual-band MILO device, two different interaction structures (i.e., SWS1 for *S*-band and SWS2 for *Ku*-band) are used. A modal matching technique has been used to analyze the beam-wave interaction inside different interaction structures for the dual-band MILO in this chapter. This technique uses the Vlasov-Maxwell's equations to explain the wave equation of RF inside the structure in the presence of the electron beam. The dispersion relation in the presence of the electron beam can be explained using the equation [Dwivedi and Jain (2012)]:

$$\det\left|(\varsigma^{I}_{n,m} - \varsigma^{II}_{n,m})\right| = 0, \quad \text{for} \quad (-\infty < n < \infty), \tag{4.11}$$

where,

$$\varsigma^{I}_{n,m} = \frac{d}{2}\frac{\sin(\beta_{(m-n)} d/2)}{(\beta_{(m-n)} d/2)} \times \left[\frac{X_0\{\gamma^{III}_m r\}\exp(j\omega t)}{\phi_n(1 - bJ_0\{\gamma_n r_b\}Y_0\{\gamma_n r_b\}J_0\{\gamma_n r\} + bJ_0^2\{\gamma_n r_b\}Y_0\{\gamma_n r\})}\right],$$

and,

$$\varsigma_{n,m}^{II} = \frac{1}{j\omega\varepsilon}\frac{d}{2}\frac{\sin(\beta_{(m-n)}d/2)}{(\beta_{(m-n)}d/2)}\frac{\lambda_n}{\phi_n}$$
$$\times\left[\frac{X_0'\{\gamma_m^{III}r\}\exp(j\omega t)}{(1-bJ_0\{\gamma_n r_b\}Y_0\{\gamma_n r_b\}J_1\{\gamma_n r\}+bJ_0^2\{\gamma_n r_b\}Y_0\{\gamma_n r\})}\right],$$

here, $d(=L-T)$ is the disc spacing, γ_n is the radial propagation constant in beam present region, rb is the beam outer radius, J_0 is the first kind and Y_0 is the second kind Bessel function with zero order. The term $X_0\left\{\gamma_m^{III}r\right\}$ is defined as:

$$X_0\left\{\gamma_m^{III}r\right\}=Y_0\left\{\gamma_m^{III}r_a\right\}J_0\left\{\gamma_m^{III}r\right\}-J_0\left\{\gamma_m^{III}r_a\right\}Y_0\left\{\gamma_m^{III}r\right\},$$

and, prime of the expression defined the derivative term.

The solution of the preceding expression (4.11) gives real as well as imaginary roots. The real root gives the information of device dispersion relation whereas the imaginary roots of the solution give the temporal growth rate (TGR). The TGR for a different band is calculated using the equation given as [Dwivedi and Jain (2012)]:

$$\delta\omega=-\omega+\left(\frac{\gamma_n}{j\varepsilon}\phi_n\frac{X_0'\{\gamma_m^{III}r\}}{X_0\{\gamma_m^{III}r\}}\right)^{0.5}$$
$$\times\left(\frac{(1-bJ_0\{\gamma_n r_b\}Y_0\{\gamma_n r_b\})J_0\{\gamma_n r\}+bJ_0^2\{\gamma_n r_b\}Y_0\{\gamma_n r\}}{(1-bJ_0\{\gamma_n r_b\}Y_0\{\gamma_n r_b\})J_1\{\gamma_n r\}+bJ_0^2\{\gamma_n r_b\}Y_1\{\gamma_n r\}}\right)^{0.5} \quad (4.12)$$

To identify the range of oscillation with the light line and beam line along with slow space charge wave line is determined using the following expression defined respectively as [Lemke (1989)]:

$$\omega=\beta_0 c, \quad (4.13)$$

$$\omega=\beta_0 v_e, \quad (4.14)$$

and,

$$\omega=\beta_0 v_{sc}. \quad (4.15)$$

here, $v_{sc}=v_e(1-\omega_p/(\omega\gamma_0^{1.5}))$, with, ω_p = plasma frequency.

4.4 PIC SIMULATION OF DIFFERENT BI-FREQUENCY MILO

A simulation study of bi-frequency MILO is mainly considered to validate the theory of generation of two different frequencies using a single device. Simulation of an HPM device is done in two steps. In the first step, the device design parameters obtained through design expressions are validated through cold simulation (beam-absent). The cold simulation gives information about the operating frequency and desired mode of operation supported by the structure. After the cold simulation, beam present simulation (also called PIC simulation) is performed to evaluate the RF generation capacity through the device. If the generated RF signal and efficiency obtained through simulation is not satisfactory then the optimization of design parameters is done through a different optimization technique. Here, a detailed simulation study of bi-frequency MILO generating frequencies of 3.4 GHz and 3.65 GHz (i.e., in *S*-band), and generating frequencies of 2.1 GHz and 14 GHz (i.e., in *S* and *Ku* dual-band) has been performed.

4.4.1 *S*-Band Bi-Frequency MILO

A detailed PIC simulation of bi-frequency MILO which is generating RF in *S*-band with frequencies 3.4 GHz and 3.65 GHz has been performed followed by cold simulation. The desired two frequencies can be generated by an azimuthal partition of slow-wave structure which is obtained through designing an anode cylinder with two different inner radii *rw1* (0°–180°) and *rw2* (180°–360°). This two different azimuthal cavity generates two different and stable frequencies. The schematic diagram of bi-frequency MILO with azimuthal partitioning is shown in Figure 4.1. The design methodology for bi-frequency MILO is similar to the conventional MILO design methodology. In bi-frequncy MILO as shown in figure 4.1, The first two discs in axial direction forms choke cavity, then three discs forms interaction cavity, last disc used as extractor disc. Cathode is designed with radius rc which emits the electrons for interaction. Collector is designed to collect the electrons after beam-wave interaction. Stub is used for impedance matching. The design parameters for *S* -band bi-frequency MILO is calculated and presented in table 4.1 using the above described expression given in section-4.2. The simulation starts with firstly performing beam absent simulation to obtain the operating frequency (using dispersion characteristics) at desired modes (using field pattern). Once the desired frequency and mode is

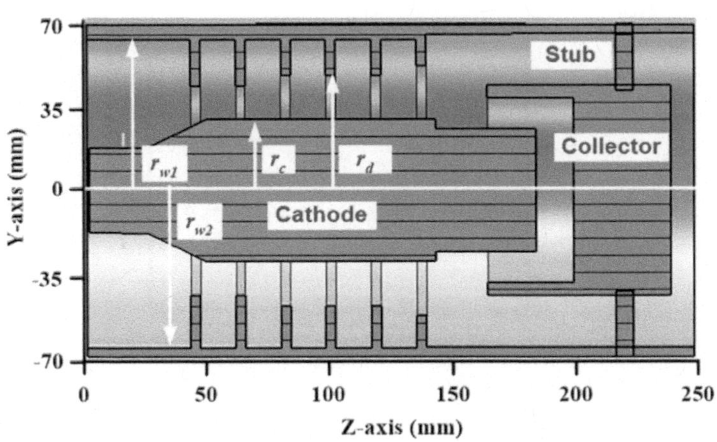

FIGURE 4.1 Azimuthal partition *S*-band bi-frequency MILO schematic diagram.

obtained for the device, the beam present simulation is carried out to obtain the RF output power from the device.

4.4.1.1 Cold Simulations (Beam Absent Simulation)

The device design parameters (shown in table 4.1) are obtained through previously explained analysis, and the cold simulation (i.e., simulation in absence of electron beam) of the unit cavity structure for *S*-band bi-frequency MILO is performed. The cold test of the interaction structure is performed using EIGENMODE solver of 'CST Microwave Studio'. The dispersion curve of the interaction structure is obtained for different modes to define the operating frequency of the device which is shown in Figure 4.2. Azimuthal partitioning of the interaction cavity always generates asymmetric modes inside the structure such as HEM_{11} mode (i.e., 2nd and 3rd mode), HEM_{21} mode (i.e., 4th and 5th mode) [Wang *et al.* (2009)]. Here, mode 1 and mode 3 are desired modes for interaction. Figure 4.3 shows the magnetic field distribution

TABLE 4.1
Design Parameters of *S*-Band Bi-Frequency MILO

Particulars	Specifications
Voltage	490 kV
Current	45 kA
Periodicity (L)	18.5 mm
Radius of Extractor Disc	52 mm
Radius of SWS Disc (r_d)	48 mm
Radius of Choke Disc	44 mm
Radius of cathode (r_c)	31 mm
Thickness of Disc (T)	4 mm
Radius of Anode	64 mm and 66 mm

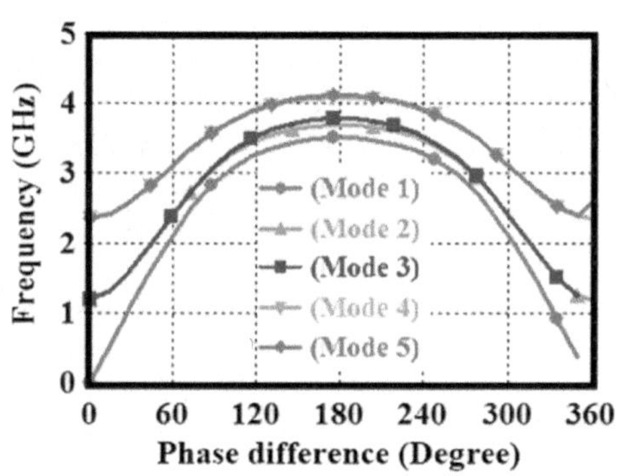

FIGURE 4.2 Dispersion curve for *S*-band bi-frequency MILO with r_{w1} = 64mm and r_{w2} = 66mm.

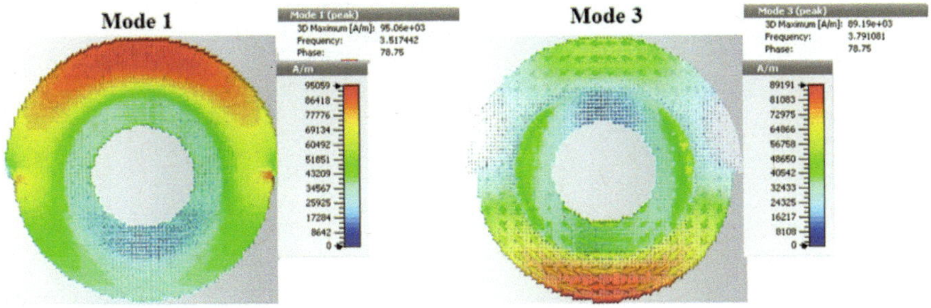

FIGURE 4.3 Magnetic field distribution vector of operating modes of the *S*-band bi-frequency MILO.

vector of two modes; one is mode 1 corresponding to cavity depth *rw1*, and the other is mode 3 corresponding to cavity depth *rw2*. It can be seen clearly that two modes are supported by two different azimuthal partitions, not a complete TE or TM mode. Here, the magnetic field distribution vector of mode 1 mainly concentrated in the azimuthal cavity of 0°–180° and that of mode 3 is concentrated in the azimuthal cavity of 180°–360°. Thus, two modes which generate two frequencies in the device are mainly mode 1 and mode 3 and the corresponding frequencies are 3.51 GHz and 3.79 GHz.

4.4.1.2 PIC Simulation (Beam Present Simulation)

The PIC simulation of *S*-band bi-frequency MILO is performed using the PIC solver of 'CST Particle Studio'. The operating DC voltage of 490 kV and current of 45 kA is used for the simulation. This high current basically generates self-magnetic insulation inside the device which results in the electrons flow in the axial direction. The axial electron flow mainly takes part in beam-wave interaction and due to different interaction cavity depth in azimuthal direction the normal TM or TE modes change to hybrid modes. The hybrid mode generation inside the device can be observed through phase space of the electron particles at a different time instant as shown in Figure 4.4. The DC voltage of 490 kV applied at the input generates 45 kA of current i.e. the input power ~22.05 GW can be seen in Figure 4.5. This device generates RF signal in two different modes as discussed earlier. The RF signal generated in two different desired modes (i.e., mode 1 and mode 3) is shown in Figure 4.6.

The frequency spectrum of the generated RF signal (which is observed at the output port) can be obtained through FFT of the signal shown in Figure 4.7. The frequency spectrum shows that the signal has two frequencies which are 3.4 GHz and 3.65 GHz.

As the RF signal generated at two different modes, so the combined effect of the RF signal can be seen in Figure 4.8. It can be observed from the figure that the overall peak RF output power of 2.86 GW and average RF power of 1.43 GW is obtained at the output which results in overall device efficiency of 6.5%.

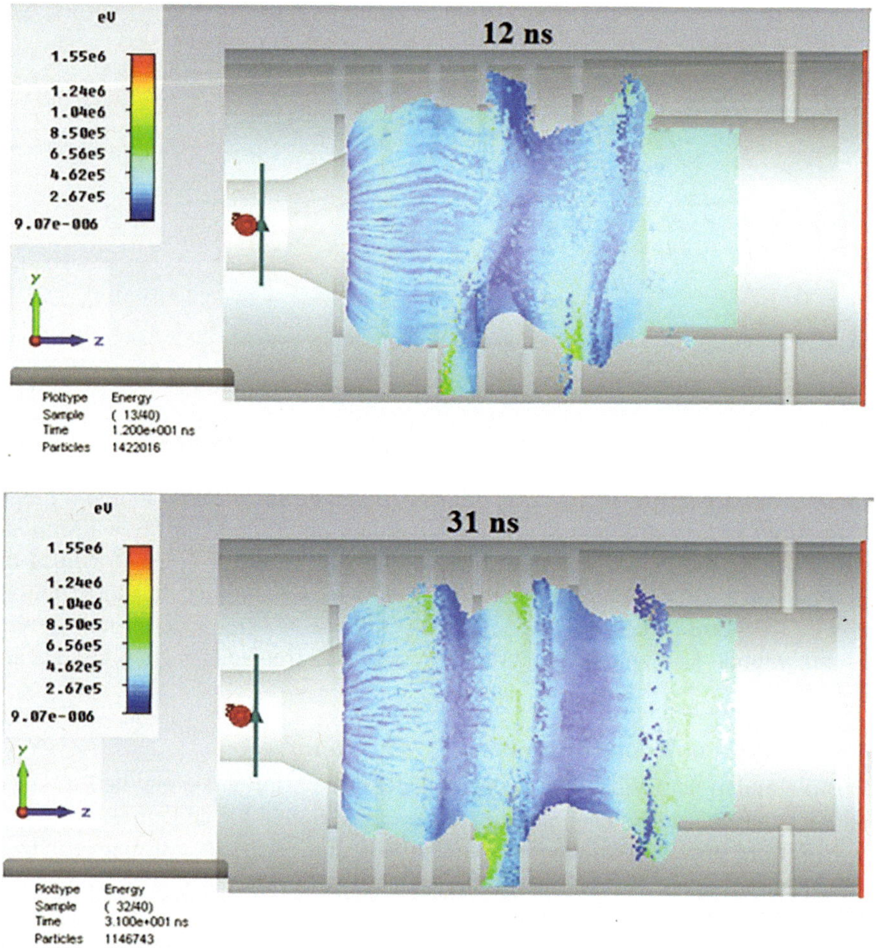

FIGURE 4.4 Phase space of particles (electrons) for *S*-band bi-frequency MILO at a different time instant of 12 ns and 31 ns.

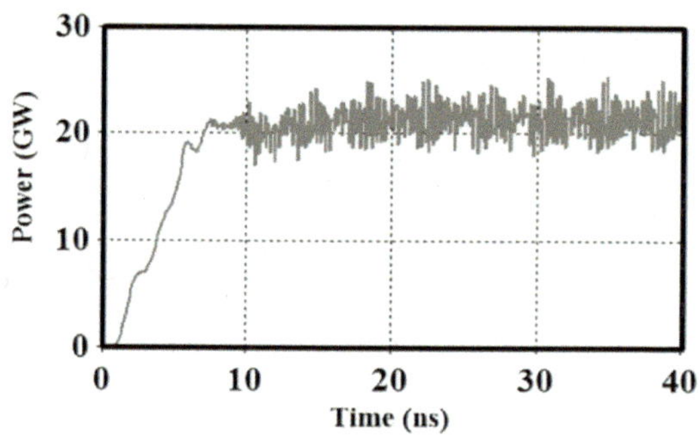

FIGURE 4.5 Temporal input power at 490 kV of voltage and 45 kA of current (also known as wave-particle power transfer).

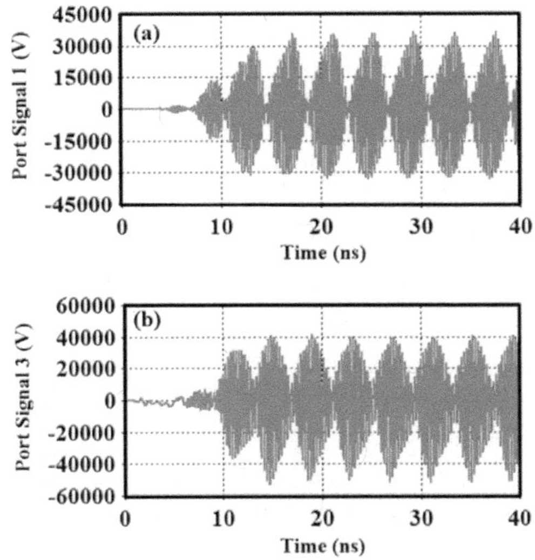

FIGURE 4.6 RF output port signal of (a) mode 1 and (b) mode 3.

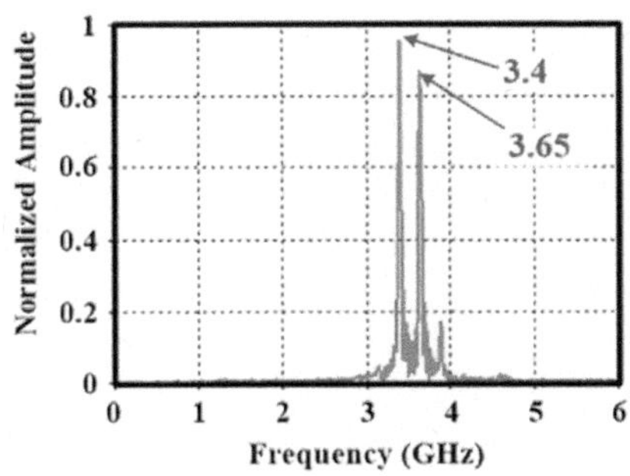

FIGURE 4.7 Frequency spectrum of the RF signal obtained at the output port.

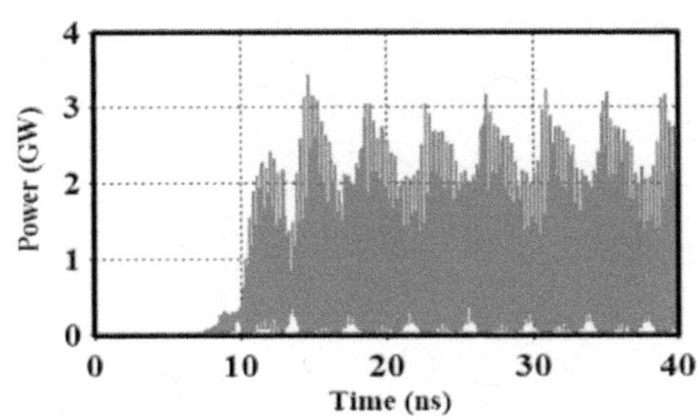

FIGURE 4.8 Combined RF output power of the two frequencies.

4.4.2 *S*/*Ku* Dual-Band MILO

A dual-band MILO which generates frequencies in two different bands (i.e., in *S*-band and *Ku*-band) has been designed using two different slow-wave structures. The dual-band MILO schematic diagram is shown in Figure 4.9. The different components of the MILO device are: SWS1 as an anode structure and C1 as a cathode structure for the *S*-band generation section. Similarly, SWS2, C2, and C3 were used as anode structure, cathode structure (i.e., for the emission of electrons), and the cathode (i.e., for load section), respectively for the *Ku*-band generation section. The self-magnetic insulation occurs inside the device in two phases. In the first phase, magnetic insulation occurs through load current (which starts at the C3 section of cathode) towards the *Ku*-band generation section. In the second phase, the *Ku*-band generation section acts as a load for the *S*-band generation section and creates magnetic insulation inside the whole structure. The RF signals which are generated through efficient beam-wave interaction at the two sections (i.e., in the *S*-band section and *Ku*-band section) can be extracted out through the two different extractor parts. To eliminate the leakage of RF from one section to another section, one segregation cavity is designed between SWS1 and SWS2. Similarly, two different stub structures i.e., Stubl and Stub2 are used for DC current feedbacking in corresponding bands.

The dual-band MILO device design methodology follows the structural design parameter explained in the section 4.2. The calculation of design parameters uses the aforementioned design expression given in Eqs. (4.1)–(4.10) and the design parameters are presented in table 4.2. Two different structure parameters are used for generating frequencies in a different band. The region of operation for the device is explained using Hull cut-off criteria and Buneman-Hartree criteria shown in Figure 4.10. The electron beam presence dispersion relation explains the RF generation and propagation inside the different coaxial interaction structures (SWS1 and SWS2) for dual-band MILO device. The imaginary roots of Eqs. (4.12) explain the instability occurs inside the different slow-wave structure due to small plasma perturbation inside the device and gives the temporal growth rate of the RF generated due to the instability.

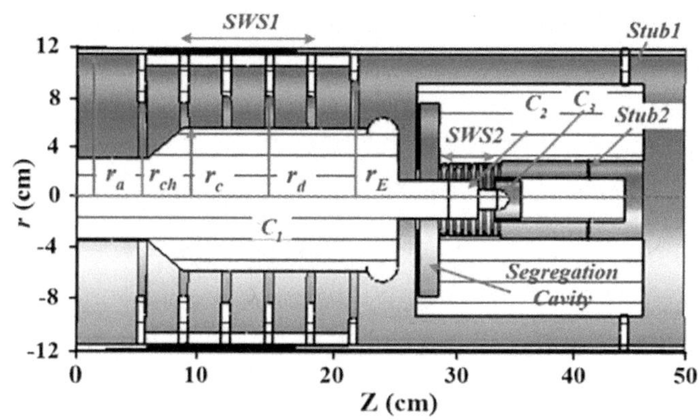

FIGURE 4.9 Dual-band MILO schematic diagram.

In the dual-band MILO device, two different interaction structures (i.e., SWS1 for *S*-band and SWS2 for *Ku*-band) are used.

With the design and beam parameters given in table 4.2, the dispersion curve and temporal growth rate are numerically calculated for different interaction structure. Figure 4.11 shows the beam present dispersion curve and temporal growth rate for two bands (i.e., for *S*-band and *Ku*-band). The numerically obtained dispersion curve is further verified using the simulation and the results found in good agreement for both the band (i.e., *S*-band and *Ku*-band). The corresponding beam line and slow space charge line explain the beam-wave interaction range whereas the light line explains that the operation of the device is restricted in the slow-wave region. It can be seen from Figure 4.11(a) that the instability region for *S*-band lies between phase difference 1.89 rad to 3.17 rad with the maximum temporal growth rate of 0.23 (1/nsec). Similarly, Figure 4.11(b) explains the instability region for *Ku*-band

TABLE 4.2
Design Parameters of *S*/*Ku* Dual-Band MILO

Particulars	Specifications	
	***S*-band**	***Ku*-band**
Voltage (*V0*)	610 kV	
Current (*Ia*)	58 kA	
Radius of Cathode (*rc*)	60 mm	15.74 mm
Radius of Anode (*ra*)	111 mm	30 mm
Radius of SWS disc (*rd*)	86 mm	26 mm
Radius of Extractor disc (*rE*)	99 mm	28 mm
Radius of Choke disc (*rch*)	80 mm	24 mm
Thickness of Disc (*T*)	7 mm	2 mm
Periodicity (*L*)	35 mm	6 mm

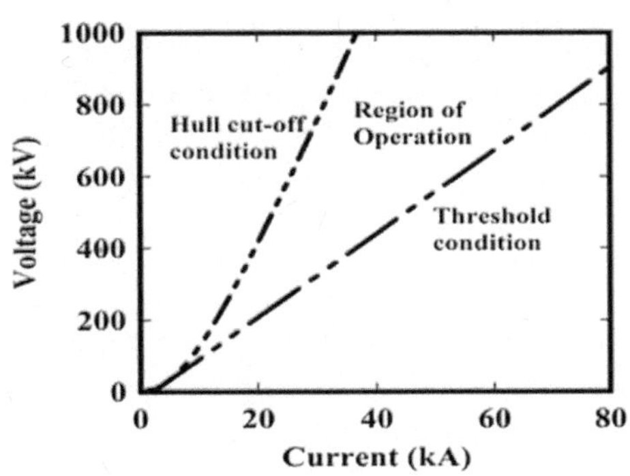

FIGURE 4.10 Region of operation curve defining Hull cut-off condition and Buneman Hartree threshold criteria for dual-band MILO.

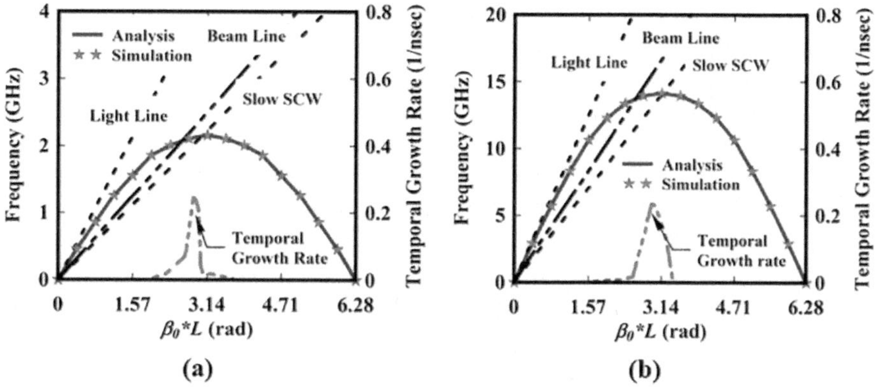

FIGURE 4.11 Dispersion curve and its temporal growth rate for fundamental mode (i.e., TM_{01} mode) (a) of *S*-band and (b) of *Ku*-band.

lies between phase difference 2.5 rad to 3.3 rad with the maximum temporal growth rate of 0.24 (1/nsec). The oscillation frequencies close to π-mode of operation for *S*-band observed ~2.12 GHz and for *Ku*-band ~14.2 GHz which is close to desired frequencies of operation.

4.4.2.1 Cold Simulation (Beam Absent Simulation)

The cold simulation as explained in the preceding section is performed for the validation of the obtained design parameters (given in table 4.2). The dispersion curve shown in the Figure 4.11 is obtained through Eigenmode solver of 'CST Microwave Studio'. The two dispersion curves for two different bands show the operating π-mode frequencies (i.e., 2.2 GHz in *S*-band and 14.1 GHz in *Ku*-band).

The segregation cavity is used mainly to stop the interference of the generated RF signal at two frequencies. The design of the segregation cavity is explained in section 4.3. Figure 4.12 shows the role of the use of the segregation cavity. It can be seen from Figure 4.12(a) that, the RF signal generated at the SWS1 section (i.e., for the *S*-band generation section) is propagated through extraction section 1 and stopped by the segregation cavity to propagate towards the extortion section 2. Similarly, Figure 4.12(b) shows that the RF signal generated at the SWS2 section (i.e., for the *Ku*-band generation section) is propagated through extractor section 2 and stopped by the segregation cavity to propagate towards the SWS1 section (i.e., *S*-band generation section).

4.4.2.2 PIC Simulation (Beam Present Simulation)

After cold simulation, the PIC simulation of the dual-band MILO is performed followed by optimization to obtain the RF signal in two different bands. The beam parameter for simulation is selected as: DC pulse beam voltage of 610 kV with 10 ns of rise time, 130 ns of hold time, and 10 ns of fall time and beam current of 58 kA. The input DC pulse voltage and generated current are shown in Figure 4.13. With

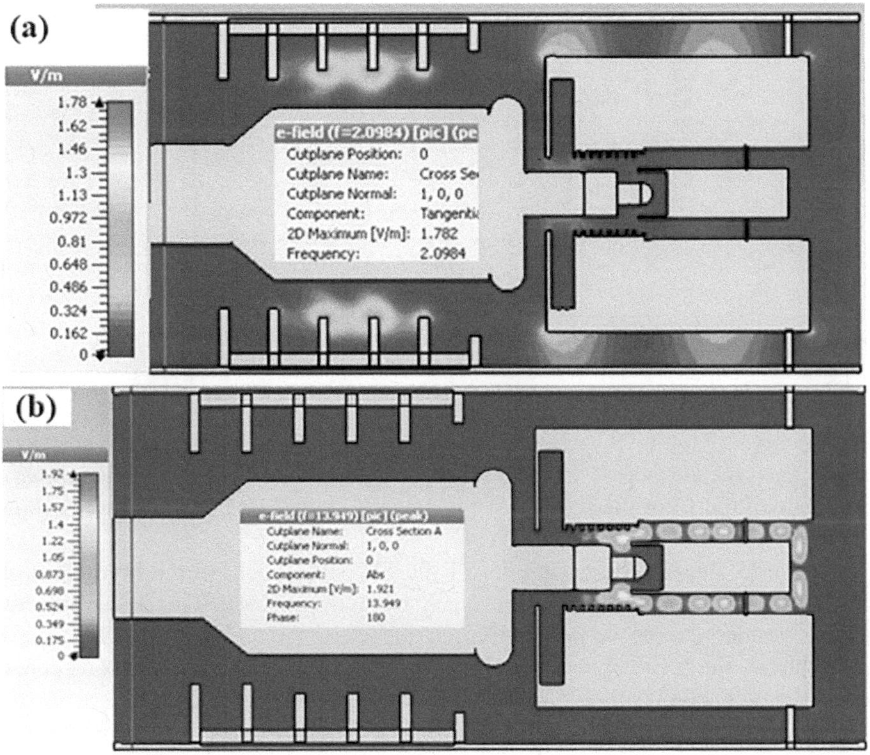

FIGURE 4.12 Role and effect of segregation cavity for different frequency bands (a) for *S*-band and, (b) for *Ku*-band.

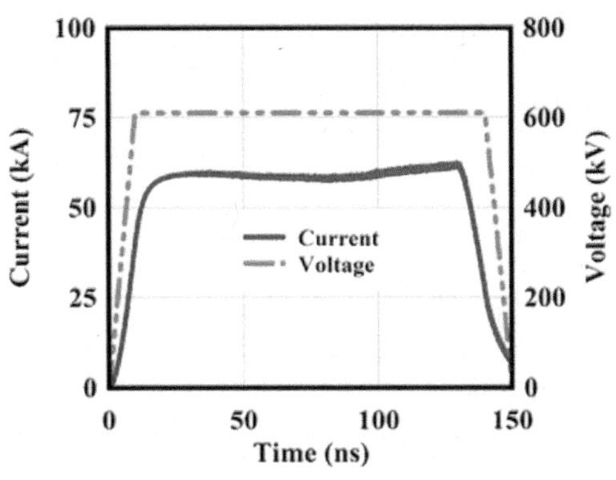

FIGURE 4.13 Applied input DC voltage (in kV) pulse with 10 ns rise time, 130 ns hold time, and 10 ns fall time with the generated current (in kA).

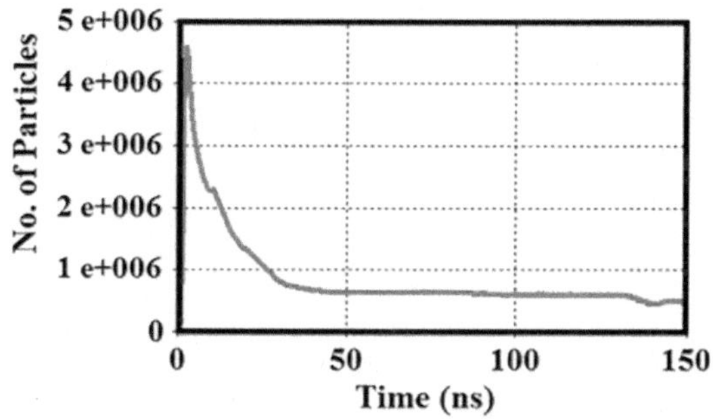

FIGURE 4.14 Number of emitted particles with respect to time in ns.

the application of input DC pulse voltage between anode and cathode, the electron is emitted from the surface of the cathode through explosive emission. The number of electrons emitted from the surface of the cathode at different time instant is shown in Figure 4.14.

Figure 4.15 shows different phases of the PIC simulation. In the first phase, the electrons emitted from the cathode surfaces start moving towards the anode structure and are considered to be the pre-oscillation phase and shown in Figure 4.15(a). In the second phase, the self-magnetic insulation is taking place and this phase is considered as the magnetic insulation phase which is shown in Figure 4.15(b).

After magnetic insulation, the third phase starts in which an electron sheath is confined between anode and cathode and flows in axial direction. Due to the instability generated inside the device, the electron sheath between anode and cathode generates oscillation which is shown in Figure 4.15 (c). In the last phase, also called the non-linear phase, shown in Figure 4.15 (d), the spoke formation takes place which is mainly responsible for beam-wave interaction. The TM_{01} mode is excited inside the azimuthal symmetric cavity structure. To track the axial motion of the electrons, an axial momentum tracker is applied during the PIC simulation with 10 ns of tracking time steps. The absolute value of normalized momentum with respect to axial distance Z is shown in Figure 4.16. It can be clearly seen from Figure 4.16 that, the diocotron instability occurs inside the device forms π-mode (i.e., 180° phase shift between consecutive cavity) of operation near both the interaction structure for efficient beam-wave interaction.

After the efficient beam-wave interaction, the RF signal generated inside the device is stored in the cavity (i.e., in the *S*-band section and *Ku*-band section) in the form of the standing wave. This stored energy was extracted out using different extractor sections (i.e., extractor section-1 for *S*-band and extractor section-2 for *Ku*-band). The extractor section mainly changes the standing wave into the traveling wave and helps in matching the impedances between the interaction section and the output section. To track the generated RF signal, a waveguide port is placed at the output section-1 (i.e., for *S*-band) and output section-2 (i.e., for *Ku*-band). The RF signal and its frequency spectrum received at the waveguide port at section-1 are shown in Figure 4.17. Figure 4.17(a) shows the RF signal amplitude with respect to time whereas Figure 4.17(b) shows the FFT (Fast Fourier Transform) of the received signal. It can be seen from Figure 4.17(b) that the received RF signal

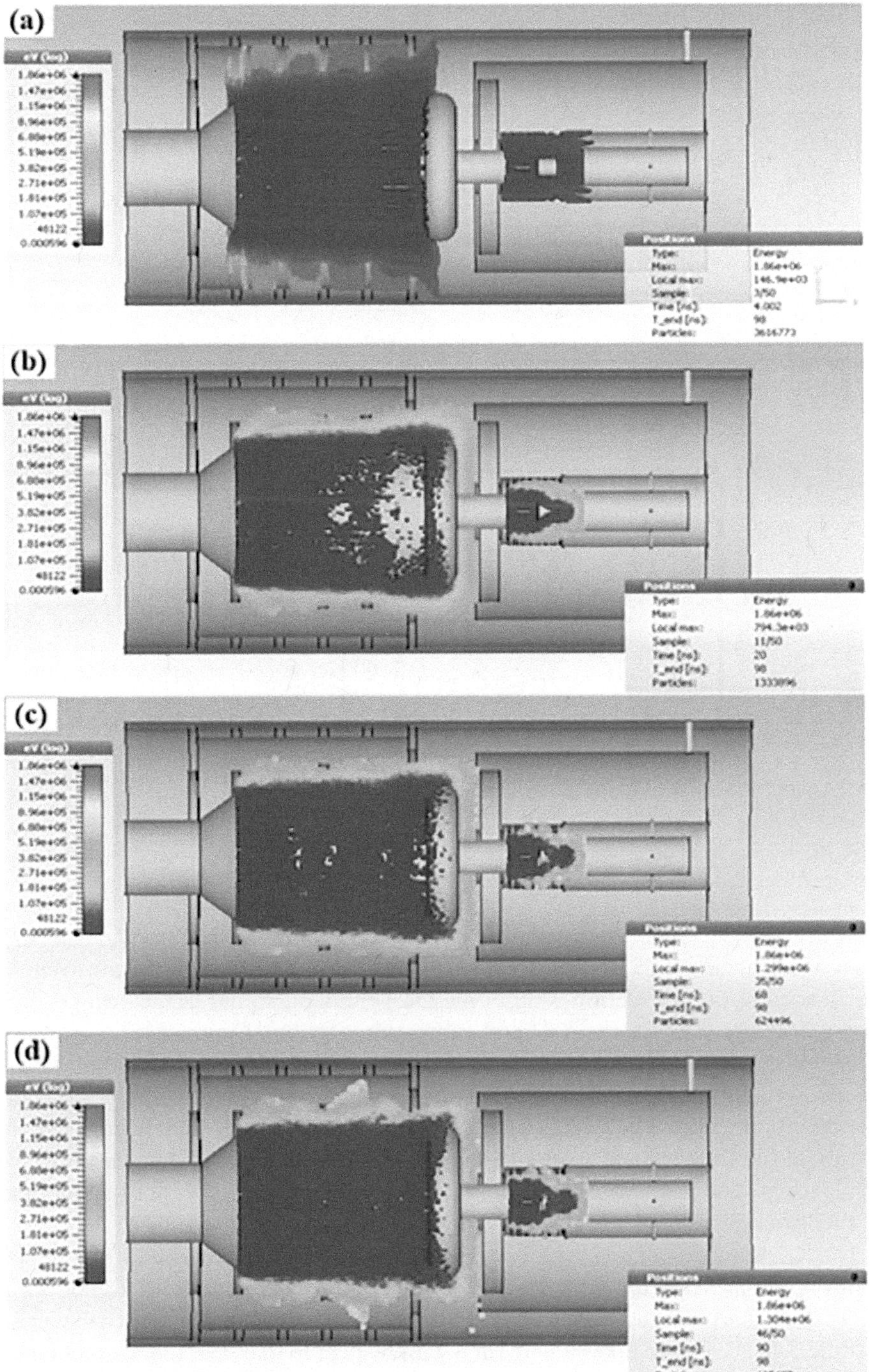

FIGURE 4.15 Four phases of interaction between electron beam and RF wave within the device (a) Phase 1 (i.e., pre-oscillation phase); (b) Phase 2 (i.e., magnetic insulation phase); (c) Phase 3 (i.e., start oscillation phase); and (d) Phase 4 (i.e., non-linear phase).

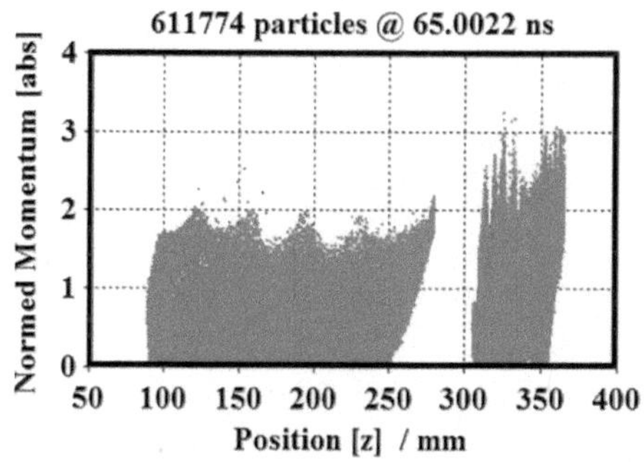

FIGURE 4.16 Normalized momentum at different axial positions at 65.02 ns time of oscillation.

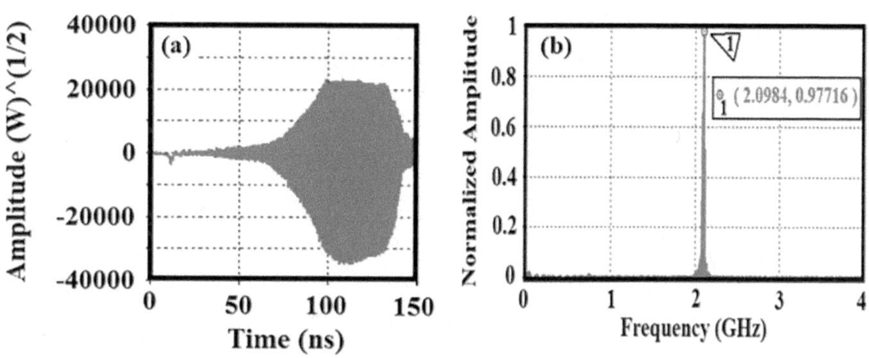

FIGURE 4.17 *S*-band (a) received RF signal at the output port (b) frequency spectrum of corresponding RF output signal found through Fast Fourier Transform (FFT).

at section-1 having 2.0984 GHz frequency (i.e., in *S*-band). Similarly, a waveguide port placed at output section-2 observes the generated signal at the second interaction structure (i.e., for the *Ku*-band generation section). Figure 4.18(a) shows the RF signal amplitude with respect to time whereas Figure 4.18(b) shows the FFT of the received signal. It can be seen from Figure 4.18(b) that the received RF signal at section-2 having 13.947 GHz frequency (i.e., in *Ku*-band). The drop in the frequency at the higher band is due to the beam loading effect and corner effect of disc loading. The beam loading effect and disc corner effect does not impact much for the lower band.

The RF power calculated for *S*-band signals is shown in Figure 4.19. It can be seen from Figure 4.19 that the peak RF power generated for *S*-band is ~1.2 GW but its pulse is very narrow. The reason behind this problem is that the instability that arises near the *S*-band section takes a long time with respect to that near the *Ku*-band interaction section. The RF power calculated for *Ku*-band signals is shown in Figure 4.20. It can be seen from Figure 4.20 that the peak RF power generated for *Ku*-band is ~3.1

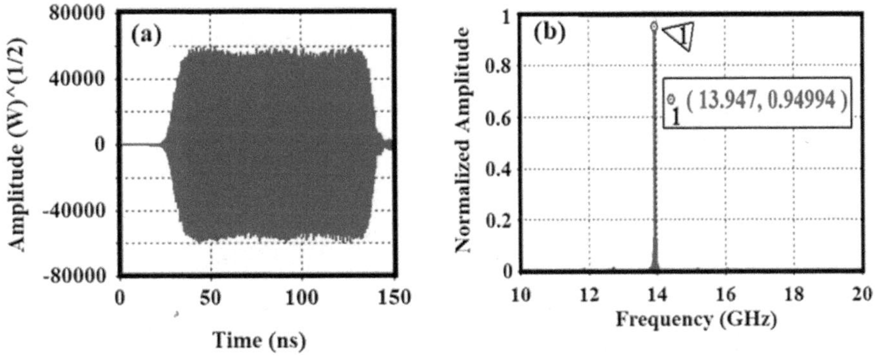

FIGURE 4.18 *Ku*-band (a) received RF signal at the output port (b) frequency spectrum of corresponding RF output signal found through Fast Fourier Transform (FFT).

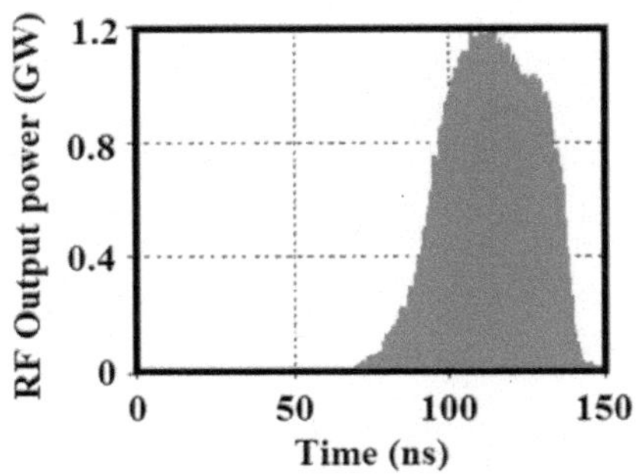

FIGURE 4.19 RF output peak power (in GW) received at the output port for *S*-band.

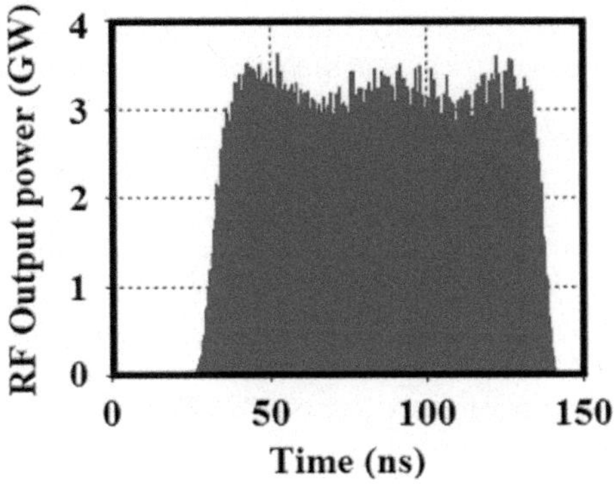

FIGURE 4.20 RF output peak power (in GW) received at the output port for *Ku*-band.

GW. Thus, the combined effect of two RF signals generated in *S*-band and *Ku*-band with application of 610 kV of DC voltage and 58 kA of current provide overall 12.1% of the efficiency.

4.5 CONCLUSION

In this chapter, *S*-band bi-frequency MILO along with dual-band (*S*-band and *Ku*-band) MILO which generates RF through different interaction structures are designed and investigated using analysis and validated through simulation. The design expression is derived for calculating the different design parameters for different bi-frequency MILO. The beam wave interaction process has been explained and the temporal growth rate calculated numerically for dual-band MILO. The numerical calculation is used to find the dispersion curve and validated using Eigenmode simulation for different interaction structures for *S*-band bi-frequency MILO and for *S*-band and *Ku*-band related to dual-band MILO, respectively. Finally, the PIC simulation was performed for different bi-frequency MILO. *S*-band bi-frequency MILO generates RF signal at two frequencies and combined RF peak power of ~2.86 GW, and average power of ~1.43 GW is obtained at the output with the application of 490 kV and 45 kA. The overall efficiency of ~6.5% is obtained through the *S*-band bi-frequency MILO. Finally, the PIC simulation for dual-band MILO is carried out by applying 610 kV voltages and 58 kA current. The device generates 3.1 GW RF power in *Ku*-band and 1.2 GW RF power in *S*-band with the overall efficiency of ~12.1%.

REFERENCES

Benford, J., Swegle, J. A. and Schamiloglu, E., *High Power Microwaves*, 2nd ed. Boca Raton, FL: CRC Press, ch. 3, pp. 43–103, 2007.

Chen, D. B., Wang, D., Meng, F. B., Fan, Z. K., Qin, F. and Wen, J., "Bifrequency HPM generation in a MILO with azimuthal partition," *IEEE Transactions on Plasma Science*, vol. 37, no. 10, pp. 1916–1920, 2009.

Cousin, R., "Compréhension des mécanismes régissant le fonctionnement d'un tube hyperfréquence de type MILO," Ph.D. thesis, Ecole Polytechnique X, France, 2005.

Davis, J. G. and Gibson, A. A. P., "Higher order mode impedances and cut-off frequencies of overmoded coaxial waveguides," *International Journal of Electronics*, vol. 93, no. 5, pp. 335–346, 2006.

Dwivedi, S. and Jain, P. K., "Beam-wave interaction analysis of a magnetically insulated line oscillator," *Physics of Plasmas*, vol. 19, no. 8, pp. 082110, 2012.

Dwivedi, S. and Jain, P. K., "Design expressions for the magnetically insulated line oscillator," *IEEE Transactions on Plasma Science*, vol. 41, no. 5, pp. 1549–1556, 2013.

He, J., Cao, Y., Zhang, J., Wang, T. and Ling, J., "Design of a dual-frequency high-power microwave generator," *Laser and Particle Beams*, vol. 29, no. 4, pp. 479–485, 2011.

Ju, J. C., Fan, Y. W., Shu, T. and Zhong, H. H., "Proposal of a gigawatt-class L/Ku dual-band magnetically insulated transmission line oscillator," *Physics of Plasmas*, vol. 21, no. 10, pp. 103–104, 2014.

Ju, J. C., Fan, Y. W., Zhong, H. H. and Shu, T., "A novel dual-frequency magnetically insulated transmission line oscillator," *IEEE Transactions on Plasma Science*, vol. 37, no. 10, pp. 2041–2047, 2009.

Kumar, A., Dwivedi, S. and Jain, P. K., "MILO performance improvement study—an equivalent circuit approach," *IEEE Transactions on Plasma Science*, vol. 47, no. 10, pp. 4642–4649, 2019.

Lemke, R. W., "Linear stability of relativistic space-charge flow in a magnetically insulated transmission line oscillator," *Journal of Applied Physics*, vol. 66, no. 3, pp. 1089–1094, 1989.

Lemke, R. W., Calico, S. E. and Clark, M. C., "Investigation of a load-limited, magnetically insulated transmission line oscillator (MILO)," *IEEE Transactions on Plasma Science*, vol. 25, no. 2, pp. 364–374, 1997.

Tang, Y., Meng, L., Li, H., Zheng, L., Wang, B. and Yin, Y., "An X-band dual-frequency coaxial relativistic backward-wave oscillator," *IEEE Transactions on Plasma Science*, vol. 40, no. 12, pp. 3552–3559, 2012.

Wang, D., Chen, D., Qin, F. and Fan, Z., "Improved bifrequency magnetically insulated transmission line oscillator," *IEEE Transactions on Plasma Science*, vol. 38, no. 1, pp. 19–25, 2009.

Xiao, R., Sun, J., Chen, C., Zhang, Y. and Shao, H., "High efficiency annular magnetically insulated line oscillator-transit time oscillator with three separate frequencies in three bands," *Journal of Applied Physics*, vol. 106, no. 3, pp. 033308, 2009.

Yang, Z., Liang, Z., Zhang, B., Li, J., Ma, W., Hu, S. and Liu, S., "A Cherenkov oscillator operating at two different wave bands," *International Journal of Infrared and Millimeter Waves*, vol. 20, no. 1, pp. 83–92, 1999.

Zhang, X., Li, Y., Li, Z., Zhong, H. and Qian, B., "Preliminary experimental investigation of a complex dual-band high power microwave source," *Review of Scientific Instruments*, vol. 86, no. 10, pp. 104703, 2015.

5 Microwave and Millimeter-Wave Radar Imaging
Challenges and Applications

Vineet Singh, Akhilendra Pratap Singh, Arjun Kumar, Prabhakar Tripathi and Rajanish Kumar Singh

5.1 INTRODUCTION

This chapter takes readers on an insightful tour of the field of advanced imaging systems, covering foundational ideas, image reconstruction methods, and several technologies with useful applications in the millimeter- and microwave-wave domain. At the end of the chapter, the thorough investigation is brought together and deep insights that go beyond theoretical frameworks and into the practical factors necessary for successful system development are provided. This chapter emphasizes how important it is to comprehend the basic ideas behind microwave and millimeter-wave imaging, as it sets the stage for the discussion of image resolution and quality that follows. This conversation functions as a useful tool for researchers and practitioners alike, as well as a helpful manual for everyone involved in the complex process of imaging design.

This chapter is notable for its thorough examination of a variety of technologies, including through-wall imaging, ground-penetrating radar, biomedical imaging, and millimeter-wave imaging-based non-destructive testing. Each technology is carefully investigated, exposing its distinct possibilities and difficulties. Through-wall imaging faces signal propagation complexities, subsurface conditions and resolution are addressed by ground-penetrating radar, and resolution and sensitivity are continuously sought after by biomedical imaging. This chapter functions as a strong and useful resource, offering a fundamental comprehension of fundamental ideas, delving deeply into difficulties, and revealing possible uses in diverse technical settings. Its important contribution reverberates in innovation and advancements, especially in the areas of exploration, monitoring, and diagnostics in a variety of sectors, highlighting the critical role that millimeter-wave and microwave imaging play in forming scientific disciplines and industries in the face of ongoing technological advancement.

DOI: 10.1201/9781032656021-5

5.2 FUNDAMENTALS OF MICROWAVE AND MILLIMETER IMAGING

Microwave imaging, a branch of remote sensing and radar technology, involves the use of electromagnetic (EM) waves in the microwave frequency range either to get a signature of the object or generate detailed images of the scenes. The fundamental concepts underlying in microwave imaging are crucial to understanding how its basic structure functions and achieves its various applications. The microwave and millimeter wave imaging techniques are based on the principle of radar (Radio Detection and Ranging), a type of active sensor which is modified as per the requirement of the application type. Radar-based microwave and millimeter-wave imaging systems use the electromagnetic field theory, emphasizing three key properties: conductivity (σ), dielectric permittivity (ε), and dielectric permeability (μ). Conductivity decides the medium's ability to facilitate charge movement, while dielectric permittivity governs the charge displacement in response to an electric field, leading to energy storage. Magnetic permeability outlines how intrinsic magnetic moments respond to a magnetic field (Jol, 2009; Sadiku, 2001). These principles shape the effectiveness of imaging systems, with practical applications often focusing on nonmagnetic mediums, where conductivity (σ) and dielectric permittivity (ε) play decisive roles. A lower conductivity enhances wave penetration for improved imaging applications (Jol, 2009).

Microwaves interacting with objects or surfaces undergo scattering or reflection, which provides valuable information for imaging (Wang, 2023). The resulting scattering patterns and reflection characteristics contribute to creating detailed images, helping in the identification of object shapes and material properties. Scattering in radar engineering comprises the redirection of electromagnetic waves due to irregularities in the propagation medium, playing a crucial role in target detection and characterization. Radar waves interacting with an object cause scattering, and the resulting scattered waves provide information about the target's size, shape, composition, and orientation (Mahafza, 2005). Three types of scattering based on object size are Rayleigh, Mie, and nonselective scattering, each dependent on the target's size relative to the wavelength. Microwave and millimeter-wave imaging primarily involve Mie and nonselective scattering, depending on the frequency used, which is essential for target recognition and tracking (She, 2001). Reflection, another key concept, refers to the bouncing back of radar waves upon encountering a boundary. The efficiency of wave reflection is influenced by the geometry and nature of the reflecting surface. Smooth surfaces reflect radar waves more efficiently than rough ones. In radar systems, the strength of the reflected signal, known as radar cross-section (RCS), is critical for target detection and tracking (Mahafza, 2005).

5.3 TECHNIQUES FOR IMAGE RECONSTRUCTION

Image reconstruction techniques are important for microwave imaging systems as they transform raw data into final and meaningful visual representations. These techniques are applied to collected data to generate detailed information from scattered or reflected microwaves because they preserve media interface information as reflections, thereby increasing the overall quality of the reconstructed images. In microwave imaging applications such as medical diagnostics or industrial inspections, precise reconstruction is essential for identifying object shapes, detecting anomalies, and interpreting material properties. The efficacy and dependability of microwave imaging systems are enhanced by sophisticated image reconstruction algorithms and techniques, which enable better decision-making across a range of industries, from non-destructive testing in industrial settings to healthcare. Three cutting-edge methods that have been explored in the imaging field are discussed here:

5.3.1 Diffraction Tomographic

As an inverse scattering algorithm, tomography is used in Through Wall Imaging (TWI) to describe the electrical characteristics of walls and in Ground Penetrating Radar (GPR) to describe the ground. In a study detailed in reference (Zhang and Hoorfar, 2012), a three-dimensional (3D) diffraction tomographic algorithm designed for real-time TWRI was introduced.

$$I(x,y,z) = \int dk \int dk_x \int dk_y \frac{j64\pi E_s\left(k_x,k_y,k\right)\exp\left(-j\sqrt{4k^2-k_x^2-k_y^2}\,z_r\right)}{k^2 n_0 \alpha_r \overline{F}\left(\frac{k_x}{2},\frac{k_y}{2},k\right)\left[\alpha_T \overline{F}\left(\frac{k_x}{2},\frac{k_y}{2},k\right)\right]\left(4k^2-k_x^2,k_y^2\right)} exp\left(jk_x x + jk_y y\right) \tag{5.1}$$

where the dyadic function $\overline{F}$ is given by

$$\overline{F} = \frac{T^{TM}}{k^2 k_\rho^2 k_z}\begin{bmatrix} k_x^2 k_z^2 & k_x k_y k_z^2 & k_x k_z k_\rho^2 \\ k_x k_y k_z^2 & k_y^2 k_z^2 & k_y k_z k_\rho^2 \\ k_x k_z k_\rho^2 & k_y k_z k_\rho^2 & k_\rho^4 \end{bmatrix} + \frac{T^{TE}}{k_\rho^2 k_z}\begin{bmatrix} k_y^2 & -k_x k_y & 0 \\ -k_x k_y & k_x^2 & 0 \\ 0 & 0 & 0 \end{bmatrix}$$

$E_s(k_x, k_y, k)$ is spatial Fourier transform of backscattering signal, denoted as Es and received by the receiver at the mth position, and originates from the transmitter at the same m^th^ position. In this representation, k_n represents the free space wavenumber of the nth operating frequency within the range of f_{min} to f_{max}, with a frequency step of Δf. α_T indicates transmitting antenna polarization direction, and η_0 is the wave impedance in the freespace, $\eta_0 = 120\pi$. T_{TE} and T_{TM} are the transmission coefficients through the layered medium for TE and TM waves This innovative method proved successful in detecting the presence of a target concealed behind a wall. In a complementary approach, Zhang presented a method for imaging buried targets beneath multi-layered media using a multi-frequency multi-monostatic measurement configuration. The manuscript utilized the Born approximation to linearize the backscattering wave, along with Green's function and Fourier transform techniques to render images of the targeted scene. To optimize data processing efficiency in Zhang's work, two techniques—coherent summation and 2D inverse fast-Fourier-transform—were proposed. In linear inverse scattering algorithms, the coherent summation technique included summing over all receiver locations coherently. This method simplified data processing by prioritizing simplicity over a non-linear solution. Reconstructing every pixel in the transverse plane at every down-range pixel was another method that was put forth. With the use of the diffraction tomography algorithm, a 2D inverse fast Fourier transform was used to complete this reconstruction. This strategy sought to improve data processing speed and effectiveness. These developments in tomography-based algorithms demonstrate the broad range of applications for which they may be used, from real-time TWI position recognition to complex multi-layered media scenarios where buried targets need to be imaged. The accomplishments of these methods highlight how well they may be applied to problems in a variety of domains, such as security, subsurface research, and remote sensing.

5.3.2 3-D Beamforming

While diffraction tomography stands as a successful imaging method, its primary application has been in two dimensions, and the algorithms associated with it often focus on single-layer homogeneous walls. However, the limitations of LISP in scenarios involving different wall-target configurations have led to the exploration of alternative methods. Numerical studies have demonstrated that a three-dimensional (3D) or adaptive beam former can yield high-quality focused images under such circumstances. In the study outlined in reference (Zhang and Hoorfar, 2014), Synthetic Aperture Radar (SAR) was employed to represent the backscattering signal from the scanning region.

$$E_s\left(r_{rm},r_{tm},k\right)=\int G\left(r_{rm},r,k_n\right)G\left(r,r_{tm},k_n\right)\sigma\left(r\right)dr \tag{5.2}$$

The backscattering signal, denoted as Es and received by the receiver at the mth position, originates from the transmitter at the same m^th^ position. In this representation, k_n represents the free space wavenumber of the nth operating frequency within the range of f_{min} to f_{max}, with a frequency step of Δf. The layered medium Green's functions, denoted as G, model the wave propagation process from the transmitter to the target and from the target to the receiver, taking into account the presence of the wall. $\sigma\left(r\right)$ represents the reflectivity of the target, and r, with their respective indices, denotes position vectors. The indices "t" and "r" distinguish between the transmitter and the receiver. To solve equation (5.2), the Green's function is approximated, constituting another inverse scattering algorithm.

$$I\left(r\right)=\int_{k_{min}}^{k_{max}}dk_n\int_{L_{x\,min}}^{L_{x\,max}}\int_{L_{y\,min}}^{L_{y\,max}}dr_{rm}E_s\left(r_{rm},r_{tm},k_n\right)G^*\left(r_{rm},r,k_n\right)G^*\left(r,r_{tm},k_n\right) \tag{5.3}$$

The discussed backscattering signal representation captures the intricate relationship of multiple parameters, such as the wavenumber, Green's functions, reflectivity, and position vectors. With the use of this all-inclusive model, the scattering phenomena inside a scanning zone can be analyzed while accounting for the complicated traits of the medium, which includes the presence of walls, and both the transmitter and receiver positions. For the purpose of solving the preceding equation, the Green's function approximation is an essential step that falls into the larger class of inverse scattering techniques. Using the measured scattering data, the objective of inverse scattering is to obtain information about the medium or scattering item. The Green's function is an essential element in this context, offering understanding of the interactions and wave propagation process within the medium. The fact that SAR was selected as the imaging method for this investigation shows how well it works to rebuild images of the scanned area by recording backscattering signals. SAR is well known for its capacity to produce images with a high degree of resolution by simulating a wide aperture with the motion of the radar antenna. Improved imaging capabilities are made possible by this synthetic aperture approach, especially in situations involving complicated geometries or numerous layers. This research recognizes that typical methods are limited when dealing with scenarios that are larger than two dimensions or involve non-homogeneous walls, which is why 3D or adaptive beam formers are the focus of this study. The project intends to enhance the quality and usefulness of focused images in situations when traditional approaches might not be sufficient by investigating alternate beam formation techniques.

5.3.3 Compressed Sensing

Compressive sensing serves as a signal processing technique designed to reconstruct a signal using fewer measurements or signal samples than conventional methods. When applied to radar systems, this approach presents advantages such as reduced current consumption and sampling time compared to standard radar systems. Researchers are increasingly favoring the adoption of compressive sensing in radar imaging to improve acquisition times. Recent findings in compressive sensing (CS) indicate the possibility of reconstructing a K-sparse signal $x = \Psi s$ of length N from O(KlogN) measurements (Yoon and Amin, 2008). CS involves acquiring non-traditional linear measurements, $y = \Phi x$, in the form of randomized projections. The signal vector, x, which has a sparse representation in a transform domain Ψ, can be reconstructed from $M = C\ (\mu^2\ (\Phi,\ \Psi)\ \log \mathrm{N})$ K compressive measurement exactly with high probability by solving a convex optimization problem of the following form min $\|x\|_1$, subject to $y = \Phi\Psi x$ **(5.4)**

This optimization problem can be efficiently solved using linear programming, where C is a small constant and $\mu(\Phi,\ \Psi)$ represents the mutual coherence between measurement matrix Φ and the transform domain Ψ.

Radar imaging akin to many sensing systems, demands extensive sampling, resulting in prolonged data acquisition times. To mitigate the challenges associated with the substantial sampling requirements and extended data acquisition times in TWRI, researchers have proposed the application of compressive sensing. The significance of compressive sensing in radar systems lies in its ability to effectively reconstruct signals with a reduced number of samples, leading to diminished current consumption and sampling times. This is particularly advantageous in TWRI, where the need for extensive sampling could otherwise impose significant constraints on system efficiency. In practical terms, the compressive sensing method allows the sampling of the space of interest in radar image, providing a means to overcome the challenges posed by extensive sampling requirements and protracted data acquisition times. By adopting this approach, researchers aim to streamline the data acquisition process in radar, making it more efficient and less resource-intensive.

5.4 DIFFERENT TECHNOLOGIES BASED ON MMW AND MICROWAVE IMAGING, THEIR USES AND CHALLENGES

5.4.1 Through the Wall Imaging

Through Wall Imaging (TWI) is a non-invasive system for detecting and recognizing targets beyond walls or barriers without physical access, utilizing Radio Frequency waves. The inherent trade-off between wall penetration and target resolution in RF waves has prompted the exploration of Ultra-Wideband (UWB) radar for TWI. UWB radar, with its large absolute signal bandwidth, offers high-resolution imaging and improved range measurement accuracy, surpassing conventional narrowband radar. It demonstrates immunity to passive interferences and enhances target classification through superior range resolution. Challenges exist, especially with various wall types, yet UWB radar systems like impulse radar, stepped frequency continuous wave (SFCW) radar, and noise radar address these concerns. Impulse radar provides sub-nanosecond range resolutions, while SFCW radar offers advantages in terms of low-speed converters, high dynamic range, and extended distance coverage, making it a suitable choice for TWI systems.

In Singh's (2023) explanation of the image formation process, a simplified flow is depicted using a Stepped Frequency Continuous Wave (SFCW) radar imaging system. Figure 5.1 illustrates the assembly of an experimental SFCW through-wall imaging system, incorporating a vector network analyzer (VNA) and a pyramidal horn antenna operating in the C-band. The handheld VNA, serving as the SFCW system's core, facilitates simultaneous transmission and reception of electromagnetic waves on its port. Antennas are strategically positioned to cover the entire wall length, creating a monostatic system across the C-band. The VNA-designed system records reflection coefficients (S_{11}) concurrently in the frequency domain, and an anechoic chamber houses the complete experimental setup to minimize external environmental effects. Table 5.1 provides a summary of the experiment's various parameters.

The usage of microwave through-wall imaging systems holds significant importance across various applications. In the realm of security and law enforcement, these systems enable personnel to detect and locate individuals hidden behind walls or barriers, providing crucial insights in scenarios like hostage situations or building

TABLE 5.1
Typical Values of SFCW Radar Parameters

Radar parameters	Value
Frequency range	3.5GHz–5.5GHz
Bandwidth	2GHz
Number of frequency points	201
Power Transmitted	–3dbm
Range Resolution	7.5cm
Antenna Type	Horn
Polarization	Vertical

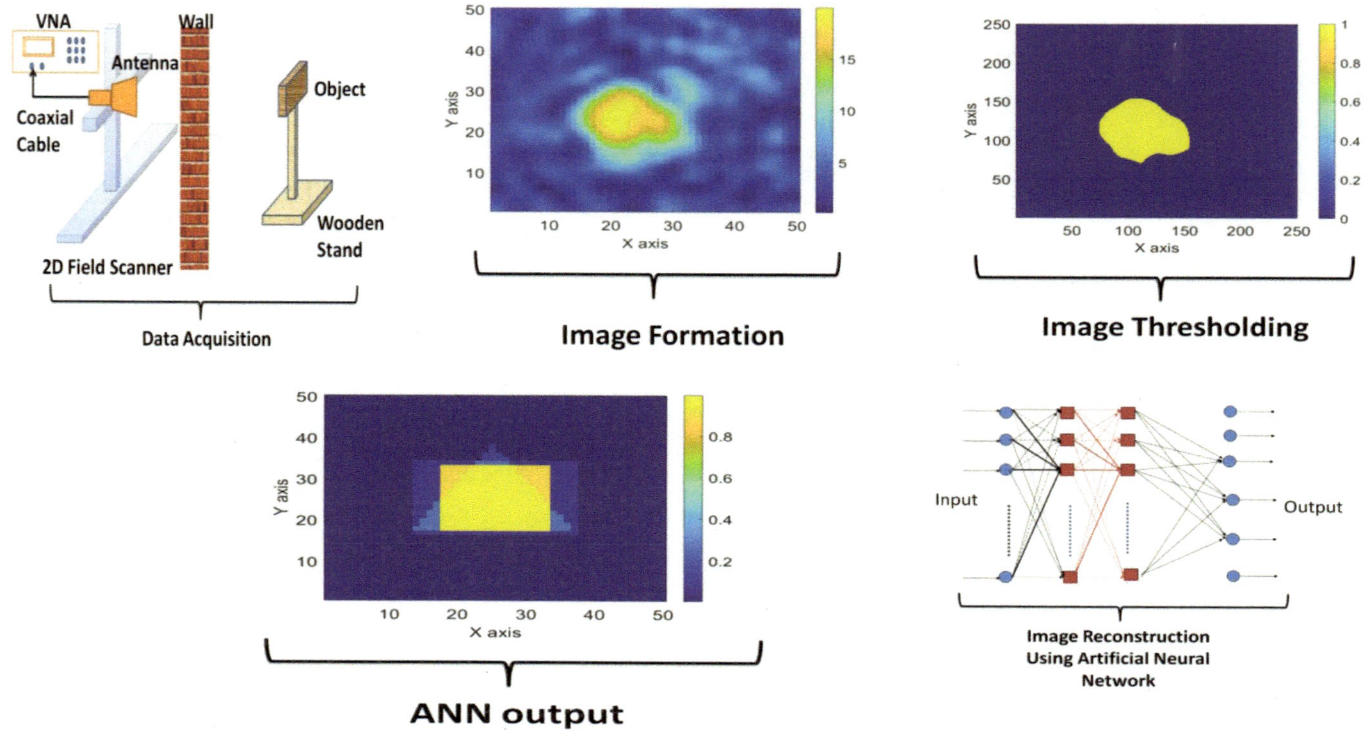

FIGURE 5.1 Block diagram of image formation process through a SFCW radar imaging system [Akhilendra Pratap Singh, "A Shape-based Approach for Recognition of Hidden Objects Using Microwave Radar Imaging System," *Progress In Electromagnetic Research* C, vol. 133, pp. 135–149, 2023. DOI: 10.2528/PIERC23041402., reproduced courtesy of The Electromagnetics Academy].

clearances. In military operations, through-wall imaging aids in situational awareness, allowing forces to assess potential threats and navigate complex environments with enhanced precision. Moreover, in search and rescue operations, microwave through-wall imaging proves invaluable for locating individuals in disaster-stricken areas where conventional visibility is limited. Microwave through-the-wall radar imaging faces several challenges that impact its effectiveness in various applications. One significant challenge is the attenuation and scattering of microwaves as they pass through different wall materials, leading to a loss of signal strength and reduced image quality. The complexity of wall structures, such as multi-layered walls or walls with varying compositions, poses difficulties in accurately imaging targets behind them. Additionally, the presence of clutter, reflections, and interference from surrounding objects can complicate target detection and recognition. The performance of through-the-wall imaging systems is also affected by environmental factors like weather conditions, which may impact signal propagation. Another challenge involves the need for real-time processing and interpretation of complex data, demanding sophisticated algorithms for accurate image reconstruction. As the technology advances, addressing these challenges is crucial for maximizing the potential of microwave through-the-wall radar imaging in fields such as security, surveillance, and emergency response.

5.4.2 Ground Penetrating Radar

Ground Penetrating Radar (GPR) is a geophysical imaging technique that employs radar pulses to capture subsurface information. This technology has found widespread applications across various fields due to its non-invasive nature and ability to provide detailed subsurface images (Huston et al., 2000). At its core, GPR also operates on the principles of radar technology, leveraging the propagation of electromagnetic waves (Jol, 2009). Here a general schematic diagram of GPR is shown in Figure 5.2 which contains all the necessary blocks of the design. To acquire data, the system sends modulated pulses of radiofrequency waves into the ground through an antenna. As these waves encounter subsurface materials with different dielectric properties, such as changes in composition or moisture content, a portion of the energy reflects back to the surface (Dojack, 2012). The GPR system records both the time delay and strength of these reflected signals, either directly or following specific processing, to generate a detailed cross-sectional image of the subsurface (Jol, 2009). The GPR has attracted the different engineering domains which need it as a prime instrument for their studies and surveys and its design is modified as per the domain requirements. Since GPR falls under the microwave imaging-based technique and itself contains a very large span of research and application and requires a separate chapter for discussion but here within the confines of this chapter, we narrow our focus to highlight specific application areas of GPR. Our objective is to provide upcoming researchers with an overview of the demand within this field, a brief discussion on the application areas of GPR and their potential uses is done.

GPR is an irreplaceable tool for geologists, providing insights into the Earth's composition and structure. It aids in identifying subsurface layers, distinguishing bedrock from soil, and mapping geological formations (Benson, 1995; Busby

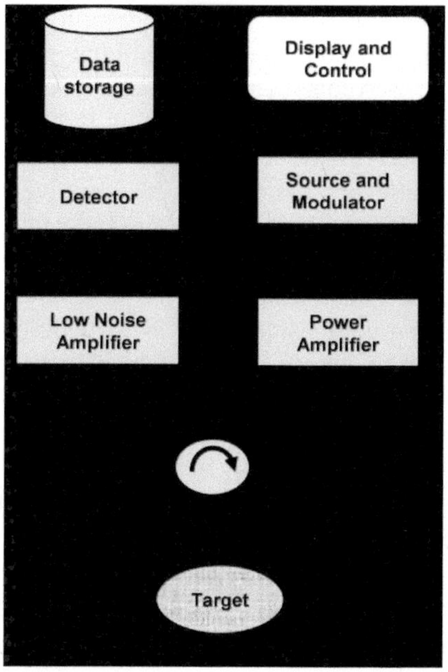

FIGURE 5.2 GPR schematic.

et al., 2004). In archaeology, GPR has revolutionized how researchers explore historical sites. By revealing buried structures, artifacts, and even ancient landscapes, GPR allows archaeologists to plan excavations strategically and non-destructively (Goodman and Piro, 2013; Dojack, 2012). Civil engineers rely on GPR for subsurface mapping in construction projects. It helps locate utilities, assess pavement conditions, and identify potential hazards, enhancing project planning and minimizing the risk of damage to existing infrastructure (Jol, 2009). Environmental scientists use GPR to study soil characteristics, delineate groundwater tables, and assess potential contamination. GPR contributes valuable information for environmental impact assessments and remediation efforts (Knight, 2001). GPR is a go-to technology for urban planning and utility management. By accurately locating underground utilities such as pipes and cables, GPR minimizes the risk of accidental damage during excavation (Metwaly, 2015). GPR aids forensic investigators in locating buried evidence, clandestine graves, or concealed structures. Its non-destructive nature is critical in preserving crime scenes and gathering crucial information (Barone and Maggio, 2019). In the realm of civil infrastructure, GPR plays a dominant role in inspecting and evaluating the condition of bridges, dams, tunnels, and other structures (Jol, 2009). It helps identify subsurface defects and weaknesses that may impact structural integrity. GPR has proven its worth in military applications by detecting tunnels, buried objects, and hidden structures. It enhances situational awareness and assists in planning and executing military operations (Sato, 2009).

While Ground Penetrating Radar (GPR) provides a rich source of information, it is not without its intricacies. The depth of penetration faces limitations in specific soil types, and variables such as electrical conductivity can exert influence, impacting resultant accuracy. The uncertain composition of soil, coupled with varying moisture content, poses a significant challenge for GPR surveys. Addressing the issue of higher contrast in subsurface targets, concealing lower reflective buried targets, remains a current center point for researchers and practitioners alike. Additionally, the uneven surface of the survey area proves critical, disrupting the calibration of the GPR system and presenting yet another formidable challenge. The implementation of image focusing algorithms on GPR data is particularly intricate due to the substantial variation in backscattered signals resulting from abrupt changes in pathloss during the survey. Ongoing research and technological advancements, however, continue to address these challenges, pushing the limitations of GPR capabilities.

5.4.3 Microwave Imaging in Biomedical Field

Biomedical microwave imaging is an emerging medical technology utilizing microwaves to create images of biological tissues. This technology operates based on the interaction of microwaves with tissues and the subsequent analysis of scattered or reflected signals (Aldhaeebi et al., 2020). The dielectric properties of tissues, particularly permittivity and conductivity related to water content, are crucial in this process (Etoz and Brace, 2018). A typical microwave imaging system comprises an antenna array, a transmitter-receiver, and a switch. When microwaves interact with biological tissues, they undergo processes such as absorption, reflection, and scattering, influenced by the tissues' dielectric properties (Aldhaeebi et al., 2020). Variations in permittivity and conductivity create contrast in microwave imaging, enabling the differentiation of tissues. Microwave imaging systems use transmitter and receiver antennas strategically placed to focus and collect scattered signals (AlSawaftah et al., 2022). Received signals are processed to extract information about tissue dielectric properties, and sophisticated algorithms reconstruct images based on the collected data. Tomographic techniques are often employed to create three-dimensional images, providing insights into spatial tissue distribution.

The interaction between biological tissues and the signals from biomedical imaging systems yields important information about their structural and functional properties. The technology holds promise in various medical fields, including breast cancer detection, brain imaging, and monitoring physiological changes (Aldhaeebi et al., 2020; Mobashsher et al., 2016; Lin, 1992). Ongoing research aims to improve resolution, sensitivity, and overall performance, making it a valuable tool in medical imaging. Specific applications include excelling in breast cancer detection and monitoring, especially in imaging dense breast tissues and facilitating real-time therapeutic monitoring (AlSawaftah et al., 2022). In brain imaging, microwaves enable stroke detection and show potential for studying various neurological conditions (Mobashsher et al., 2016). Cardiovascular imaging involves non-invasive assessment of abnormalities and real-time monitoring of blood flow (Lauteslager et al., 2022). The technology aids in wound healing assessment and diabetic foot ulcer detection, and contributes to chronic wound management and proactive diabetes intervention (Owdaet al., 2020). In orthopedic applications, microwave imaging assesses bone density and monitors joint disorders for treatment planning (Zhou et al., 2010). Overall, microwave-based biomedical imaging holds the promise of revolutionizing diagnostics and patient care across diverse medical disciplines, with ongoing research and technological progress set to expand its integration into routine clinical practice.

With its non-invasive diagnostic capabilities and ability to completely change medical diagnostics, microwave imaging has emerged as a game-changing technology in biomedical applications. However, there are a number of complex issues with its integration that call for a thorough comprehension in order to advance technology. The first difficulty is finding the best compromise between penetration depth and resolution. Low energy levels are used for safety, but it's still difficult to get enough penetration through a variety of tissues and keep good resolution (Lin, 2016). In addition, biological tissues are inherently complex, necessitating the development of strong algorithms to take into consideration fluctuations in dielectric characteristics and improve the dependability of microwave-based biomedical imaging (Chandra et al., 2015). Significant challenges are also presented by artifacts and signal noise, which are introduced by a variety of elements such as patient movement, technological interference, or reflections. To increase the overall quality and accuracy of microwave-based biomedical imaging, sophisticated signal processing methods and noise reduction tactics are essential. Another obstacle is the computational complexity of picture reconstruction, which has prompted continuous research to improve efficiency and simplify algorithms to make microwave imaging more useful for real-time applications in clinical settings (Chandra et al., 2016). Patient safety is of the utmost importance during microwave imaging procedures, necessitating a careful balancing act between acquiring adequate imaging data and following safety protocols. To guarantee consistent procedures and encourage moral application in healthcare, it is imperative to establish explicit regulatory frameworks for microwave imaging in biomedical applications.

5.4.4 Non-Destructive Testing in Industry

Millimeter wave imaging stands out as a superior technology for non-destructive testing (NDT) across diverse applications, presenting compelling advantages over microwave imaging. One key strength lies in its higher spatial resolution, achieved through operating at higher frequencies with shorter wavelengths (Appleby and Anderton, 2007). This feature enables millimeter wave imaging to excel in the detection of smaller defects, providing complicated details of surface structures crucial for comprehensive assessments. Additionally, its proficiency in surface-level imaging, attributed to limited penetration depth, makes it ideal for applications focused on analyzing materials near the surface without the need for deep penetration. Millimeter wave imaging in non-destructive testing involves the generation of electromagnetic waves using a transmitter, with the choice of frequency crucial for specific applications. As these waves interact with materials, their sensitivity to dielectric properties enables the detection of anomalies. Reflection and scattering occur when encountering variations in the material, providing information about internal structures (Kharkovsky and Zoughi, 2007). Processing of the gathered A, B, and C-scan data results in high-resolution images that help with material characterization and flaw identification. This technology is applied across industries, including aerospace, automotive, and electronics, for inspecting composite materials, assessing weld quality, and ensuring structural

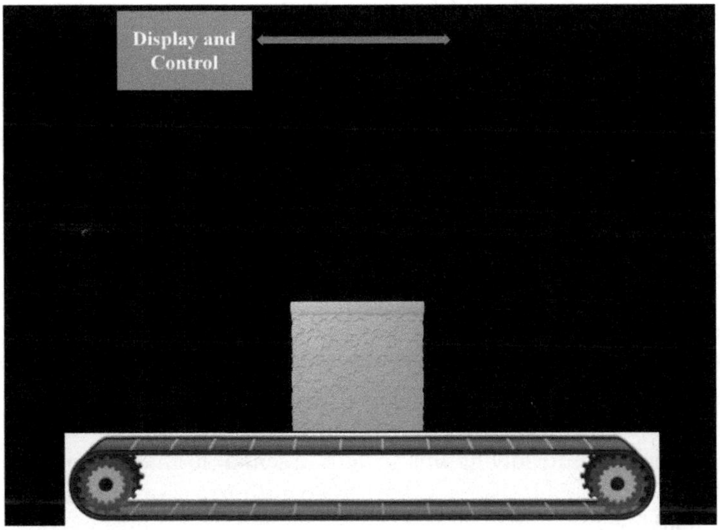

FIGURE 5.3 Millimeter wave imaging scanner for packed item quality testing.

integrity. A millimeter wave imaging scanner for nondestructive quality testing of packed items moving on a conveyor belt is shown in Figure 5.3.

Millimeter wave imaging-based non-destructive testing (NDT) is a versatile and powerful technology with widespread applications across various industries, providing invaluable solutions for quality control, safety, and reliability assessments. In the aerospace sector, it plays a crucial role in inspecting composite materials used in aircraft structures, ensuring the detection of internal defects or delaminations without causing damage (Gupta et al., 2021). In automotive manufacturing, millimeter wave imaging is utilized for Weld Quality Assurance (WQA), assessing weld quality without compromising the integrity of components, ensuring structural soundness, and complying with safety standards. In electronics manufacturing, particularly for Printed Circuit Board (PCB) Inspection, millimeter wave imaging assists in examining PCBs for defects, ensuring the integrity of connections and components without the need for physical contact (Kief and COSMIAC, 2018). In security and defense applications, it contributes to Hidden Threat Detection, enhancing security screening processes in airports and public spaces (Corredoura et al., 2006). For Material Science Research, researchers employ millimeter wave NDT to study and characterize materials, exploring properties, composition, and structural integrity without altering the material's state (Kharkovsky and Zoughi, 2007). It is used in pipeline inspection in the oil and gas sector, where it enables the non-invasive detection of structural problems such as thinning, corrosion, and other concerns (van Berlo et al., 2021; Ghasr et al., 2014). Within the food industry, it is employed for Quality Control in Packaging, inspecting packaging materials for defects or contamination, ensuring the safety and quality of packaged products (Farina et al., 2019; Vipiana et al., 2020). Overall, millimeter wave imaging-based non-destructive testing emerges as a pivotal technology, offering non-invasive solutions for evaluating the integrity and quality of diverse materials and structures across a spectrum of industries.

Millimeter wave imaging-based non-destructive testing (NDT) presents several challenges that require careful consideration for effective implementation. Among the notable issues are resolution constraints, where obtaining high resolution is essential for accurately identifying and characterizing tiny anomalies, particularly with materials that have complicated structures or minute imperfections. Challenges also include material penetration, where certain materials limit the depth of inspection and hinder the penetration of millimeter waves, necessitating solutions for thorough evaluations of thick or dense materials. Environmental interference poses another challenge, with external factors like weather conditions impacting performance; developing techniques to mitigate these influences is vital for accurate imaging. Moreover, because of the massive volume of data created during millimeter wave NDT, complicated data processing methods are required for real-time or almost real-time analysis, which presents a considerable challenge. The equipment cost is a barrier that prohibits millimeter wave imaging from being widely used, particularly in smaller enterprises. Resolving this issue is essential to making millimeter wave imaging financially feasible. Human exposure to millimeter waves raises safety problems that should be carefully considered, with a focus on establishing precise limits and standards (Lin, 2016; Wu et al., 2015). Creating established procedures is crucial since the lack of industry standards makes it difficult to guarantee consistency and dependability across various applications. Furthermore, the effective application of millimeter wave imaging-based nondestructive testing (NDT) necessitates education and training, calling for qualified operators capable of deciphering intricate imaging data. The last layer of complication is regulatory compliance, which highlights the necessity for broad adoption by requiring adherence to standards and approvals for the usage of millimeter wave technology across industries. In order to ensure the efficacy and safety of millimeter wave imaging-based non-destructive testing in a variety of industrial situations, it is imperative that these problems be addressed.

5.5 CONCLUSIONS

In the concluding remarks, this chapter consolidates a comprehensive exploration of microwave and millimeter-wave imaging systems, offering insights into fundamental concepts, image reconstruction techniques, and various technologies with practical applications. The significance of microwave and millimeter-wave imaging is established, emphasizing the importance of understanding fundamental concepts for effective system development. The discussion on image resolution and quality provides valuable insights for practitioners and researchers, serving as a practical guide for designing imaging systems.

The chapter explores a variety of technologies and applications, including through-wall imaging, ground-penetrating radar, biomedical imaging, and millimeter-wave imaging-based non-destructive testing. Every technology has different potential and challenges, which are covered in detail in the chapter. Through-wall imaging encounters complexities in signal propagation, while ground-penetrating radar confronts challenges related to resolution and subsurface medium conditions. Biomedical imaging requires continuous improvement in resolution and sensitivity, critical for various medical applications. Millimeter-wave imaging-based non-destructive testing addresses challenges associated with varying material properties during materials inspection. All things considered, this chapter proves to be a useful tool, offering a basis in basic ideas, tackling difficulties, and investigating possible uses in a variety of technological contexts. It highlights the critical role that microwave and millimeter-wave imaging is expected to play in influencing the future of companies and scientific disciplines as technology continues to progress. It also contributes to breakthroughs and innovation in diagnostics, monitoring, and exploration in numerous sectors.

REFERENCES

Aldhaeebi, M.A., Alzoubi, K., Almoneef, T.S., Bamatraf, S.M., Attia, H. and Ramahi, O.M., 2020. Review of microwaves techniques for breast cancer detection. *Sensors*, 20(8), p. 2390.

AlSawaftah, N., El-Abed, S., Dhou, S. and Zakaria, A., 2022. Microwave imaging for early breast cancer detection: Current state, challenges, and future directions. *Journal of Imaging*, 8(5), p. 123.

Appleby, R. and Anderton, R.N., 2007. Millimeter-wave and submillimeter-wave imaging for security and surveillance. *Proceedings of the IEEE*, 95(8), pp. 1683–1690.

Barone, P.M. and Di Maggio, R.M., 2019. Forensic geophysics: Ground penetrating radar (GPR) techniques and missing persons investigations. *Forensic Sciences Research*, 4(4), pp. 337–340.

Benson, A.K., 1995. Applications of ground penetrating radar in assessing some geological hazards: Examples of groundwater contamination, faults, cavities. *Journal of Applied Geophysics*, 33(1–3), pp. 177–193.

Busby, J.P. , Cuss, R.J., Raines, M.G., and, Beamish, D., 2004. Application of ground penetrating radar to geological investigations. *British Geological Survey*.

Chandra, R., Zhou, H., Balasingham, I. and Narayanan, R.M., 2015. On the opportunities and challenges in microwave medical sensing and imaging. *IEEE Transactions on Biomedical Engineering*, 62(7), pp. 1667–1682. doi: 10.1109/TBME.2015.2432137.

Corredoura, P., Baharav, Z., Taber, B. and Lee, G., 2006, May. Millimeter-wave imaging system for personnel screening: Scanning 10^ 7 points a second and using no moving parts. In *Passive Millimeter-Wave Imaging Technology IX* (Vol. 6211, pp. 68–75). SPIE.

Etoz, S. and Brace, C.L., 2018. Development of water content dependent tissue dielectric property models. *IEEE Journal of Electromagnetics, RF and Microwaves in Medicine and Biology*, 3(2), pp. 105–110.

Farina, L., Scapaticci, R., Vasquez, J.T., Rivero, J., Litman, A., Crocco, L. and Vipiana, F., 2019, July. Microwave imaging technology for in-line food contamination monitoring. In *2019 IEEE International Symposium on Antennas and Propagation and USNC-URSI Radio Science Meeting* (pp. 817–818). IEEE.

Ghasr, M.T., Ying, K. and Zoughi, R., 2014, February. 3D millimeter wave imaging of vertical cracks and its application for the inspection of HDPE pipes. In *AIP Conference Proceedings* (Vol. 1581, No. 1, pp. 1531–1536). American Institute of Physics.

Goodman, D. and Piro, S., 2013. *GPR Remote Sensing in Archaeology* (Vol. 9, p. 233). Springer.

Gupta, R., Mitchell, D., Blanche, J., Harper, S., Tang, W., Pancholi, K., Baines, L., Bucknall, D.G. and Flynn, D., 2021. A review of sensing technologies for non-destructive evaluation of structural composite materials. *Journal of Composites Science*, 5(12), p. 319.

Huston, D., Hu, J.Q., Maser, K., Weedon, W. and Adam, C., 2000. GIMA ground penetrating radar system for monitoring concrete bridge decks. *Journal of Applied Geophysics*, 43(2–4), pp. 139–146.

Jol, H.M. ed., 2009. *Ground Penetrating Radar Theory and Applications*. Elsevier.

Kharkovsky, S. and Zoughi, R., 2007. Microwave and millimeter wave nondestructive testing and evaluation-overview and recent advances. *IEEE Instrumentation & Measurement Magazine*, 10(2), pp. 26–38.

Kief, C. and COSMIAC at UNM Albuquerque United States, 2018. *Environment Testing 3D Electronics Research Program*. UNM.

Knight, R., 2001. Ground penetrating radar for environmental applications. *Annual Review of Earth and Planetary Sciences*, 29(1), pp. 229–255.

Lauteslager, T., Tømmer, M., Lande, T.S. and Constandinou, T.G., 2022. Dynamic microwave imaging of the cardiovascular system using ultra-wideband radar-on-chip devices. *IEEE Transactions on Biomedical Engineering*, 69(9), pp. 2935–2946.

Lin, J.C., 1992. Microwave sensing of physiological movement and volume change: A review. *Bioelectromagnetics*, 13(6), pp. 557–565.

Lin, J.C., 2016. Human exposure to RF, microwave, and millimeter-wave electromagnetic radiation [health effects]. *IEEE Microwave Magazine*, 17(6), pp. 32–36.

Mahafza, B.R., 2005. *Radar Systems Analysis and Design Using MATLAB*. Chapman and Hall/CRC.

Metwaly, M., 2015. Application of GPR technique for subsurface utility mapping: A case study from urban area of Holy Mecca, Saudi Arabia. *Measurement*, 60, pp. 139–145.

Mobashsher, A.T., Mahmoud, A. and Abbosh, A.M., 2016. Portable wideband microwave imaging system for intracranial hemorrhage detection using improved back-projection algorithm with model of effective head permittivity. *Scientific Reports*, 6(1), p. 20459.

Owda, A.Y., Owda, M. and Rezgui, N.D., 2020. Synthetic aperture radar imaging for burn wounds diagnostics. *Sensors*, 20(3), p. 847.

Sadiku, M.N., 2001. *Electromagnetics*. New York, Oxford University Press.

Sato, M., 2009. Principles of mine detection by ground-penetrating RADAR. In *Anti-Personnel Landmine Detection for Humanitarian Demining: The Current Situation and Future Direction for Japanese Research and Development* (pp. 19–26). Springer.

She, C.Y., 2001. Spectral structure of laser light scattering revisited: Bandwidths of nonresonant scattering lidars. *Applied Optics*, 40(27), pp. 4875–4884.

Singh, A.P., 2023. A shape-based approach for recognition of hidden objects using microwave radar imaging system. *Progress in Electromagnetics Research C*, 133, pp. 135–149.

van Berlo, B., Elkelany, A., Ozcelebi, T. and Meratnia, N., 2021. Millimeter wave sensing: A review of application pipelines and building blocks. *IEEE Sensors Journal*, 21(9), pp. 10332–10368.

Vipiana, F., Crocco, L. and LoVetri, J., 2020. Electromagnetic imaging and sensing for food quality and safety assessment [guest editorial]. *IEEE Antennas and Propagation Magazine*, 62(5), pp. 16–17.

Wang, L., 2023. Microwave imaging and sensing techniques for breast cancer detection. *Micromachines*, 14(7), p. 1462.

Wu, T., Rappaport, T.S. and Collins, C.M., 2015, June. The human body and millimeter-wave wireless communication systems: Interactions and implications. In *2015 IEEE International Conference on Communications (ICC)* (pp. 2423–2429). IEEE.

Yoon, Y.-S. and Amin, M.G., 2008. "Compressed sensing technique for high-resolution radar imaging", *International Society for Optics and Photonics*, Apr. 2008. doi: 10.1117/12.777175.

Zhang, W. and Hoorfar, A., 2012. Three-dimensional real-time through-the-wall radar imaging with diffraction tomographic algorithm. *IEEE Transactions on Geoscience and Remote Sensing*, 51(7), pp. 4155–4163.

Zhang, W. and Hoorfar, A., 2014. "Three-Dimensional Synthetic Aperture Radar Imaging Through Multilayered Walls," *IEEE transactions on antennas and propagation*, 62(1), pp. 459–462, Jan. 2014, doi: https://doi.org/10.1109/tap.2013.2287274.

Zhou, T., Meaney, P.M., Pallone, M.J., Geimer, S. and Paulsen, K.D., 2010, August. Microwave tomographic imaging for osteoporosis screening: A pilot clinical study. In *2010 Annual International Conference of the IEEE Engineering in Medicine and Biology* (pp. 1218–1221). IEEE.

6 High Isolation and Low RCS Tunable Wide Band Graphene-Based THz MIMO Antenna for Nano Communication with 6G/IOT Application

Gaurav Saxena, Harshit Srivastava, Ishan Vishwakarma and Dipanshu Yadav

6.1 INTRODUCTION

The increasing need for wireless high-speed connectivity systems has led to the exploration of new frequency bands, including the terahertz (THz) band. The THz band, which ranges from 0.1 THz to 10 THz, is considered a promising frequency band for future wireless communication systems, including the 6G communication system. However, the use of the THz frequency is difficult because of the limited availability of suitable materials and devices that can operate in this frequency range [1]. Metamaterials and metasurfaces are known for their unique and exceptional properties in altering electromagnetic (EM) waves, making them highly useful in the development of various THz (terahertz) devices such as filtration systems, polarized converters, and absorbers. Among these devices, the THz metasurface absorber (MSA) is particularly significant, as it has the potential to create more robust and versatile gadgets for a wide range of applications, including identification, image processing, spectrum recognition of hazardous compounds, and other functional uses. As a result, the research and development of high-performance MSAs have gained popularity in recent years. MSAs can achieve high absorption values by dissipating EM wave energy through interference cancellation or by converting the energy of the incoming THz pulse into thermal or alternative energy sources loss [2–3]. In recent years, researchers have developed and investigated numerous designs of metasurface absorbers (MSAs) with diverse resonant structures to achieve single, dual, or multi-band, as well as large-scale absorption. However, many of these MSA designs suffer from limitations such as restricted bandwidth, poor absorption performance, and fixed absorption efficiency, which hinder their practical applications.

DOI: 10.1201/9781032656021-6

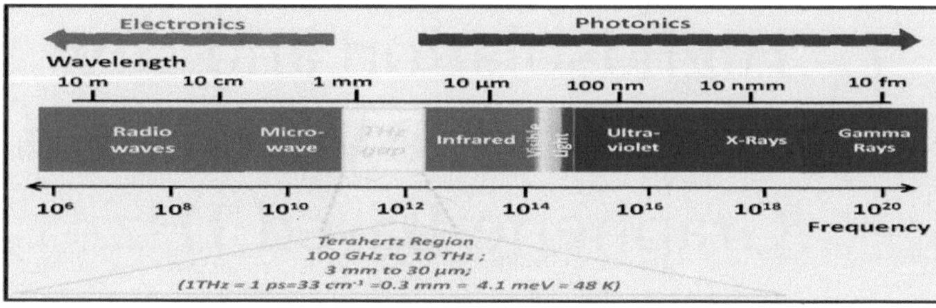

FIGURE 6.1 Frequency spectrum of THz Communications.

As a result, there is an urgent need for the development of high-performance MSAs using novel materials to enhance their potential for practical use [4]. Graphene, a unique material composed of a single layer of carbon atoms, possesses exceptional mechanical, chemical, and electrical properties that can be adjusted. Graphene has shown potential uses in infrared and adjustable terahertz plasmonic devices, as it can support different plasmon resonances in the THz and infrared frequencies [5–6]. A UWB-MIMO antenna system is designed that has gain between 2 and 6.5 dB which is good for wireless communication systems like linked objects and 5G [7].

Graphene, a unique material composed of an individual carbon atom layer, has exceptional properties that can be adjusted. It has shown promising applications in infrared and tunable terahertz plasmonic devices. The use of graphene in wideband MIMO antenna systems for THz applications can help overcome challenges associated with conventional antennas, such as limited bandwidth and high loss [8].

Some of the drawbacks of conventional antenna systems, including their high complexity, high loss, and narrow bandwidth, can be mitigated by using graphene in wideband MIMO antenna systems for THz applications [9–10]. The scarcity of accessible bandwidth is driving research approaching commercial uses in millimeter-wave technology; in the near future, the next frontier in communication is expected to be in the terahertz (THz) range. To meet current standards, wireless communication must have a high data rate and enhanced channel capacity [11]. As such, significant absorptions, attenuation, route, and multipath fading present additional challenges for THz-range MIMO communication systems in terms of data transfer. Because of this, the MIMO system is a fantastic choice for handling these issues [12].

A MIMO antenna's component spacing affects how the antenna parts interact with one another. Closer elements produce more mutual coupling, which boosts correlation coefficients and lowers efficiency. For exchange of coupling (mutual coupling) to be less than 15 dB, the distance between the elements must be at least $\lambda/2$ or $\lambda/4$ of the frequency [13].

Graphene, due to its high conductivity, enables the use of thinner and smaller antenna elements, which can be integrated into compact and flexible devices. Additionally, the high mechanical strength and flexibility of graphene make it possible to use it in conformal and wearable antenna systems, which can be used in a wide range of applications, including wireless communication, sensing, and imaging [14–15]. Despite the many potential benefits of graphene-based wideband MIMO antenna systems for THz applications, there are also several challenges associated with the design and implementation of these systems. One of the main challenges is the lack of suitable fabrication methods that can produce high-quality graphene films with controlled properties. Additionally, the lack of standard measurement techniques and characterization methods for graphene-based antenna systems can impede the ability to compare the performance of different designs [16–17]. The performance of the antenna is improved by enhancing isolation and diversity parameters. A modified array antenna decoupling-surface (MAADS) is used to minimize the transfer of current-density between antennas. An inverted neutralization line parasitic and two paired-feed looped antennas with linked capacitor embedded parts are acceptable decoupling structures that are added to enhance isolation. Planar metasurface is also utilized to improve decoupling and reduce radar cross-section in important frequency bands. The components of the antenna are set orthogonally to each other to further enhance isolation [18–19]. Graphene-based antennas have shown promise for 6G-IoT and bio-medical sensing applications. These antennas can be used to create programmable beam-steering capabilities via Reconfigurable Intelligent Surfaces (RIS). RIS are a type of metamaterial that can be programmed to reflect or absorb electromagnetic waves in a desired way. This allows for the precise control of the radiation pattern of an antenna, which is essential for applications such as IoT and bio-medical sensing. This approach offers tunable and reconfigurable properties through the use of graphene, enabling beam scanning and beam forming without the need for traditional switches such as varactor diodes or P-I-N diodes [20–22]. Additionally, nano communication in 6G/IoT and life sciences-medical sensing uses has been suggested using CSRR packed graphene influenced EBG oriented extremely isolated wideband THz dielectric oscillator MIMO antennas [23]. These antennas have the potential to improve isolation and diversity parameters in MIMO systems, and can be used for sensing applications such as haemoglobin and urine detection. The use of graphene in these antenna designs allows for improved performance and adaptability in the THz frequency range, overcoming limitations faced by traditional copper-based antennas [24–25].

In this research, we present a comprehensive study of a graphene-based metasurface absorber and antenna for the THz frequency range (0.1–2 THz) that offers high isolation and reduced radar cross-section (RCS). The MSA is composed of a diamond-shaped structure with concentric circles at the centre, which is designed to be polarization insensitive to both waves that are both transversely electric and magnetic. A MIMO antenna is designed that contains two hexagonal elements with graphene patch which is closely coupled and both the elements are orthogonally oriented with a 180° angle. To improve the isolation a decoupling framework is positioned in between the two components, in the ground layer. The low RCS is obtained with the help of a metasurface array, by placing this array below the proposed antenna design.

6.2 GRAPHENE CONDUCTIVITY

The relationship between the energy of electrons and their wavevector in graphene is described by the dispersion relation, which is influenced by the conductivity of graphene, $\sigma(\omega)$. Consequently, the resonant properties of graphene are primarily given by its conductivity. When dealing with sizes greater than 100 nm, the impact of the edges becomes negligible, allowing us to describe the conductivity of graphene using the Kubo formula. The overall conductivity is made up of two components, specifically σ(total) = σ(inter) + σ(intra).

The interband contribution represented in equation (6.1) [6]:

$$\sigma_{\text{int}er} = \frac{e^2}{4h}\left(H\left(\frac{\omega}{2}\right)+i\frac{4\omega}{\pi}\int_0^{4\omega}\frac{H(\varepsilon)-H\left(\frac{\omega}{2}\right)}{\omega^2-4\varepsilon^2}\partial\varepsilon\right) \tag{6.1}$$

$$H(\varepsilon)=\frac{\sinh\left(\frac{h_\varepsilon}{K_BT}\right)}{\cosh\left(\frac{\mu_c}{K_BT}\right)+\cosh\left(\frac{h_\varepsilon}{K_BT}\right)} \tag{6.2}$$

The intraband contribution represented in equation (6.3)[6]:

$$\sigma_{\text{int}ra}(\omega)=\frac{2e^2K_BTi}{\pi h^2\left(\omega+i\tau^{-1}\right)}\ln\left(2\cosh\left(\frac{\mu_c}{2K_BT}\right)\right) \tag{6.3}$$

Where, μc denotes chemical capacity, e denotes electron's charge, τ denotes relaxation time, T denotes temperature, $\hbar$ denotes reduced Plank's constant, and kB denotes Boltzmann constant. At terahertz band, the intraband conductivity, which is similar to the Drude model, dominates over interband conductivity and this is due to the simple fact that electron energy (EF) has a higher energy than the photons ($\hbar\omega$) in this frequency range. This leads the total conductivity represented by equation (6.4) [6]:

$$\sigma(\omega)=\frac{2e^2K_BTi}{\pi h^2\left(\omega+i\tau^{-1}\right)}\ln\left(2\cosh\left(\frac{\mu_c}{2K_BT}\right)\right) \tag{6.4}$$

For frequencies in the optical range and above the terahertz band, the contribution of intraband conductivity decreases and approaches a value of $\sigma o = \pi e^2/(2h)$ [6]. This is because at these high frequencies, the electrons in graphene are not able to respond quickly enough to changes in the electric field, leading to a decrease in intraband conductivity. Also, in numerical simulations with plasma frequency, graphene is regarded as a thin material with a tiny thickness Δ represented by equation (6.5) [6]:

$$\omega_p=\sqrt{\frac{2e^2K_BT}{\Delta\pi\varepsilon_0h^2}\ln\left(2\cosh\left(\frac{\mu_c}{2K_BT}\right)\right)} \tag{6.5}$$

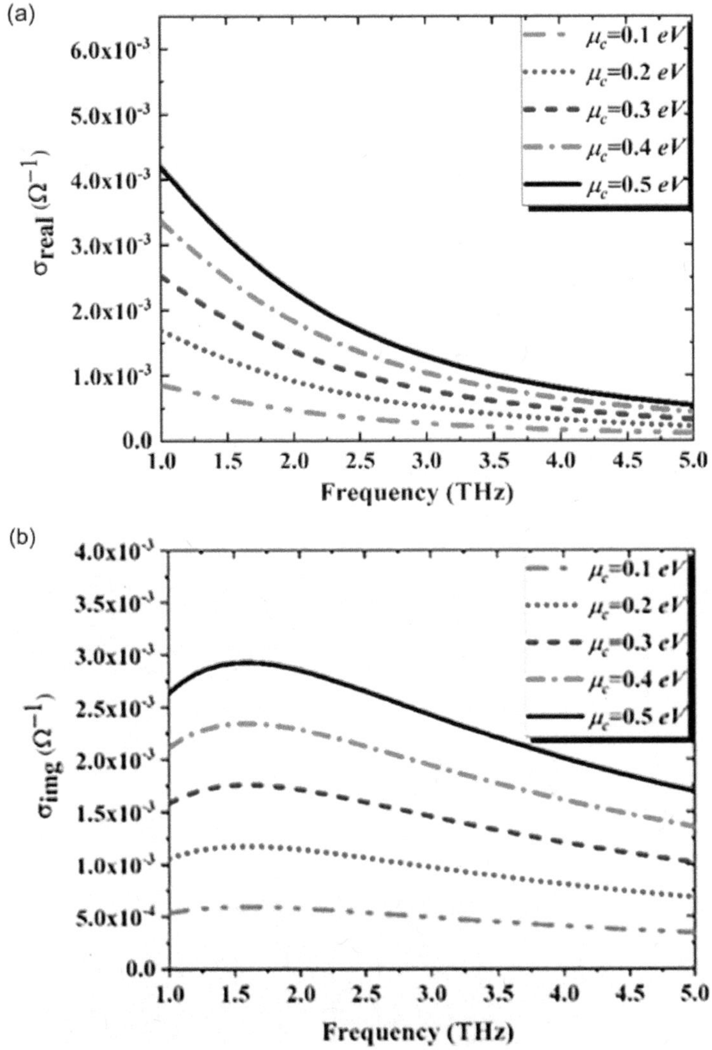

FIGURE 6.2 At T = 300 K, the graphene conductivity curve changes in relation to the chemical potential (μc) for $\tau = 0.1$ ps. (a) real (b) imaginary.

6.3 METASURFACE DESIGN PROCEDURE AND RESULT DISCUSSION

The top, bottom, and dimensions of the proposed metasurface, as seen in Figure 6.3, are given in the table that goes with it. Its base is completely coated in a 5 μm thick coating of gold with a conductivity of 4.11 × 107 S/m, and it is constructed on a 30 μm high quartz substrate with a permittivity of 4.5. The metasurface is activated by means of Floquet ports and periodic boundary conditions, and its unit cell is

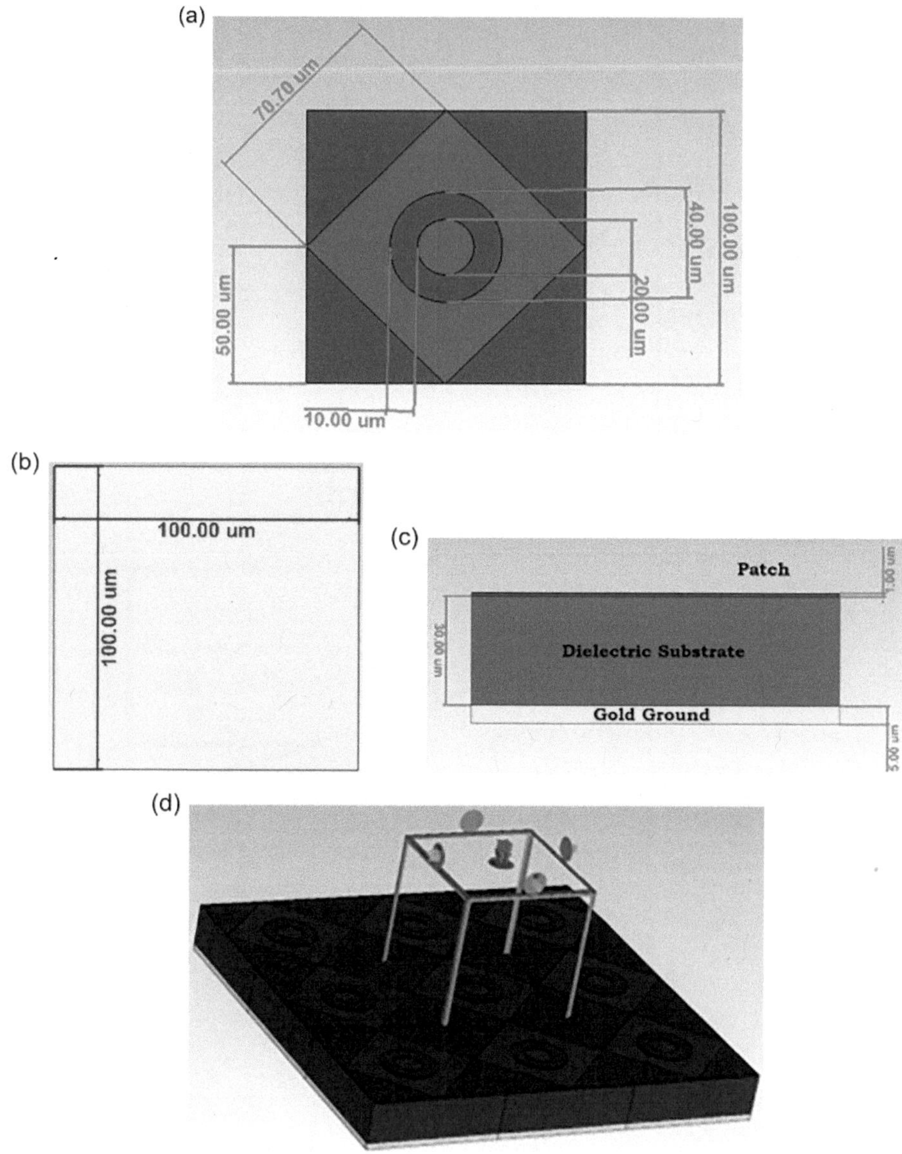

FIGURE 6.3 (a) Metasurface front view with dimension. (b) Back view of metasurface. (c) Height of each layer of metasurface. (d) 3 × 3 array at normal incident.

composed of concentric circular forms with diamond cuts. The development of the suggested absorber depends on the design of a special metasurface structure and its efficient electromagnetic response. It might be difficult for researchers to develop alternative metasurface structures, nevertheless, in order to bolster their conclusions. The size of the unit cell metasurface is $50 \times 50 \times 36\ \mu m^3$.

6.4 METASURFACE SIMULATED RESULTS

Figure 6.4a shows the simulated return loss (S11) of a single cell inside the suggested metasurface over the intended frequency range of 0.6–1.75 THz. It illustrates in Figure 6.4b the reflectivity and absorbance characteristics of the selected unit cell within the intended frequency band (0.6–1.75 GHz) are nearly zero and more than 90% respectively. It is also comprehended using Equation (6.6). Furthermore, the absorptivity can be expressed as:

$$\text{Absorptivity}\,(f) = 1 - |S11|^2 = 1 - \text{reflected power} \tag{6.6}$$

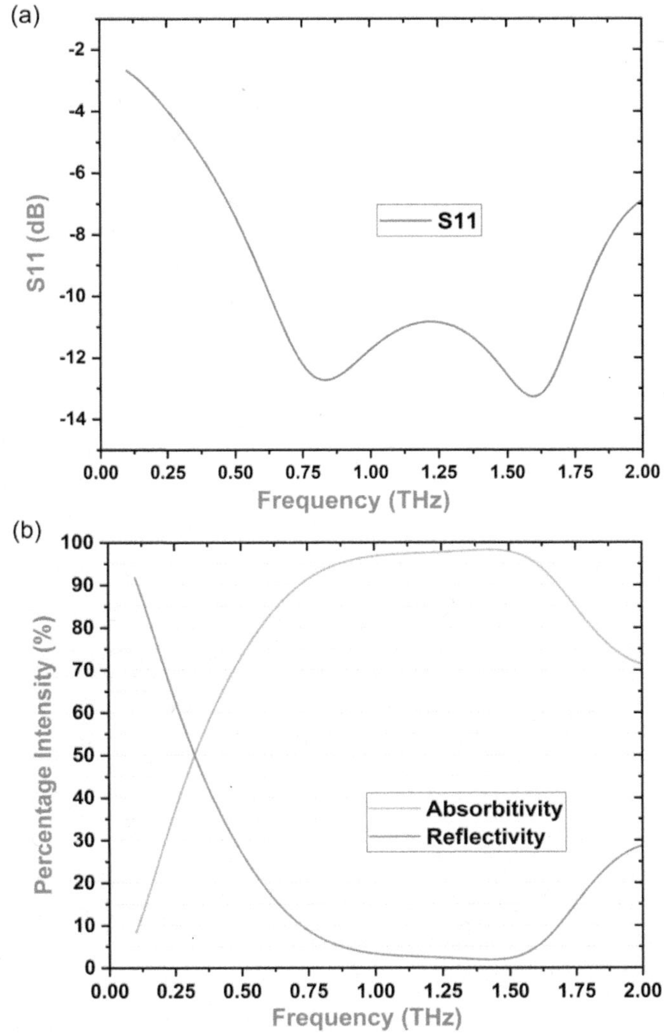

FIGURE 6.4 (a) S11 parameter of metasurface. (b) Absorptivity and reflectivity of metasurface. (c) Theta variations in S11 parameters of metasurface. (d) Phi variations in S11 parameters of metasurface.

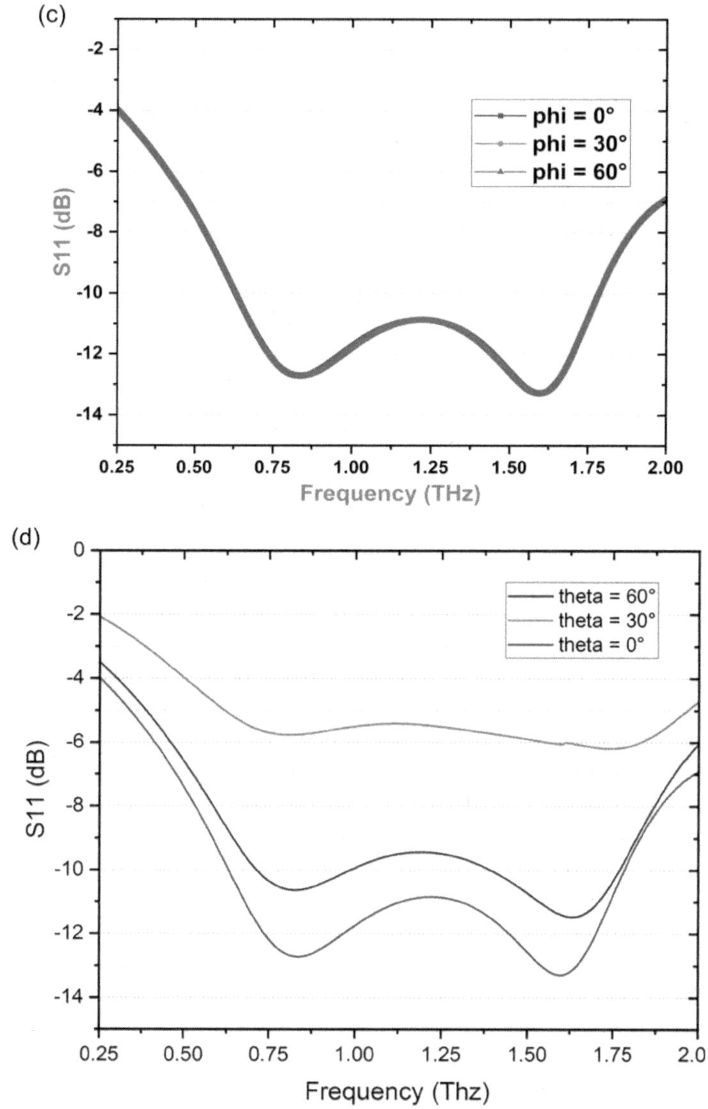

FIGURE 6.4 (Continued)

The reflection coefficient $|S11|^2$ can be decomposed into two components, $|S11, xx|^2$ and $|S11, xy|^2$, which can be used to evaluate the performance of the absorber. By using these equations, researchers can analyse the behaviour of absorbers and optimize their design for specific application.

The proposed cell configuration is designed for developing a metasurface absorber suitable for a range of defence and antenna applications, where the electromagnetic-signals may be acknowledged from various incident-angles (θ). Figure 6.4c depicts the change in return loss at different incidence angles, illustrating the dependency of return loss on the incident angle. However, Figure 6.4d demonstrates that when the phi angles vary between 0° and 60°, the return loss remains constant. This characteristic establishes the proposed structure as a polarization-insensitive metasurface, making it versatile and effective in its response to incoming electromagnetic waves regardless of their polarization state.

The distribution of electric field vectors across the structure's surface is referred to as the E-field surface of a metasurface. These electric fields are generated by the electromagnetic waves that impact the surface and are essential in governing the metasurface's behaviour and functionality. The specific pattern of the E-field surface will depend on the design and properties of the metasurface, as well as the frequency and polarization of the incident waves. In general, however, the E-field vectors will follow the contours of the metasurface, with variations in amplitude and direction across the surface. By analysing the distribution of E-field vectors, it is possible to gain insights into the behaviour of the metasurface and optimize its design for a specific application. The E-field of proposed metasurface at (i) 1.05 THz, (ii) 1.2025 THz, and (iii) 1.4404 THz are given in Figure 6.6(a,b,c). The surface current of proposed metasurface at (i) 1.05 THz, (ii) 1.2025 THz, and (iii) 1.4404 THz are given in Figure 6.7(a,b,c).

Surface current reveals the distribution of electric currents across the surface of the metasurface. This distribution can vary based on the design of the metasurface, including the size, shape, and spacing of the meta-atoms or other components.

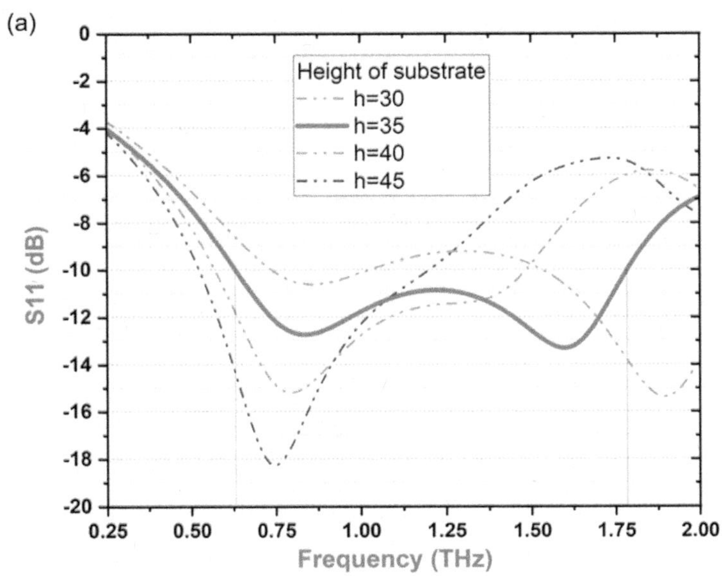

FIGURE 6.5 (a) S11 when height is varied of substrate in metasurface. (b) S11 when relaxation time of graphene is varied in metasurface. (c) S11 when graphene conductivity of graphene is varied to obtain large bandwidth.

The far-field radar cross-section (RCS) analysis of a metasurface was conducted at three different frequencies: 1.404 THz, 1.2025 THz, and 1.05 THz. The total ACS (Aperture Cross-Section) at each frequency was measured, as well as the total RCS, radiation efficiency, total efficiency, and RCS max.

At a frequency of 1.05 THz in Figure 6.8a, the total RCS was −80.68 dBm^2, the total RCS was −91.40 dBm^2, the radiation efficiency was −11.08 dB, the total efficiency was −11.40 dB, and the RCS max was −89.52 dBm^2.

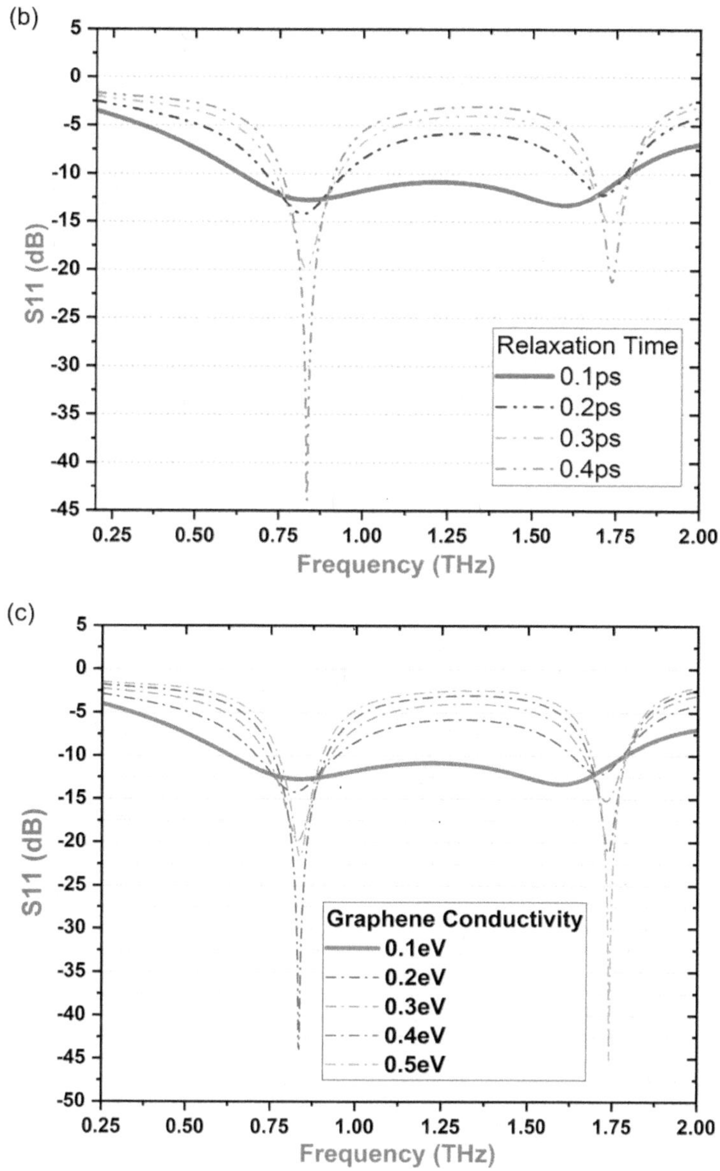

FIGURE 6.5 (Continued)

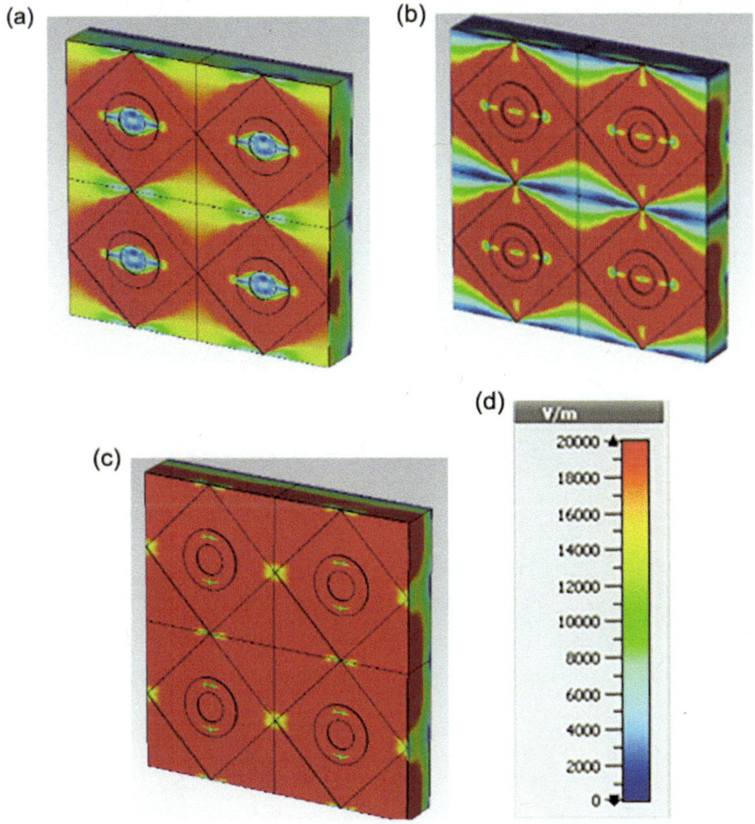

FIGURE 6.6 Electric field (a) at 1.05 THz, (b) at 1.2025 THz, (c) at 1.4404 THz, and (d) scale.

At a frequency of 1.2025 THz in Figure 6.8b, the total RCS was −80.78 dBm^2, the total RCS was -90.85 dBm^2, the radiation efficiency was −10.48 dB, the total efficiency was −10.85 dB, and the RCS max was −87.80 dBm^2.

At a frequency of 1.404 THz in Figure 6.8c, the total RCS was −80.60 dBm^2, the total RCS was −91.93 dBm^2, the radiation efficiency was −11.64 dB, the total efficiency was −11.93 dB, and the RCS max was −87.30 dBm^2.

These measurements provide valuable insights into the electromagnetic behaviour of the metasurface at different frequencies in the far-field region. They offer important information regarding the metasurface's interaction with incident waves, scattering characteristics, and efficiency in radiating and absorbing energy.

6.5 ANTENNA DESIGN

It is suggested to use a common patch antenna made of graphene for THz operation. The 30 μm thick, 2.2 relative permittivity RT5880 substrate serves as the foundation

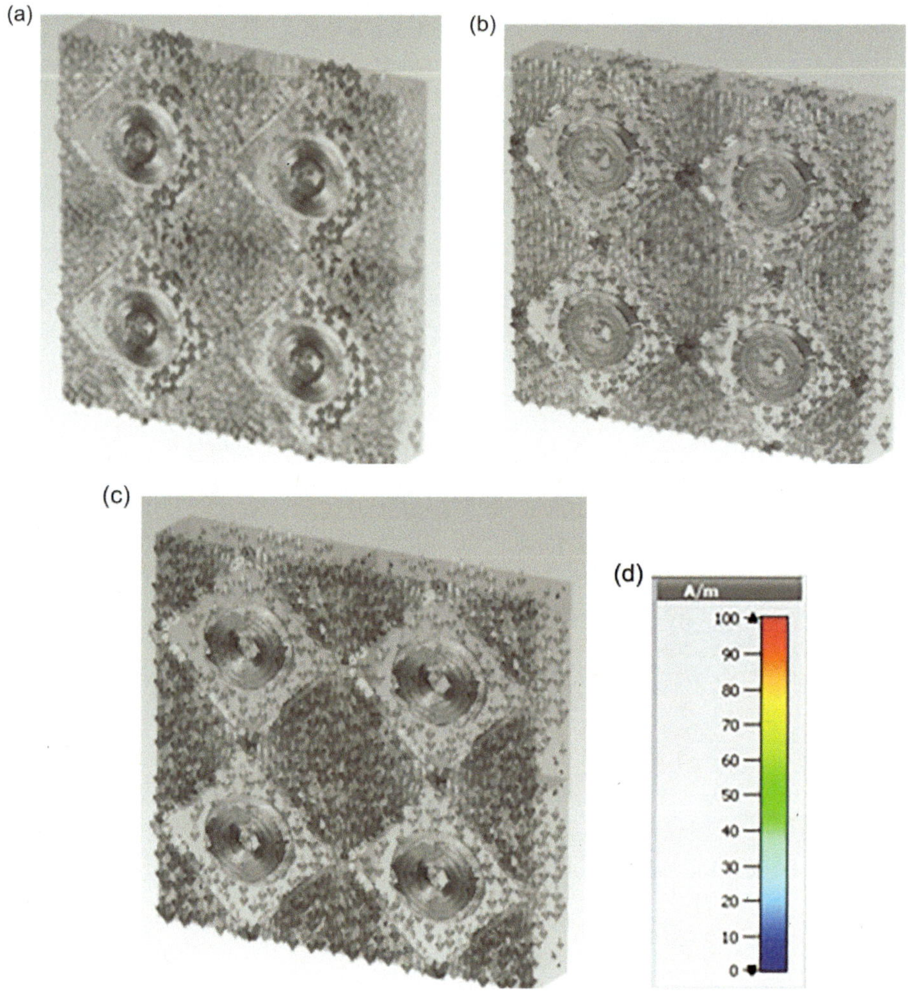

FIGURE 6.7 Surface current density (a) at 1.05 THz, (b) at 1.2025 THz, (c) at 1.4404 THz, and (d) scale.

for the antenna. The ideal patch dimensions were found using the Characteristics-Mode-Analysis (CMA) method and frequency domain solver found in CST Studio Suite. The patch is 25 μm in length and 32 μm in depth. By utilizing the special qualities of graphene, which enable the production and control in the THz frequency range of Surface Plasmon Polaritons (SPP), this design seeks to improve the performance of the antenna. Potential applications for the suggested antenna include sensing, imaging, and communication systems—all areas where high-frequency operations are necessary. The two components form a MIMO antenna with two ports, a double notch, and a closely connected patch. The microstrip is made of RT5880, the relative

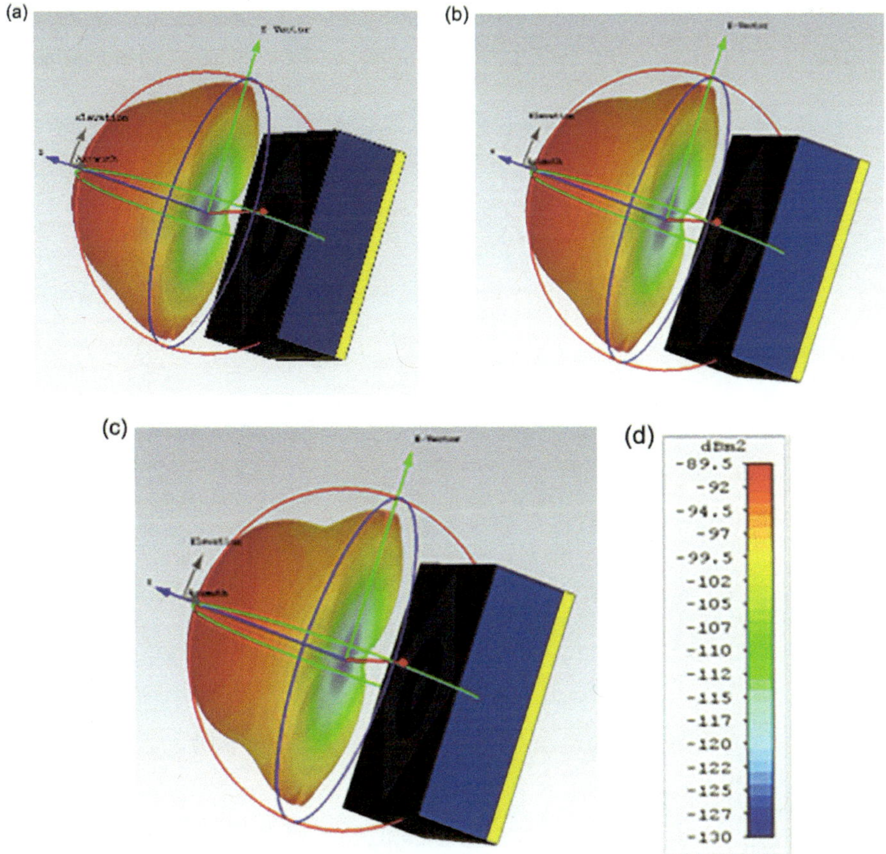

FIGURE 6.8 RCS of unit cell of metasurface (a) at 1.05 THz, (b) at 1.2025 THz, (c) at 1.4404 THz, and (d) scale.

permittivity of a dielectric substrate is 2.2, and it measures 1550 × 1000 × 150 μm^3. The antenna's gold ground plane has a C-loop-shaped decoupling structure that improves the isolation between the two components. Because of its excellent thermal conductivity, which makes it ideal for high-power applications, graphene is used to make the antenna's patches. Furthermore, the application of graphene leads to enhanced radiation properties and diversity performance at a 180° angle. The MIMO antenna is a viable option for several wireless communication applications due to its utilization of sophisticated materials and design approaches.

The graphene MIMO antenna's |S11| curve, which is displayed in Figure 6.10a, reveals how much power is reflected from the antenna at various frequencies. Better impedance matching between the antenna and the feeding line results in increased energy transfer efficiency, as indicated by a lower value of |S11|. The interdependence of elements in a MIMO (Multiple Input Multiple Output) array can impact

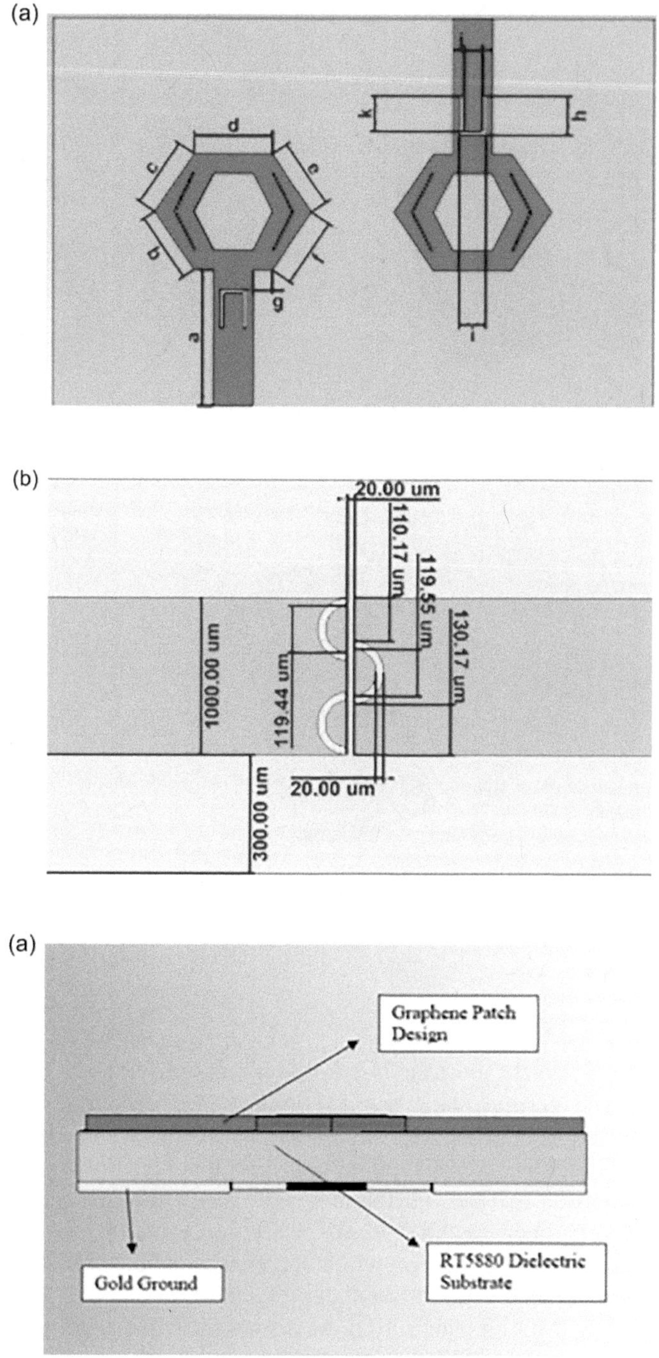

FIGURE 6.9 (a) Antenna Patch Design. (b) Back view and dimensions of decoupling structure. (c) Layered view of antenna.

TABLE 6.1
Dimensions of Patch of Proposed Antenna

Measurement	Size in μm
Feed length (a)	350 μm
Hexagonal patch side (b)	180.28 μm
Hexagonal patch side (c)	180.28 μm
Hexagonal patch side (d)	200 μm
Hexagonal patch side (e)	180.28 μm
Hexagonal patch side (f)	180.28 μm
Length from patch end to feed (g)	50 μm
Patch notch outer vertical length (h)	100 μm
Patch notch outer horizontal length (i)	68 μm
Patch notch inner horizontal length (j)	48 μm
Patch notch inner horizontal length (k)	90 μm

the performance of a single-element antenna. This mutual coupling is produced by electromagnetic contact between MIMO units, and the strength of this interaction is dependent on the distance between the elements. The primary objective of the system is to determine the optimal separation distance between adjacent MIMO elements. A high spacing between components can assure non-correlation while also increasing the total antenna size. A little distance, on the other hand, enhances the mutual coupling between the components. As a result, it is crucial to investigate mutual coupling as a critical parameter in MIMO antenna systems. The MIMO antenna's performance drastically decreases as the distance between its elements decreases because this distance causes the mutual coupling problem. To solve this, a decoupling framework is required. In the ground plane, a complementary new form with three C-looped loops is being carved out. The elevated band-stop property |S21| of the C looping-shaped structure unit cell is shown by the S-parameter curve in Figure 6.10a. The C-looped structure is reduced mutual interaction between the two portions of the graphene due to etching in the ground plane. MIMO setups due to its band rejection feature in the relevant frequency range. Therefore, this approach should be used to enhance isolation. To make this decoupling structure between patch, we firstly connect a ground line with upper and lower ground level of Ground surface of microstrip, then three half circles are looped against each other with dimensions mentioned in Figure 6.9b.

Due to two notches in microstrip patch created to get a peak notch in S11 parameter to minimize the intended band frequency to work on for various business model communication projects. The two notches are created firstly at the feed on the patch and secondly at the side of the hexagonal-shaped antenna's patch.

An antenna's ability to concentrate electromagnetic waves in a specific direction is indicated by its gain. It is commonly expressed in decibels (dB), where an isotropic radiator—one that radiates equally in all directions—is represented by a gain of 0 dB. The antenna in the suggested model has a peak gain of 10.274 dB at 0.8 THz. In comparison to an isotropic radiator operating at that frequency, this indicates that the antenna has a higher gain in the direction of the peak gain.

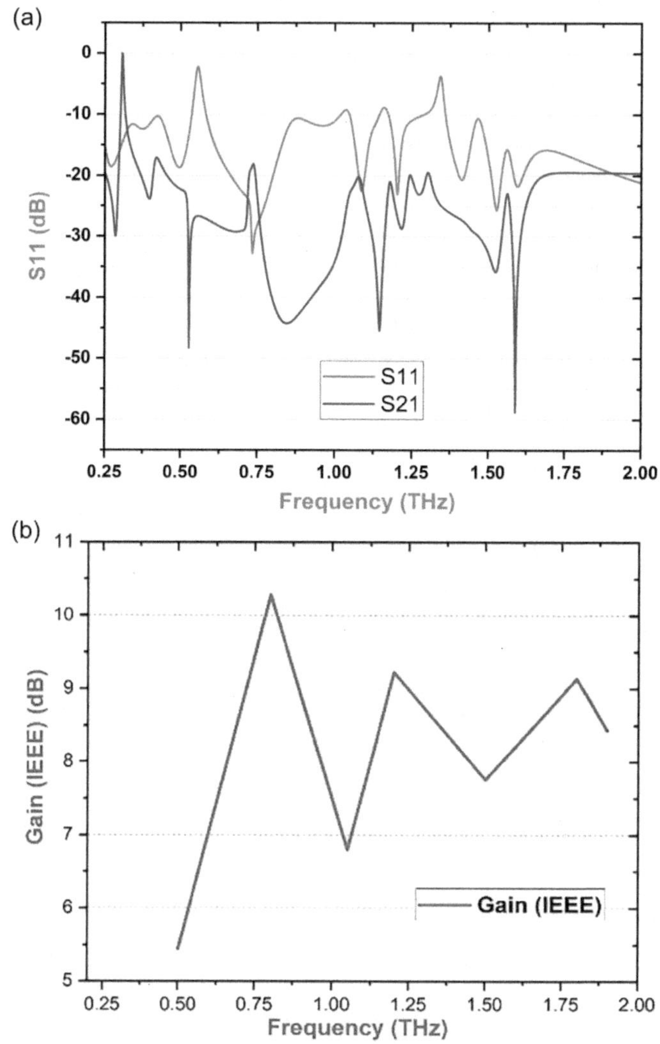

FIGURE 6.10 (a) S11 & S21 Parameter of Antenna (b) Gain of Antenna.

6.6 SIMULATED CO-CROSS-POLARIZATION OF PROPOSED ANTENNA

CST Studio Suite was used to simulate the co-cross-polarization behaviour of a proposed MIMO antenna at frequencies of 1.05 THz, 1.2 THz, and 11.5 THz. The results are shown in Figure 6.11, 6.12, 6.13. In Figure 6.11, the far-field E-field at 1.05 THz is depicted for various theta values ($0° < \theta < 360°$) with a fixed phi value of 90°. At this frequency, the cross-polarization level exhibits a side lobe level of −2.8 dB, a halved power (−3 dB) angular width of 12.9°, and a main lobe magnitude of approximately −4.76 dB. On the other hand, the co-polarization exhibits a side lobe level of −1.6 dB, a halved power (−3 dB) angular width of 22°, with the main lobe directed at 160° and

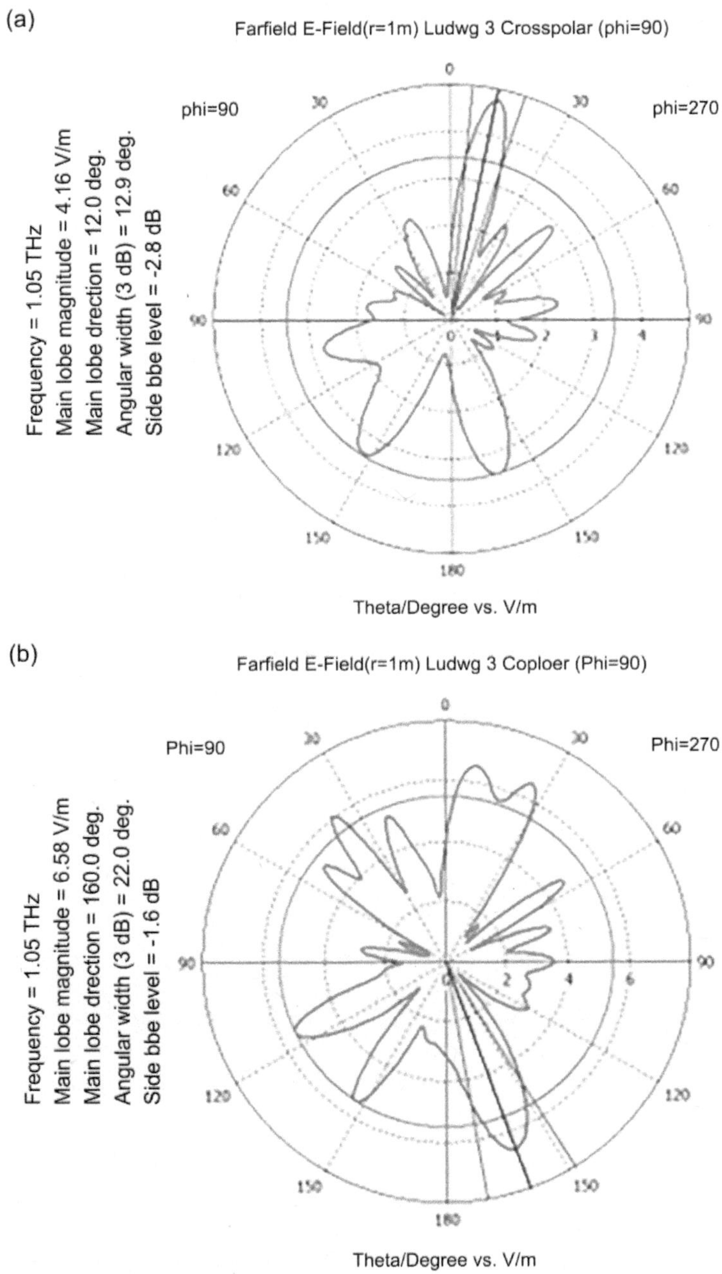

FIGURE 6.11 Simulated co-cross-polarization results of proposed antenna at 1.05 THz.

a magnitude of 6.58 dB. These characteristics indicate a clear separation between the co-cross-polarization at 1.05 THz.

Moving on to the results at 1.2 THz, as depicted in Figure 6.12, the co-cross far-field patterns are shown. In cross-polarization, the side lobe level is approximately

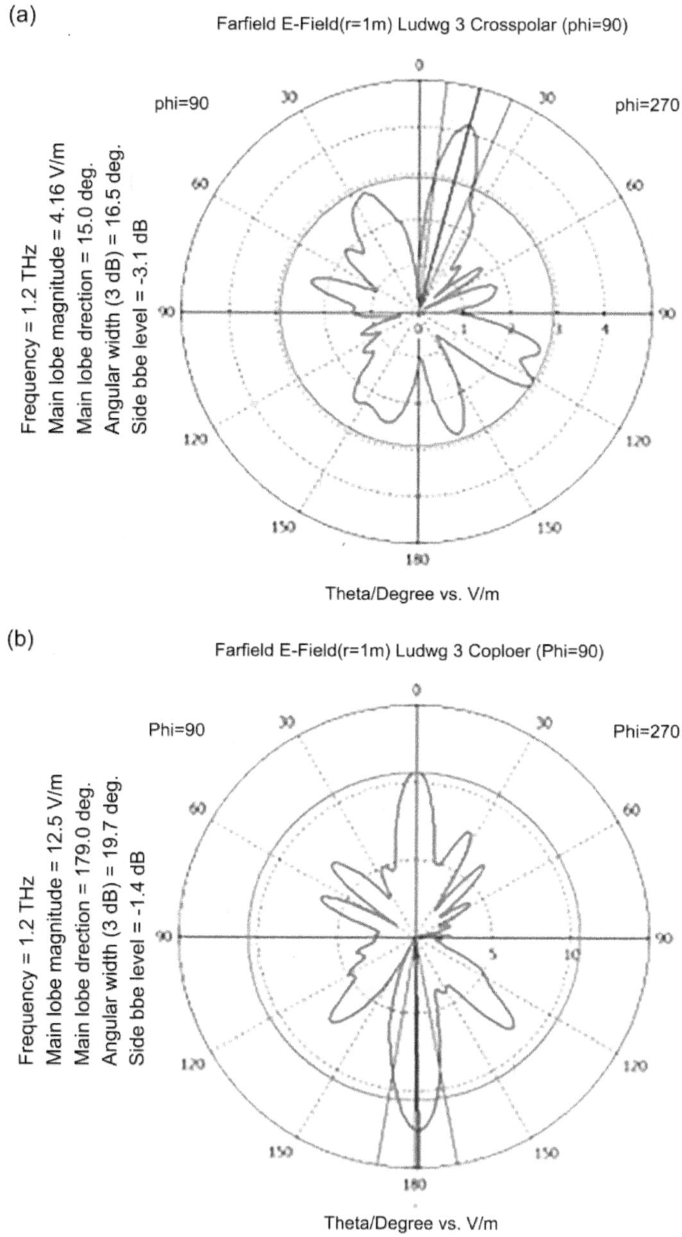

FIGURE 6.12 Simulated co-cross-polarization results of proposed antenna at 1.2 THz.

–3.1 dB, with a 3 dB angular width of 16.5°, and the main lobe is directed at 15° with a magnitude of 4.16 V/m. In co-polarization, the side lobe level is –1.4 dB, the 3 dB angular width is 19.7°, the main lobe is directed at 179°, and the magnitude is 12.5 V/m. These results also demonstrate a distinct separation between the co-cross far-field polarized patterns at 1.2 THz.

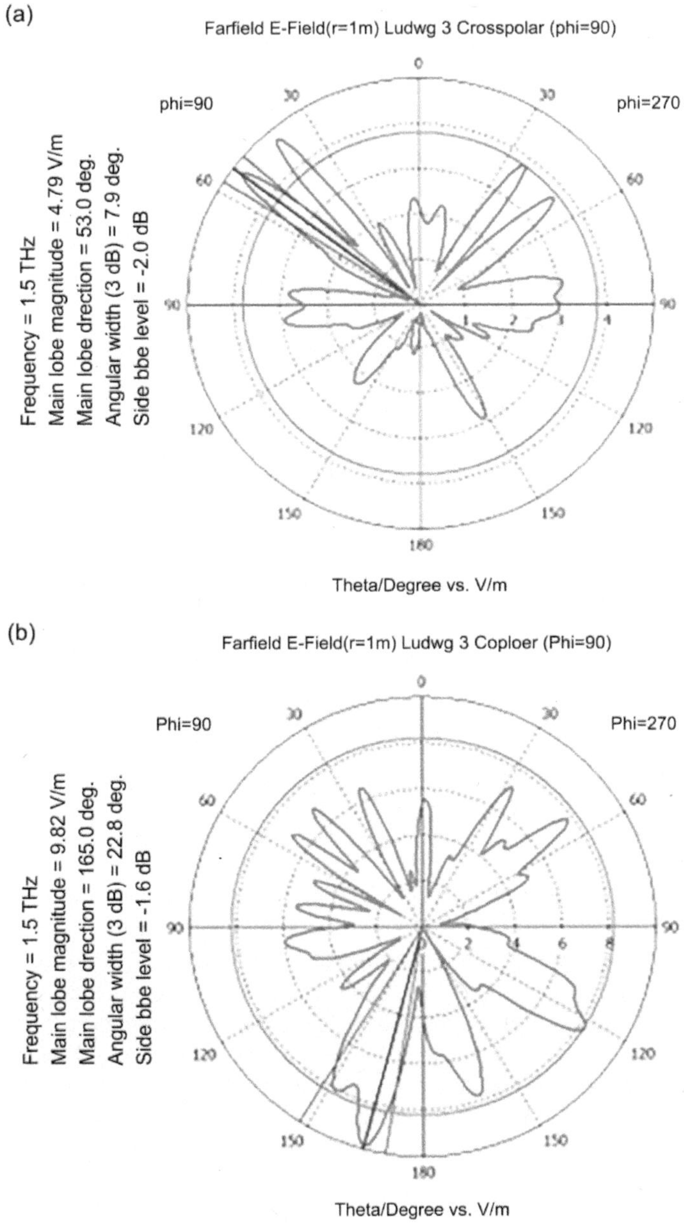

FIGURE 6.13 Simulated co-cross-polarization results of proposed antenna at 1.5 THz.

At last, the results at 1.5 THz are depicted in Figure 6.13, and the co-cross far-field patterns are shown. In cross-polarization, the side lobe level is approximately –2.0 dB, with a 3 dB angular width of 7.9°, and the main lobe is directed at 53° with a magnitude of 4.79 V/m. In co-polarization, the side lobe level is –1.6 dB, the 3 dB angular width is 22.8°, the main lobe is directed at 165°, and the magnitude is

9.82 V/m. These results also demonstrate a distinct separation between the co-cross far-field polarized patterns at 1.05 THz.

6.7 DIVERSITY MEASURES FOR THE PROPOSED ANTENNA

The S11 parameter is a crucial measurement used to evaluate the performance of an antenna. A low value of the S11 parameter, typically below −10dB, indicates efficient power transfer from the source to the antenna. In this study, the S11 parameter was obtained below −10dB for the intended frequency band, indicating that the antenna is properly appropriate for the load and transmission line. The S11 parameter graph shows two notches at frequencies of 0.55 THz and 1.34 THz, indicating that the antenna has low reflection coefficients at these frequencies. These notches correspond to the antenna's resonant frequencies, and their positions can be used to optimize the design of the antenna for specific applications. By analysing the S11 parameter graph, designers can ensure that the antenna has maximum power transfer efficiency and is optimized for the intended frequency band. *The Envelope Correlation Coefficient (ECC)* is a critical factor to consider when assessing how well MIMO antenna systems execute in relation to diversity. A MIMO antenna's ability to perform must be verified by its extremely low ECC, which shows a high level of isolation and correlation across the communication lines. Both the S-parameter approach and the pattern of far-field radiation technique may be used to compute ECC, as shown in equation 6.7 [12]. The S-parameter technique is thought to be less effective than the radiation pattern technique for calculating ECC.

$$ECC = \frac{\left|S^*_{11}\, S_{12} + S^*_{21}\, S_{22}\right|^2}{\left|\left(1 - |S_{11}|^2 - |S_{21}|^2\right)\left(1 - |S_{22}|^2 - |S_{12}|^2\right)\right|} \tag{6.7}$$

The evaluation of MIMO antenna performance involves considering the *Diversity Gain (DG)*, which quantifies the improvement in signal quality. Equation 6.8 [12] provides an estimation for determining the DG. In Figure 6.14, the simulated results for Effective Channel Capacity (ECC) and DG of the proposed MIMO antenna are displayed. The findings indicate that the MIMO antenna design demonstrates strong diversity performance across the frequency spectra of the WLAN band. This suggests that the proposed MIMO antenna offers favourable characteristics for enhancing signal quality and overall system performance.

$$DG = 10\sqrt{1 - (ECC)^2} \tag{6.8}$$

Total Active Reflection Coefficient (TARC): The Total Antenna Reflection Coefficient (TARC) serves as a crucial metric for evaluating the level of reflected power in a MIMO antenna system. The calculation involves dividing the total reflected power by the total throughput power and calculating the square root of the result. A desirable TARC value is typically greater than 10 dB, indicating minimal levels of reflected power. In the proposed system, as shown in Figure 6.16, the TARC surpasses 25 dB over the desired bandwidth. The equation (6.9) for calculating TARC in terms of S-parameters can be referenced from the pertinent literature [5].

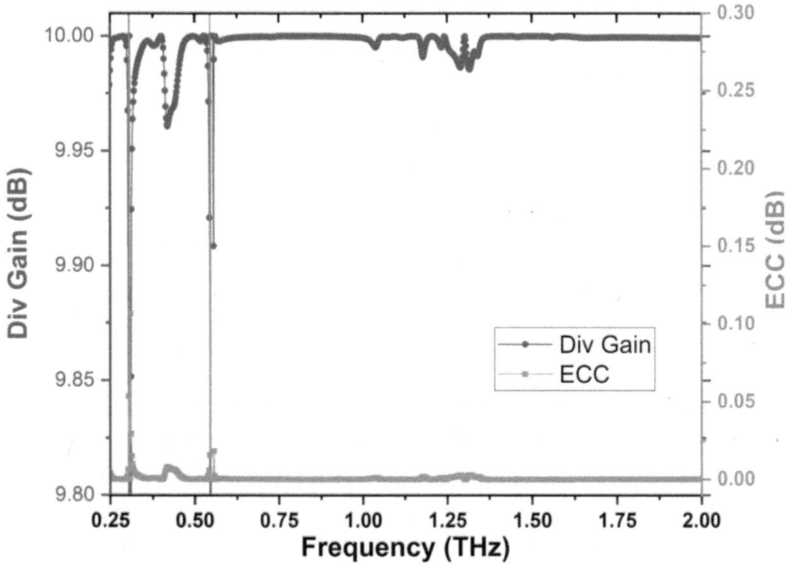

FIGURE 6.14 ECC and DG of proposed antenna.

$$TARC = \frac{\sum_{n=1}^{N} |b_i|^2}{\sum_{n=1}^{N} |a_i|^2} = \sqrt{\frac{\left(S_{ii} + S_{ij}\right)^2 + \left(S_{jj} + S_{ji}\right)^2}{n}} \tag{6.9}$$

Where, *ai* is the incident EM wave, *bi* is the reflected EM wave, *n* is the number of antenna elements.

The effectiveness of a MIMO antenna system can be evaluated using the TARC. A high TARC value indicates that the system is appropriate for the transmission line and that there is little reflected power. This is important for maximizing the efficiency of the system and for minimizing the interference with other systems.

Channel Capacity Loss (CCL): The capacity of a MIMO (Multiple Input Multiple Output) system is directly influenced by two factors: the number of antenna elements and the correlation between these elements. Increasing the number of antennas improves the overall channel capacity, leading to higher data rates in the proposed system, as shown in Figure 6.15. However, as the correlation between antenna elements becomes stronger, the Coefficient of Correlation Loss (CCL) also increases, resulting in a reduction in data rate. It is crucial to have a low CCL to accurately assess the MIMO system's ability to handle actual data rates. For a reliable transceiver link, it is recommended that the threshold value for two-port antennas remains below 0.35 bits/s/Hz. This threshold value can be determined using equations referenced as (6.10–6.12) in the relevant literature [5].

$$CCL = -\log_2\left(\psi^p\right) \tag{6.10}$$

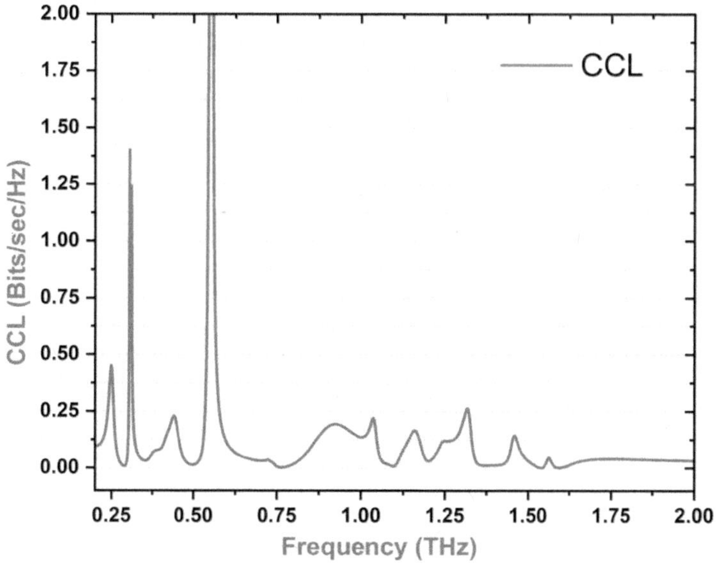

FIGURE 6.15 Channel capacity loss.

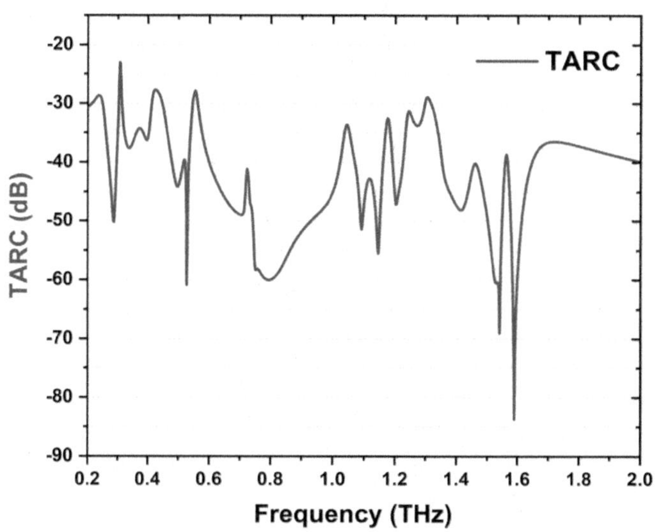

FIGURE 6.16 Total active reflection coefficient (TARC).

Where, $\psi^{p} = \begin{vmatrix} \rho_{11} & \rho_{12} \\ \rho_{21} & \rho_{22} \end{vmatrix}$ And,

$$\rho_{ii} = 1 - \left(\sum_{n=1}^{N} \left| S_{in}^{*} S_{ni} \right| \right) \tag{6.11}$$

$$\rho_{ij} = -\left(\sum_{n=1}^{N}\left|S_{in}^{*}S_{nj}\right|\right) \tag{6.12}$$

6.8 E- AND H-FIELD AND SURFACE CURRENT AND SIMULATED RESULTS DISCUSSION OF PROPOSED HEXAGONAL MIMO ANTENNA

The results of the simulations are presented in Figure 6.17, 6.18, 6.19, 6.20(a)-(b)-(c), showing the distribution of electric, magnetic fields and surface current. In Figure 12(a), the electric field distribution at three different frequencies (1.05 THz, 1.2025 THz, 1.4404 THz and 2 THz) is displayed, and the maximum electric field is observed on the excited antenna-1. At 1.05 THz, the electric field is below 423109 V/m on Graphene patch and gold ground with decoupling structure. At 1.4404 THz and 2 THz, the electric field is approximately 547979 V/m and 368665 V/m confined on Graphene patch and gold ground with decoupling structure.

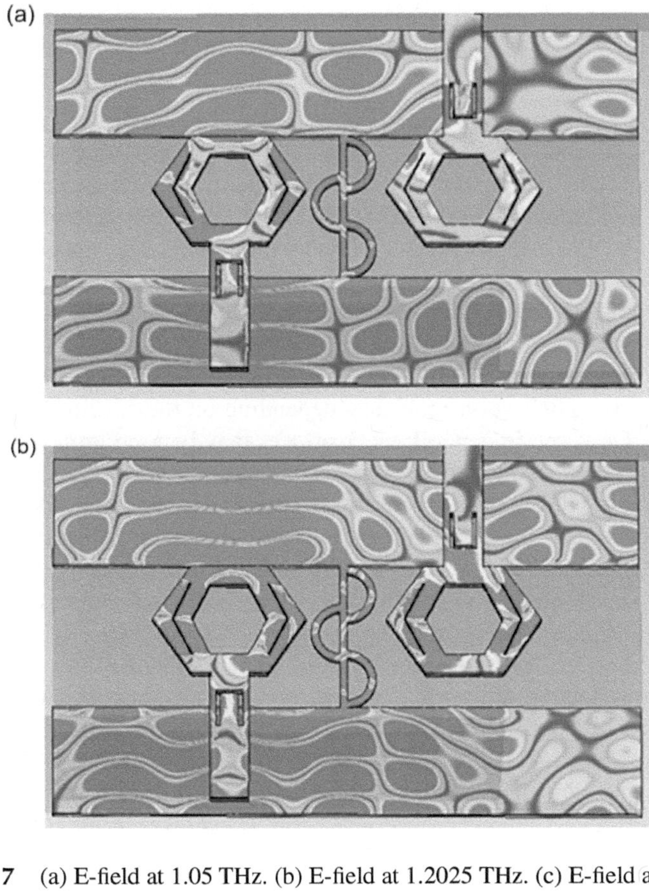

FIGURE 6.17 (a) E-field at 1.05 THz. (b) E-field at 1.2025 THz. (c) E-field at 1.4404 THz. (d) Scale.

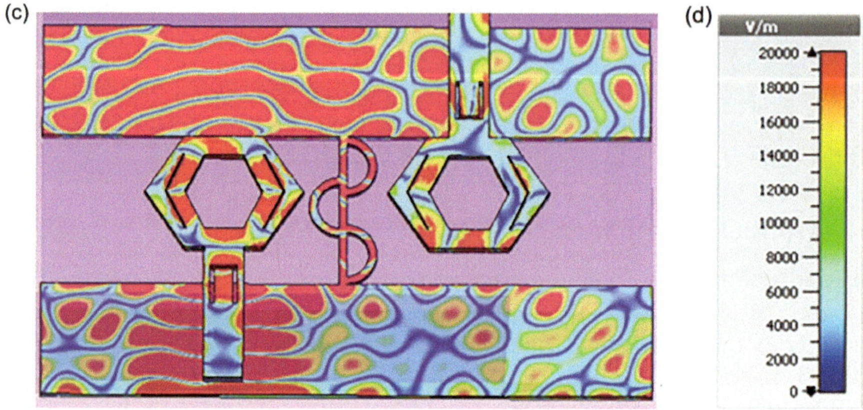

FIGURE 6.17 (Continued)

In Figure 6.17(a)-(b)-(c), the magnetic field intensity is displayed in A/m at 0.5276 THz, 1.145 THz, and 1.5896 THz, where the magnetic field is observed on the excited antenna.

6.9 ANTENNA WITH METASURFACE

In this chapter, it has been shown it is possible to use a metasurface placed below a MIMO antenna to reduce the total RCS (Radar Cross Section) of the system. In the scenario we described, it appears that a metasurface array was placed below the MIMO antenna, and that a layer of wood with a thickness of 50 μm was applied between the metasurface and the antenna. The resulting RCS of the proposed model was reduced to approximately –50dBm^2. The specific design of the metasurface array and the wood layer can have an impact on the resulting reduction in RCS, and the optimal configuration may vary depending on the specific application and requirements. However, in general, metasurfaces may be used to modify EM waves flowing through them, reducing wave scattering and reflection. By placing a metasurface below the MIMO antenna, the metasurface can effectively absorb and/or scatter a significant portion of the electromagnetic waves that would otherwise be scattered back towards the radar system, resulting in a reduced RCS.

The use of a layer of wood or other materials between the metasurface and the antenna can also have an impact on the resulting RCS reduction. By selecting materials with appropriate electromagnetic properties, it is possible to further reduce the scattering and reflection of the electromagnetic waves, resulting in a further reduction in RCS. In this chapter, we have used wood of thickness 400 μm. Overall, the use of a metasurface and other techniques can be effective in reducing the RCS of a MIMO antenna system, improving its performance and reducing the risk of detection by radar systems.

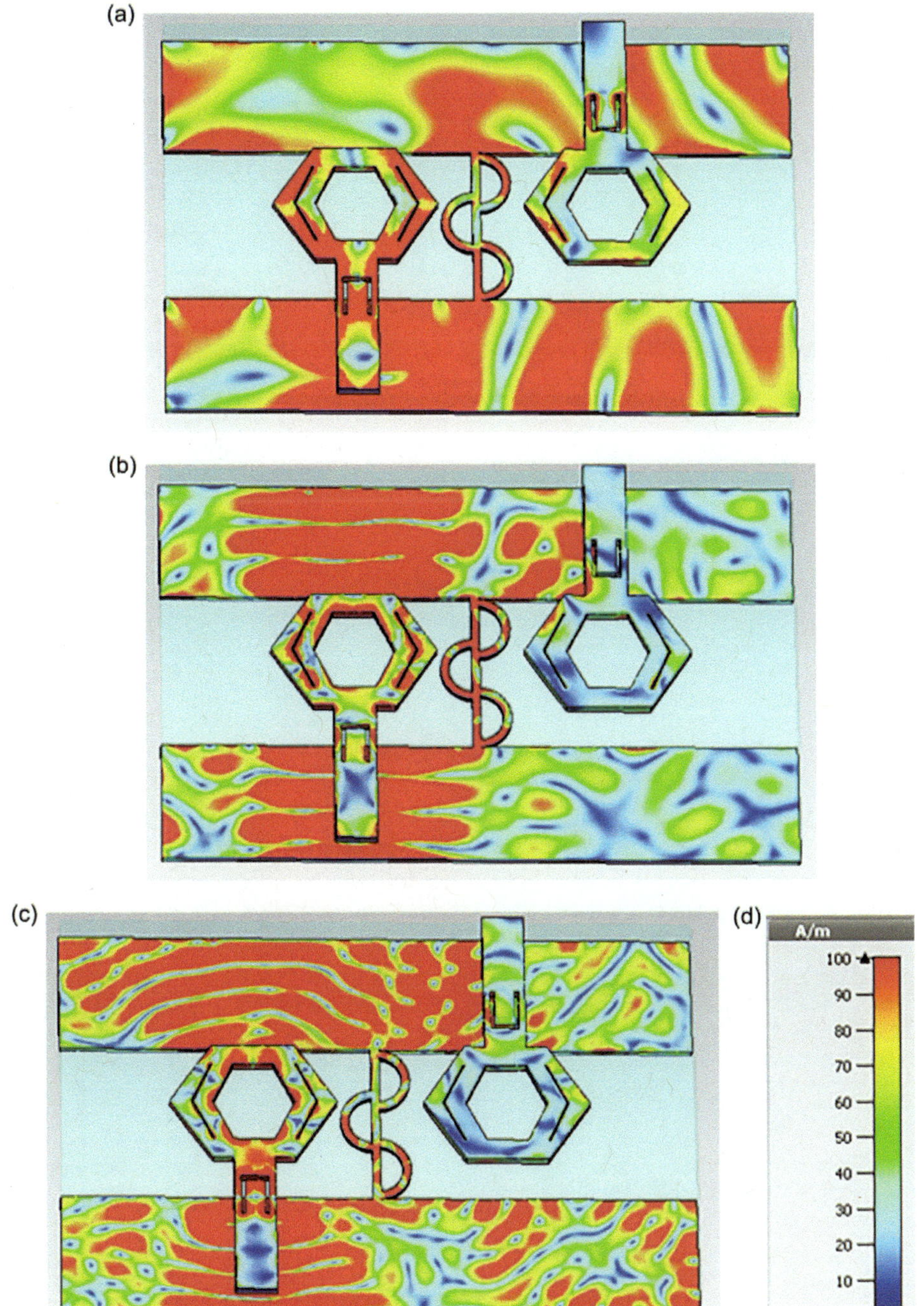

FIGURE 6.18 H-field (a) at 0.5276 THz, (b) at 1.145, (c) at 1.5896 THz, and (d) scale.

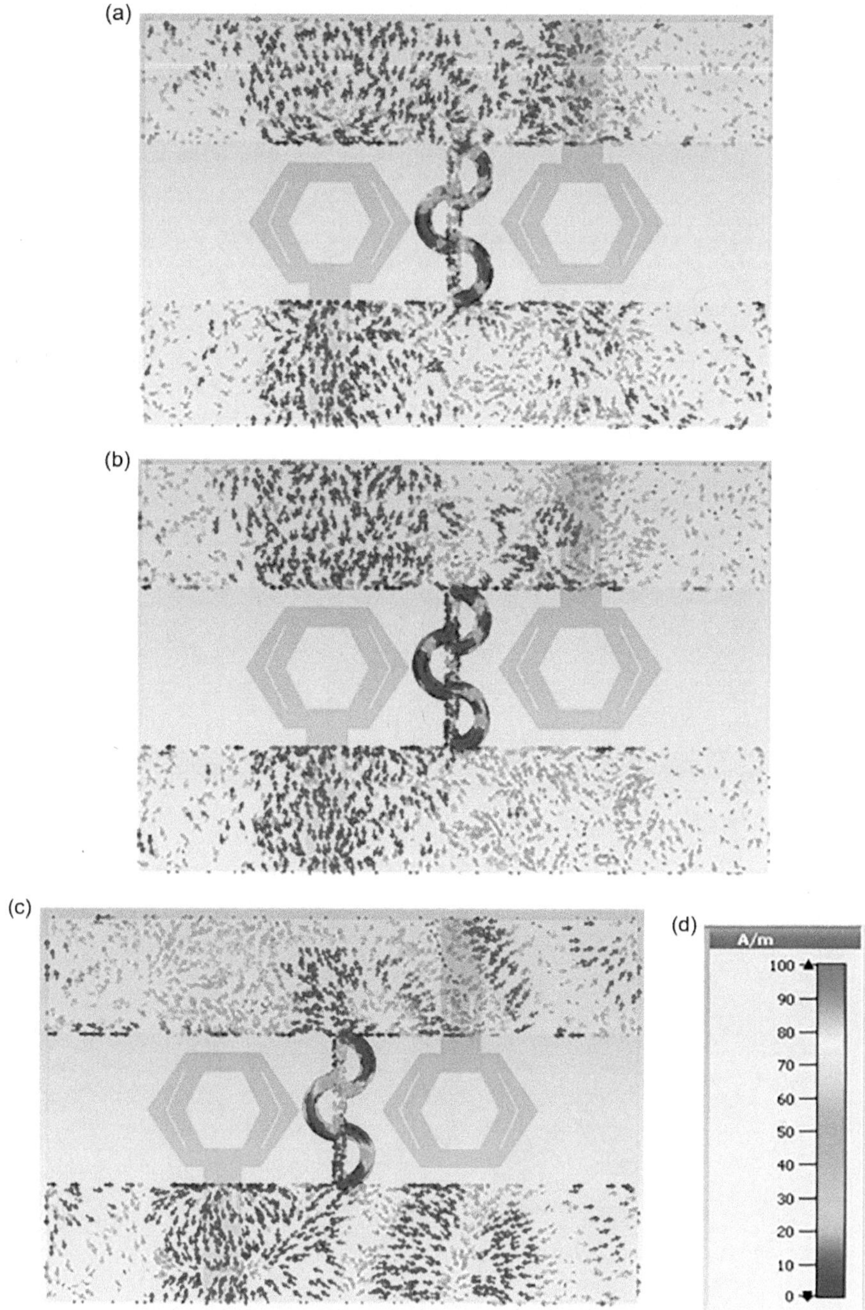

FIGURE 6.19 Surface current (a) at 0.5276 THz, (b) at 1.145 THz, (c) at 1.5896 THz, and (d) scale.

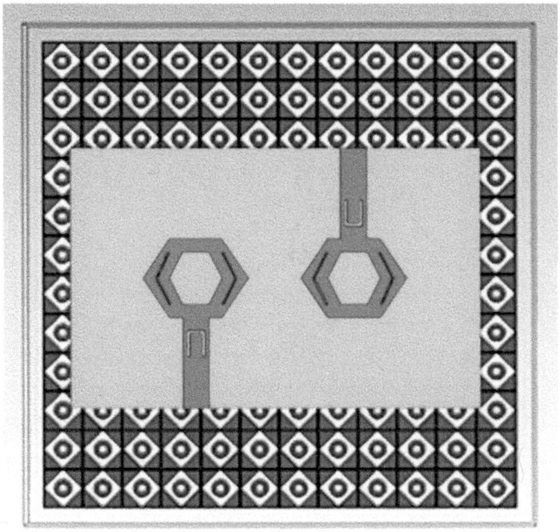

FIGURE 6.20 12 × 12 array of unit cell metasurface.

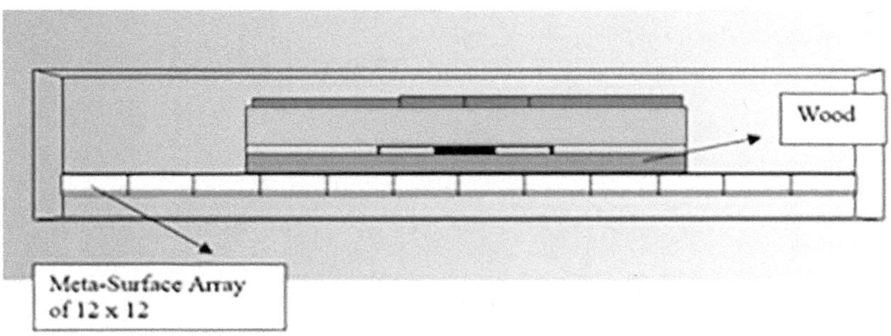

FIGURE 6.21 Layered structure of whole antenna system design.

6.10 CONCLUSION

In this study, we have successfully developed an excellent graphene patch antenna tailored for THz frequencies. This antenna exhibits outstanding efficiency, surpassing 95%, and an impressive antenna gain of 10.274 dBi. To extend the capabilities of the antenna for a bandwidth of (0.1–2.0) THz, we designed a two-element MIMO antenna. We implemented a C-loop constructed ground plane decoupling structure to boost performance and resolve interference, which led to a significant improvement in isolation between the MIMO antenna parts. Moreover, we incorporated a metasurface absorber below the antenna, positioned approximately 50 µm away. This strategic placement allowed us to achieve high isolation, surpassing 20 dB, and a notable reduction in Radar Cross Section (RCS) by 10 dBm^2. Notably, in the desired frequency band, the MIMO antenna demonstrated exceptional radiation efficiency,

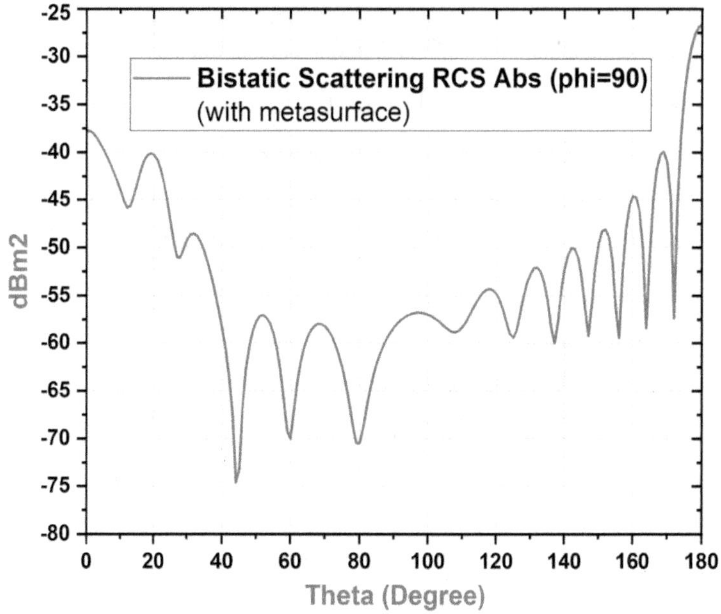

FIGURE 6.22 Bistatic scattering RCS abs (phi = 90) graph reaching –75dBm^2.

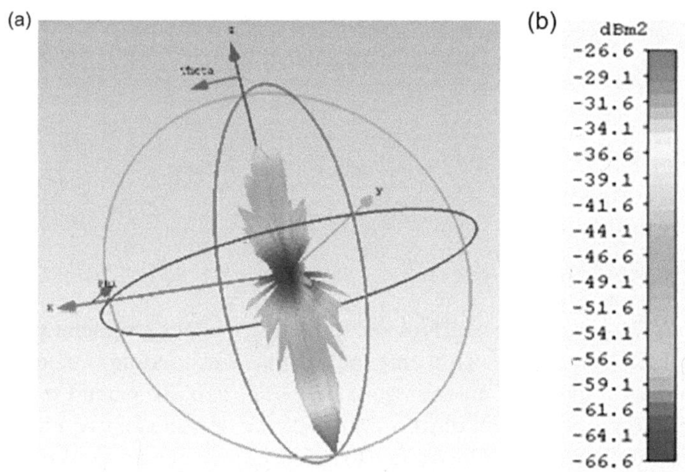

FIGURE 6.23 (a) Radiation pattern in 3D of bistatic scattering RCS with metasurface, and (b) scale.

exceeding 99%, and attained a 10.2 dBi peak gain. This conclusion demonstrates the successful development of a highly efficient and high gain graphene patch antenna for terahertz applications. The integration of a two-element MIMO design with decoupling structures and a metasurface absorber has resulted in remarkable performance enhancements, such as improved isolation and reduced RCS. These findings make the proposed antenna system well-suited for a wide range of applications, including high-frequency communication, sensing, and imaging systems.

REFERENCES

[1] Saxena, G., Awasthi, Y. K., & Jain, P. (2021). Design of metasurface absorber for low RCS and high isolation MIMO antenna for radio location & navigation. *AEU-International Journal of Electronics and Communications*, *133*, 153680.

[2] Wang, Z., Li, C., & Yin, Y. (2020). A meta-surface antenna array decoupling (MAAD) design to improve the isolation performance in a MIMO system. *IEEE Access*, *8*, 61797–61805

[3] Cheng, Y., Zhao, H., & Li, C. (2020). Broadband tunable terahertz metasurface absorber based on complementary-wheel-shaped graphene. *Optical Materials*, *109*, 110369.

[4] Kumar, P., Lakhtakia, A., & Kumar Jain, P. (2020). Tricontrollable pixelated metasurface for stopband for terahertz radiation. *Journal of Electromagnetic Waves and Applications*, *34*(15), 2065–2078.

[5] Saxena, G., Chintakindi, S., Kasim, M. A., Maduri, P. K., Awasthi, Y. K., Kumar, S., . . . Dewan, C. (2022). Metasurface inspired wideband high isolation THz MIMO antenna for nano communication including 6G applications and liquid sensors. *Nano Communication Networks*, *34*, 100421.

[6] Khan, M., Kaium, A., Ullah, M., Kabir, R., & Alim, M. A. (2020). High-performance graphene patch antenna with superstrate cover for terahertz band application. *Plasmonics*, *15*(6), 1719–1727.

[7] Sakli, H., Abdelhamid, C., Essid, C., & Sakli, N. (2021). Metamaterial-based antenna performance enhancement for MIMO system applications. *IEEE Access*, *9*, 38546–38556.

[8] Khaleel, S. A., Hamad, E. K., Parchin, N. O., & Saleh, M. B. (2022). MTM-inspired graphene-based THz MIMO antenna configurations using characteristic mode analysis for 6G/IoT applications. *Electronics*, *11*(14), 2152.

[9] Aliqab, K., Alsharari, M., Sorathiya, V., & Armghan, A. (2023). A numerical investigation of graphene-based Hilbert-shaped multi-band MIMO antenna for the terahertz spectrum applications. *Sensors*, *23*(1).

[10] Raj, U., Sharma, M. K., Singh, V., Javed, S., & Sharma, A. (2021). Easily extendable four port MIMO antenna with improved isolation and wide bandwidth for THz applications. *Optik*, *247*, 167910.

[11] Saxena, G., Awasthi, Y. K., & Jain, P. (2020). High isolation and high gain super-wideband (0.33–10 THz) MIMO antenna for THz applications. *Optik*, *223*, 165335.

[12] Sharma, P., Tiwari, R. N., Singh, P., Kumar, P., & Kanaujia, B. K. (2022). MIMO antennas: Design approaches, techniques and applications. *Sensors*, *22*(20), 7813.

[13] Ali, M. F., Bhattacharya, R., & Varshney, G. (2022). Tunable four-port MIMO/self-multiplexing THz graphene patch antenna with high isolation: Optical and Quantum Electronics, 54, 822.

[14] Zhang, J., Tao, S., Yan, X., Zhang, X., Guo, J., & Wen, Z. (2021). Dual-frequency polarized reconfigurable terahertz antenna based on graphene metasurface and TOPAS. *Micromachines*, *12*(9), 1088.

[15] Xu, Z., Dong, X., & Bornemann, J. (2014). Design of a reconfigurable MIMO system for THz communications based on graphene antennas. *IEEE Transactions on Terahertz Science and Technology*, *4*(5), 609–617.

[16] Vasu Babu, K., Das, S., Varshney, G., Sree, G. N. J., & Madhav, B. T. P. (2022). A micro-scaled graphene-based tree-shaped wideband printed MIMO antenna for terahertz applications. *Journal of Computational Electronics*, *21*(1), 289–303.

[17] Ning, B., Tian, Z., Chen, Z., Han, C., Yuan, J., & Li, S. (2021). Prospective beamforming technologies for ultra-massive MIMO in terahertz communications: A tutorial. *IEEE Open Journal of the Communications Society*, *4*, arXiv preprint arXiv:2107.03032.

[18] Hajiyat, Z. R., Ismail, A., Sali, A., & Hamidon, M. N. (2021). Antenna in 6G wireless communication system: Specifications, challenges, and research directions. *Optik*, *231*, 166415.

[19] Thummaluru, S. R., Kumar, R., & Chaudhary, R. K. (2018). Isolation enhancement and radar cross section reduction of MIMO antenna with frequency selective surface. *IEEE Transactions on Antennas and Propagation*, *66*(3), 1595–1600.

[20] Saxena, G., Kumar, S., Chintakindi, S., Al-Tamim, A., ABIDI, M. H., Saif, W. A. M., Awasthi, Y. K. (2023). Metasurface instrumented high gain and low RCS X-band circularly polarized MIMO antenna for IoT over satellite application. *IEEE Transactions on Instrumentation and Measurement*, 72, 5503210.

[21] Jarchi, S. (2021). Radiation pattern direction control of THz antenna with applying planar graphene metasurface. *Optik*, *243*, 167458.

[22] Khaleel, S. A., Hamad, E. K., Parchin, N. O., & Saleh, M. B. (2022). Programmable beam-steering capabilities based on graphene plasmonic THz MIMO antenna via reconfigurable intelligent surfaces (RIS) for IoT applications. *Electronics*, *12*(1), 164.

[23] Khaleel, S. A., Hamad, E. K., & Saleh, M. B. (2022). High-performance tri-band graphene plasmonic microstrip patch antenna using superstrate double-face metamaterial for THz communications. *Journal of Electrical Engineering*, *73*(4), 226–236.

[24] Kumar, A., Saxena, G., Kumar, P., Awasthi, Y. K., Jain, P., Singhwal, S. S., & Ranjan, P. (2022). Quad-band circularly polarized super-wideband MIMO antenna for wireless applications. *International Journal of RF and Microwave Computer-Aided Engineering*, *32*(6), e23129.

[25] Saxena, G., Chintakindi, S., Al-Tamim, A., Abidi, M. H., Saif, W. A. M., Maduri, P. K., . . . Awasthi, Y. K. (2023). CSRR loaded graphene inspired EBG based high isolation miniaturized wideband THz dielectric resonator MIMO antenna for nano/optical communication in 6G IoT and bio-medical sensing application. *Optical Communication in 6G IOT and Bio-Medical Sensing Application, 1,* 1–20.

7 Frequency Selective Surfaces/Metasurfaces *Equivalent Circuit Modelling*

Deepak Sood

7.1 INTRODUCTION

In 1785, Mr. Francis Hopkinson has noticed the presence of multiple images when he tried to see a distant street lamp through a portion of stretched silk handkerchief. Upon his request and a scientific exchange about this observation an American physicist David Rittenhouse in 1786 has found an explanation to this interesting phenomenon [1]. Rittenhouse has proved that non-continuous surfaces can diffract electromagnetic waves by filtering the white light into distinct wavelengths through a grating of fifty hairs connected between two finely threaded screws [1]. A similar experiment was also conducted by Fraunhofer for measuring the wavelength of different colors by the passing of light through diffraction grating. Further, Wood and Rayleigh have done an exhaustive work on periodic gratings and the determination of diffraction angles [2–3]. Further, the resonance effects observed by Wood were comprehensively discussed in 1965 by Hessell and Oliner [4]. Most likely, Marconi and Franklin provide the first use of gratings by designing the wire section based parabolic reflector for microwave applications [5]. The phenomenon of diffraction of electromagnetic waves by non-continuous surfaces is primitive and can be applicable to any surface designed using periodic patches or rings. Variations in structural element increases the complexity of analysis as the design is two dimensional. Scattering of incident electromagnetic fields on the two dimensional array of periodic cells is contributed by the configuration of each element, inter-element spacing and periodicity [6]. The scattered field exhibits resonances when the element size is an integer multiple of half wavelengths of the incident field. The distinct spectral responses with a distinct order of resonances can be observed with the variations in excitation wavelengths. The wave filtering capability of the frequency selective surfaces will hardly be investigated till the beginning of 20th century. Some of the early applicative work of periodic surfaces includes the use of periodic gratings for gain enhancement of antennas [7], periodic structure loaded waveguides [8], leakywave waveguides [9], frequency selective intermediate reflector [10, 11], design of Cassegrain subreflector for dish antennas and a double-frequency reflector designed for satellite Voyager 77 [12]. In the last decade the frequency selective property of these periodic screens was extensively exploited in microwave regime to design radomes [13, 14], reflect

DOI: 10.1201/9781032656021-7

and transmit arrays [15, 16], lenses [17, 18], low RCS antennas [19], filters [20], EM shielding [21], metamaterial absorbers [22], polarization converters [23], metasurfaces [24] etc. In THz regime metasurfaces are used to design modulators [25], imaging devices [26] and sensors [27]; and in optical domain they are employed for the development of photovoltaic cells [28] and diffraction gratings [29].

In order to analyze any configuration of frequency selective surfaces several classical techniques based on Finite Difference Time Domain (FDTD) and Finite Element Method (FEM) can be used, but these methods are complex [30]. However, some more efficient methods such as methods of moments (MOM) and Integral Equation methods (IEM) are also reported by the experts [31].

7.2 FSS MODELLING

Initially, basic elements such as wire grids or patches of arbitrary shapes were considered to analyze the scattering of incident wave. At first, the transmission line model was used to model scattering of an EM wave from parallel wire grid [32, 33]. The wire grid is modelled as impedance connected in shunt, and surrounding medium is modelled as infinite transmission line. Thereafter, analyses of grid structures loaded with lumped elements were reported [34]. Models to analyze the transmission and reflection behavior at the interface between the dielectrics were also developed [35]. Equivalent circuit analysis for inductive and capacitive grids with additional passive circuit elements was reported [36]. Averaging the currents flowing on the periodic structure is an alternative way to derive the impedance of the metallic grids. The basic circuit modelling equations for periodic gratings in free space were originally proposed by Marcuvitz in 1951 [37] and can be applied only for oblique angles of incidence to an inductive component illuminated by a TE-incident wave or a capacitive component illuminated by a TM-incident wave. A modal expansion method was reported [38, 39] for modelling the scattering by infinite periodic array on a dielectric substrate. The method provides a comprehensive analysis of frequency and polarization responses with physical understanding of operation but requires lengthy computing time and is not suitable for complex elements. Further, spectral domain analysis of cross dipoles and Jerusalem crosses based periodic frequency selective surfaces reported in [40] provides much better physical insights with full wave analysis, but it is computationally onerous. Therefore, the requirement of simple, effective and accurate modelling of periodic surfaces anticipates the development of equivalent circuit modelling. The equivalent circuit modelling technique was first applied to frequency selective circuits by Anderson in 1975 on self-resonant grids [41]. Surface impedance for array of patches can be obtained by using the Babinet principle [42]. To consider the substrate effects FSS capacitance is multiplied by averaged permittivity of the dielectrics when periodicity is lesser than the operating wavelength or when dielectric is thicker than half of periodicity [42]. An important parameter in the FSS analysis is the wavelength at which the grating lobes or trapped modes onset [43]. The wavelength for the grating lobe onset is:

$$\lambda_g = [\sqrt{(\varepsilon_r)} + \mathrm{Sin}\theta]; \qquad \text{For Embedded FSS} \tag{7.1}$$

$$\lambda_g = [1 + \mathrm{Sin}\theta]; \qquad \text{For Freestanding FSS (`D' represents periodicity)} \tag{7.2}$$

In the analysis of periodic structures, three different fundamental regions of operation are classified [43].

Case-I D << λ_g, quasi-static regime, the periodic surfaces can be efficiently analyzed by using homogenized theory.

Case-II D ≈ λ_g, the periodic surface element can be resonant FSS and modelled by using the circuit theory, but the values of the lumped parameters need to be retrieved by using a full-wave simulations.

Case-III D > λ_g Grating Lobe or Trapped Mode region and FSS elements need to be represented by using a multi-mode network.

7.2.1 The Averaged Modelling Technique (D << λg) (Quasi-Static Regime)

This modelling technique is based on the averaged boundary conditions (ABC). These boundary conditions are formulated earlier using the approximate Babinet principle for parallel inductive strips and square strip meshes [44]. The averaged model works properly in the quasi-static regime only when periodicity is much less than the wavelength and grids or patches have a pure inductive and capacitive behavior respectively. The inductance or the capacitance values are calculated by averaging the currents flowing on the periodic structure. For the grid structure as shown in Figure 7.1(a) if the periodicity D is much less than the wavelength (D << λg), the structure is nearly isotropic thereby its electromagnetic response poorly depends upon the choice of the plane of incidence.

When the electric field is tangential to the grid structure i.e. parallel to the strips, the electromagnetic response of the grid is inductive for low frequencies and is characterized by the grid impedance, Z_{grid}. The grid impedance Z_{grid} relates the averaged tangential component of the total electric field in the grid to the induced averaged surface current density '*J*' flowing along the strips. The derived impedances are of second order and angle dependent. It possesses different expressions for the TE and TM polarizations. The expressions for electric field and grid impedances are as:

$$E_x = j\eta_{eff}\alpha\left[J_x + \frac{1}{2k_{eff}^2}\frac{\partial^2}{\partial x^2}J_x\right] \tag{7.3}$$

$$E_y = j\frac{\eta_{eff}}{2}\alpha J_y \tag{7.4}$$

$$\alpha = \frac{Dk_{eff}}{2\pi}\ln\left(\frac{1}{\sin\left(\frac{\pi w}{2D}\right)}\right) \tag{7.5}$$

$$Z_{grid}^{TE/TM} = j\omega L_{grid} \tag{7.6}$$

$$L_{grid}^{TE} = \frac{D\mu_0}{2\pi}\ln\left(\frac{1}{\sin\left(\frac{\pi w}{2D}\right)}\right) \tag{7.7}$$

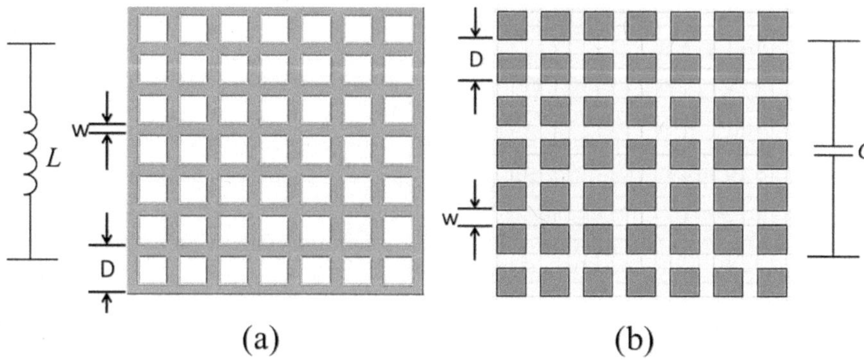

FIGURE 7.1 (a) Inductive wire grid; (b) capacitive patch array (averaged technique).

$$L_{grid}^{TM} = \frac{D\mu_0}{2\pi}\ln\left(\frac{1}{\sin\left(\frac{\pi w}{2D}\right)}\right)\left(1-\frac{k_0^2}{k_{eff}^2}\frac{sin^2\theta}{2}\right) \tag{7.8}$$

The impedance for an array of patches as shown in Figure 7.1(b) is capacitive [44]. Using the approximate Babinet principle the impedances for the array of patch are as:

$$Z_{patch}^{TE/TM} = 1/j\omega C_{patch} \tag{7.9}$$

$$C_{patch}^{TE} = \frac{D\varepsilon_0\left(\varepsilon_{r1}+\varepsilon_{r1}\right)}{\pi}\ln\left(\frac{1}{\sin\left(\frac{\pi w}{2D}\right)}\right)\left(1-\frac{k_0^2}{k_{eff}^2}\frac{sin^2\theta}{2}\right) \tag{7.10}$$

$$C_{patch}^{TM} = \frac{D\varepsilon_0\left(\varepsilon_{r1}+\varepsilon_{r2}\right)}{\pi}\ln\left(\frac{1}{\sin\left(\frac{\pi w}{2D}\right)}\right) \tag{7.11}$$

Here 'k_0' is the wave number in free space, 'w' is the gap between the squares patches, ε_{r1} and ε_{r2} are the relative dielectric permittivity of the dielectric slabs surrounding the periodic surface, and 'θ' is the incident angle. 'K_{eff}' is the wave number of the incident wave vector in the effective host medium as:

$$k_{eff}^2 = k_0\sqrt{\left(\varepsilon_{r1}+\varepsilon_{r2}\right)/2} \tag{7.12}$$

7.2.2 First Order Retrieving Method (D ≈ λg)

When periodicity is equal to wavelength i.e. under the resonant zone the simple model given by averaged technique can be improved by incorporating additional lumped elements in it. A shunt LC connection fits the frequency response for an

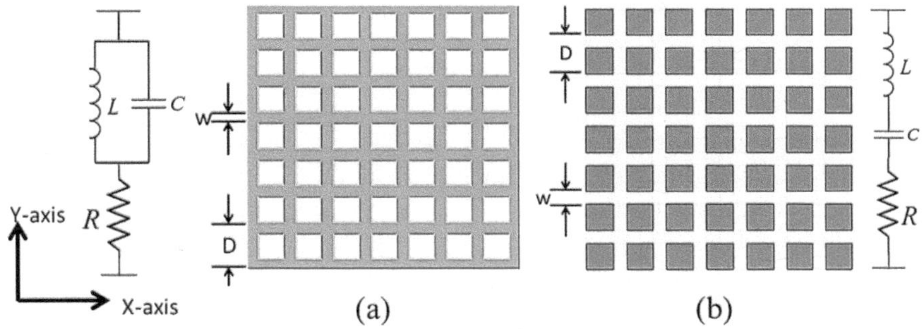

FIGURE 7.2 (a) Inductive wire grid; (b) capacitive patch array (first order retrieving).

inductive FSS grid whereas series LC circuit can fit the response of a capacitive patch FSS array as shown in Figure 7.2.

For considering the effects of losses, a series resistance in the equivalent circuit of the FSS needs to be introduced. A retrieving technique is used to determine the complex reflection coefficient through a preliminary full-wave simulation reported [45]. Thereafter, classical transmission line theory is applied to determine the impedance of the freestanding FSS given as:

$$Z_{FSS} = -\frac{Z_0^2 \left(1+\Gamma_{in}\right)}{2Z_0 \Gamma_{in}} \tag{7.13}$$

$$\Gamma_{in} = \frac{Z_0 \left(1+\Gamma_d\right) - Z_1 \left(1-\Gamma_d\right)}{Z_0 \left(1+\Gamma_d\right) + Z_1 \left(1-\Gamma_d\right)} e^{j2\beta_1 d} \tag{7.14}$$

'Γ_{in}' is the reflection coefficient of the periodic structure calculated at the FSS position. 'Z_0' is the free space impedance and 'Γ_d' represents the input reflection coefficient determined by preliminary full-wave simulation. Inductance and capacitance values are evaluated by approximating the real impedance at two frequency points 'ω_1' and 'ω_2' as:

$$\omega_1 L - \frac{1}{\omega_1 C} = Z_{\omega_1} \tag{7.15}$$

$$\omega_2 L - \frac{1}{\omega_2 C} = Z_{\omega_2} \tag{7.16}$$

After solving the preceding two equations the inductance and capacitance values are evaluated as:

$$L = \frac{Z_{\omega_1 + 1/\omega_1 C}}{\omega_1} \tag{7.17}$$

$$C = \frac{\omega_1 - \omega_2^2 / \omega_1}{\omega_2^2 Z_{\omega_1} - \omega_1 \omega_2 Z_{\omega_2}} \tag{7.18}$$

The preceding technique requires fewer computations, but it is somewhat unstable w.r.t the chosen frequency points, ω_1 and ω_2, and thereby causes some inaccuracies in the calculations. Further, a more accurate method requires the computation of the null impedance. At this null impedance point the capacitance is related to inductance as:

$$C = \frac{1}{\omega_{zero}^2 L} \tag{7.19}$$

After achieving the null condition inductance can be obtained by the iterative procedure. Further, when FSS is embedded in the dielectric layer then extra transmission line must be added in the equivalent circuit to consider the effect of dielectric and to retrieve the L and C values as represented in Figure 7.3 [46]. The reflection coefficient, Γ_{in}, mentioned in equation (7.13) is calculated as [45]:

$$\Gamma_{in} = \frac{Z_0 (1 + \Gamma_d) - Z_1 (1 - \Gamma_d)}{Z_0 (1 + \Gamma_d) + Z_1 (1 - \Gamma_d)} e^{i\beta_1 d} \tag{7.20}$$

Here Γ_d is the input reflection coefficient. When a FSS is sandwiched between two dielectrics the approximate reflection and transmission coefficients of the frequency-selective surface can be computed according to the following transmission-line relations [45]:

$$\Gamma = \frac{AZ_0 + B - CZ_0^2 - DZ_0}{AZ_0 + B + CZ_0^2 + DZ_0} = S_{11} \tag{7.21}$$

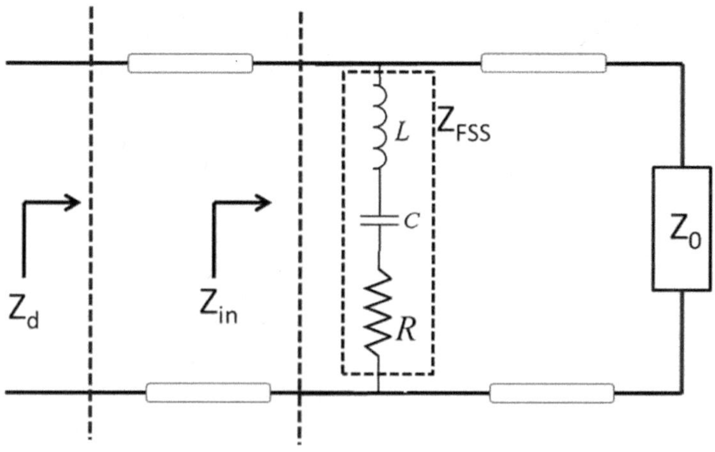

FIGURE 7.3 A transmission-line equivalent of a dielectric embedded FSS [46].

$$\tau = \frac{2Z_0}{Z_0 + B + CZ_0^2 + DZ_0} = S_{21} \tag{7.22}$$

Here, A, B, C, D coefficients represent the terms of the ABCD matrix of the FSS system as reported in [45]. After evaluation of the reflection coefficient using equations (7.20) and (7.21), simple parallel circuit calculations as shown in Figure 7.3 can be done to extract the frequency-selective surface's impedance.

The equivalent parameters inductance and capacitance can be extracted for a particular FSS, but there are certain parameters such as periodicity, dielectric effects and angle of the incident wave that affect these values (L and C). The periodicity greatly affects the frequency response of a FSS system. If all the dimensions of a FSS unit cell are rescaled by some factor then a shift in resonance frequencies can be observed [46]. Therefore, the frequency behavior of this scaled FSS can be analyzed by rescaling the equivalent inductance and capacitance values. Further, the thickness and permittivity of the dielectrics are the two main factors that affect the performance of a FSS array [42, 46]. For the thick dielectrics on both sides of FSS array the resonance frequency is reduced by $\sqrt{\varepsilon_r}$. For single side the reduction in resonance frequency will be by a factor of ($\sqrt{(\varepsilon_r + 1)}$ [42]. With a fixed permittivity, as dielectric thickness decreases the FSS resonance shifts to higher frequencies. It is quite difficult to formulate the relationship between capacitance, permittivity and dielectric thickness [46]. However, for the patch array embedded between two dielectrics the capacitance as a function of permittivity and thickness is reported [45]. An accurate formula that exactly fits the effective permittivity as a function of dielectric thickness is given [45] as:

$$\varepsilon_{eff} = \varepsilon_r^{av} + \left(\varepsilon_r^{av} - 1\right)\left[\frac{-1}{e^N(x)}\right]; \quad x = 10(\mathrm{d/D}) \tag{7.23}$$

$$\varepsilon_r^{av} = \frac{\left(d_1\varepsilon_{r1} + d_2\varepsilon_{r2}\right)}{\left(d_1 + d_2\right)} \tag{7.24}$$

Here, d_1 and d_2 represent thicknesses of two dielectrics in which FSS array is embedded and 'N' is an exponential factor that is different for different cell shapes. The capacitance of the dielectric embedded FSS can be computed with a multiplication of the free-standing value by the effective permittivity. For the FSS embedded within two dielectric slabs [45] with different permittivities (ε_{r1}, ε_{r2}) and thicknesses (d_1, d_2) the averaged value of the effective permittivity is represented in equation (7.24). Further, as the angle of incident wave varies the impedance of FSS gets changed. For oblique angles the impedance of periodic surfaces is expressed in matrix form as reported [63].

$$Z = \begin{bmatrix} Z_{TE,TE}^{FSS}\left(k_x, k_y\right) & Z_{TM,TE}^{FSS}\left(k_x, k_y\right) \\ Z_{TE,TM}^{FSS}\left(k_x, k_y\right) & Z_{TM,TM}^{FSS}\left(k_x, k_y\right) \end{bmatrix} \tag{7.25}$$

Here the wave numbers $k_x = k_0 \, Sin(\theta)Cos(\phi)$ ***and*** $k_y = k_0 \, Sin\,(\theta)\,Sin(\phi)$. The off-diagonal terms represent the coupling between the modes and are considered negligible w.r.t the dependence on θ for symmetric unit cells. Hence, the FSS impedance for oblique incidence is a two-term diagonal matrix. For different FSS elements the dependency of FSS capacitance on incident angle was analyzed using iterative procedure in [45]. The capacitances of patch and ring arrays exhibit similar angular dependencies, but for different widths of cross elements the capacitance varies differently as the angle of incidence increases. If the periodicity is less than half wavelength the expression of patch array capacitance is equally valid for both normal and oblique incidence of wave. Applying Babinet's theorem to the wire grid impedance patch impedance is evaluated. With an assumption of averaging of currents flowing on it and same angular dependence the normal-incidence impedance for a generic FSS with a correction factor 'α' is given as:

$$C_{TE}(\theta) = C_0 \left[1 - \frac{k_0^2}{k^2} \frac{Sin^2\theta}{\alpha} \right] \tag{7.26}$$

$$C_{TM}(\theta) = C_0 \tag{7.27}$$

C_0 represents the capacitance computed in the free-space for normal incidence. The value of α is equal to 2, for the large square elements (patch and ring).

7.2.3 Multimode Approach (D >> λg)

The frequency response of FSS above the visible region is non-linear due to the generation of higher order mode. The generation of higher modes causes grating lobe propagation due to the reflection or transmission of energies that occurs in other directions as per the Rayleigh Theory. If FSSs are required to be operated in the frequency region between resonant zone and the grating lobe propagation zone and above the grating lobe zone then two additional impedances are required to be connected in series in the equivalent circuit comprising inductances and capacitances as shown in Figure 7.4.

Inductive elements contribute the excitation of TE harmonics, whereas the capacitive components contribute for the TM harmonics. Impedances that represent TE and TM higher order modes are as follows [47]:

$$Z_L(\omega) = \sum_{h=1}^{N_{TE}} A_h^{TE} Z_{TE,h}^{in}(\omega) \tag{7.28}$$

$$Z_C(\omega) = \sum_{g=1}^{N_{TM}} A_g^{TM} Z_{TM,g}^{in}(\omega) \tag{7.29}$$

Here, A_h and A_g represent the degree of excitation of the harmonics, the indices h and g are referred to a particular couple of (m,n) pair. Further, h and g are referred to a particular couple of (m,n) pair and $Z_{in}TE$ and $Z_{in}TM$ are the shunt connection between the impedances seen from the left and the right of the FSS respectively.

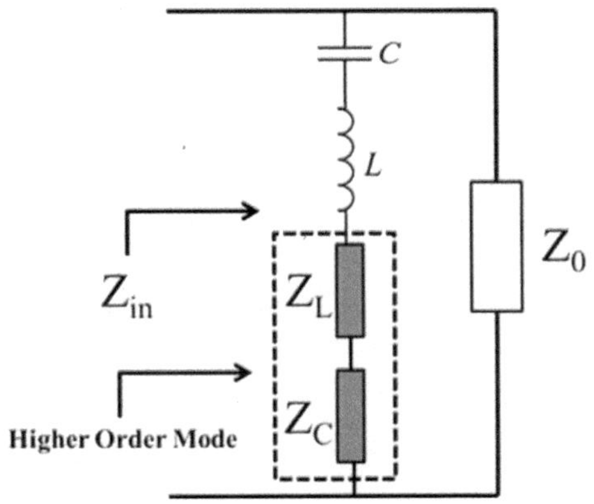

FIGURE 7.4 ECM model for single resonant FSS for region between resonant and grating lobe zone.

7.3 EQUIVALENT CIRCUITS FOR GENERIC FSSs

After the retrieval of equivalent circuit parameters, the S_{11} and S_{21} parameters of FSS are computed with the use of equations (7.21) and (7.22). A, B, C, D parameters are calculated using ABCD matrix of a FSS system consists of 'N' number of dielectrics as [64]:

$$\begin{bmatrix} A & B \\ C & D \end{bmatrix} = [M_1]\ldots\ldots\ldots\ldots[M_n][M_{FSS}][M_{n+1}]\ldots\ldots\ldots[M_N] \tag{7.30}$$

Here M represents scattering matrix of each layer and n signifies the nth layer of dielectric. Hence,

$$[M_n] = \begin{bmatrix} Cos(k_{zn}d_n) & jZ_nSin(k_{zn}d_n) \\ \dfrac{j\ Sin(k_{zn}d_n)}{Z_n} & Cos(k_{zn}d_n) \end{bmatrix} \tag{7.31}$$

and

$$[M_{FSS}] = \begin{bmatrix} 1 & 0 \\ 1/Z_{FSS} & 1 \end{bmatrix} \tag{7.32}$$

Here $k_{zn} = \sqrt{\varepsilon_r k_0^2 - k_t^2}$ is the normal component of wavenumber

(7.33)

$$k_t = k_0\, Sin\theta \tag{7.34}$$

The equivalent circuit of non-resonant elements like strip and patch consists of capacitive element and single resonant elements like dipole, loop, cross and tripole are represented by L and C in series combinations while double resonant elements such as double square loop or double cross elements are represented by two LC combinations in parallel as reported in [46]. However multi-resonant type elements such as concentric loops are represented by multiple LC combinations in parallel. The four basic types of FSS element groups include Group-1 (centrally connected such as dipole, tripole, square-spiral, Jerusalem crosses), Group-2 (looped shapes, e.g., circular, square, hexagonal loops), Group-3 (solid interiors or patch shapes of different shapes), and Group-4 (combinations of all of the above). The equivalent circuits for the basic generic FSS are represented in Table 7.1.

For double square loop FSS structure as shown in Figure 7.7 for which the equivalent circuit is represented in Table 7.1 the lumped circuit parameters can be extracted accurately by using the following set of equations [37, 48]:

$$X_1 = \omega L_1 = 2.0\frac{X_{1p}X_{2p}}{X_{1p}+X_{2p}}\left(\frac{d_1}{p}\right) \tag{7.35}$$

$$X_2 = \omega L_2 = F\left(p, 2w_{2,}\lambda\right).\left(\frac{d_2}{p}\right) \tag{7.36}$$

$$X_{1p} = F\left(p, w_1, \lambda\right) \tag{7.37}$$

$$X_{2p} = F\left(p, w_2, \lambda\right) \tag{7.38}$$

$$B_1 = \omega C_1 = 0.75.B_{1p}.\left(\frac{d_1}{p}\right) \tag{7.39}$$

$$B_1 = \omega C_2 = \frac{B_{1p}B_{2p}}{B_{1p}+B_{2p}}.\left(\frac{d_2}{p}\right) \tag{7.40}$$

$$B_{1p} = 4\varepsilon_r F\left(p, g_1, \lambda\right) \tag{7.41}$$

$$B_{2p} = 4\varepsilon_r F\left(p, g_2, \lambda\right) \tag{7.42}$$

$$F_1\left(p, w, \lambda\right) = \frac{p}{\lambda}\left[\ln\left(Cosec\frac{\pi w}{2p}\right) + G_1\left(p, w, \lambda\right)\right] \tag{7.43}$$

$$F_2\left(p, g, \lambda\right) = \frac{p}{\lambda}\left[\ln\left(Cosec\frac{\pi g}{2p}\right) + G_2\left(p, g, \lambda\right)\right] \tag{7.44}$$

7.4 EQUIVALENT CIRCUIT MODELLING OF DUAL AND MULTIBAND METASURFACES

The technique of equivalent circuit modelling suggested by Marcuvitz [37] was first applied to the frequency selective surfaces by Anderson [41] in 1975. This technique is not very fit for multilayer FSS geometries as it doesn't support the effects of the

TABLE 7.1
Equivalent Circuit Diagrams for Different FSS.

S. No.	FSS Types	Unit Cell Design	Equivalent Circuit	References
1	Square Loop	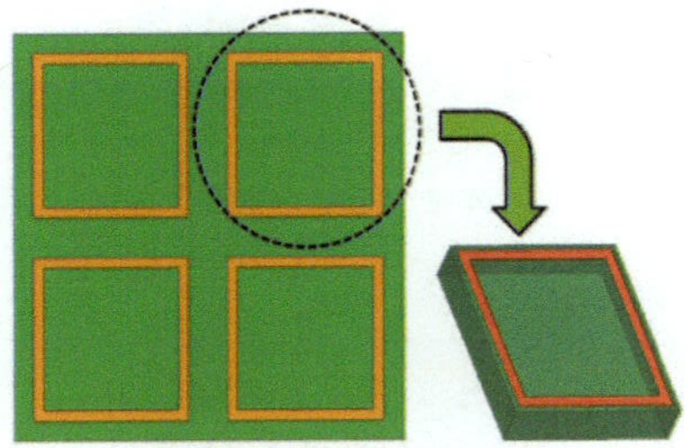	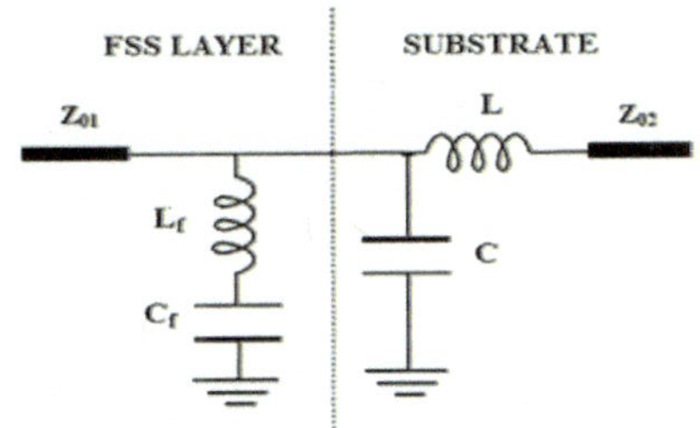	[48]
2	Grid		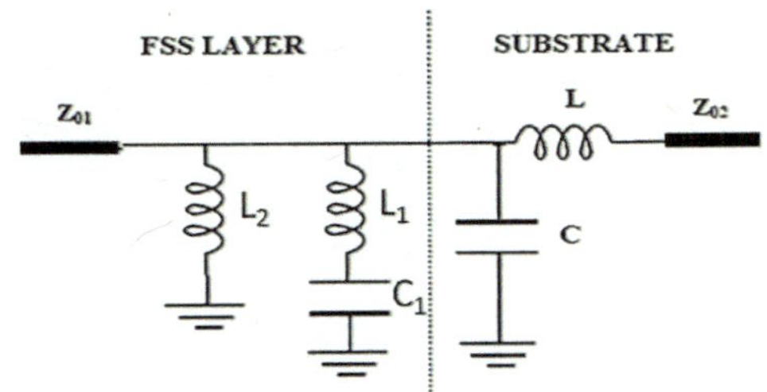	[48]
3	Double Square Loop	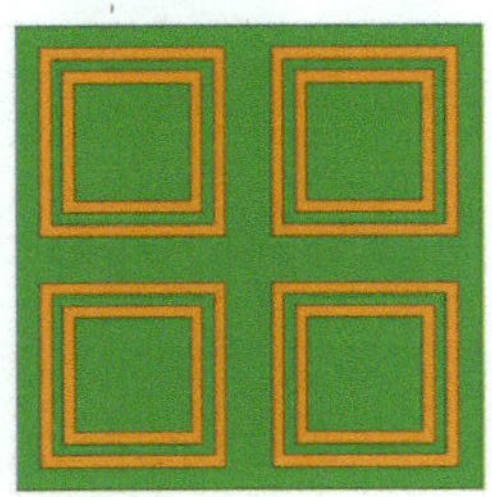	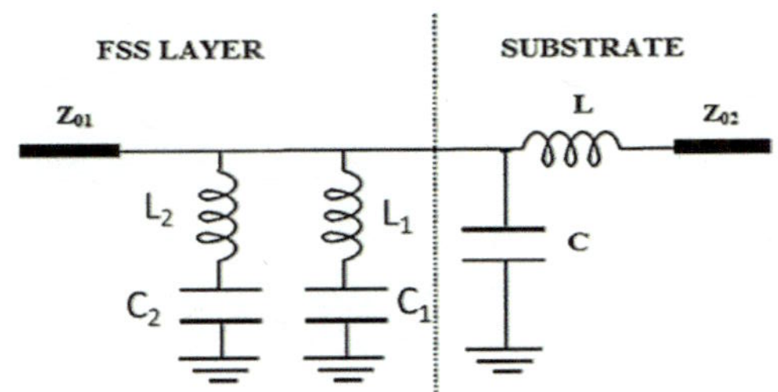	[48]

(*Continued*)

TABLE 7.1 (*Continued*)
Equivalent Circuit Diagrams for Different FSS.

S. No.	FSS Types	Unit Cell Design	Equivalent Circuit	References
4	Jerusalem Cross	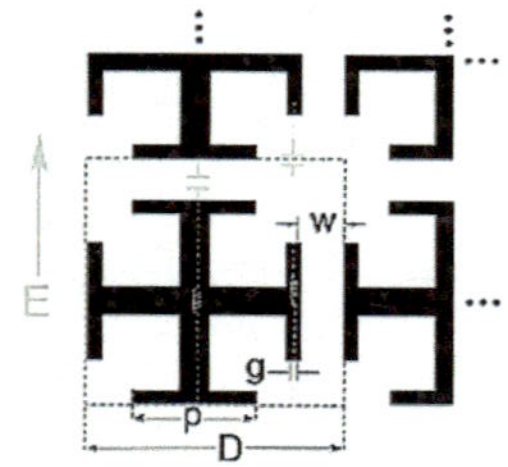	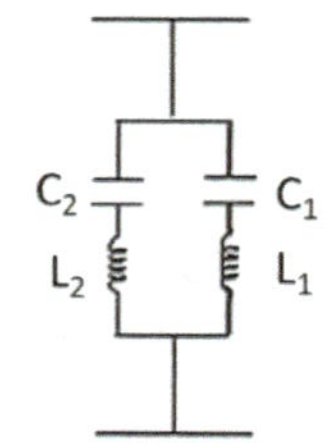	[45]
5	ELC Resonator	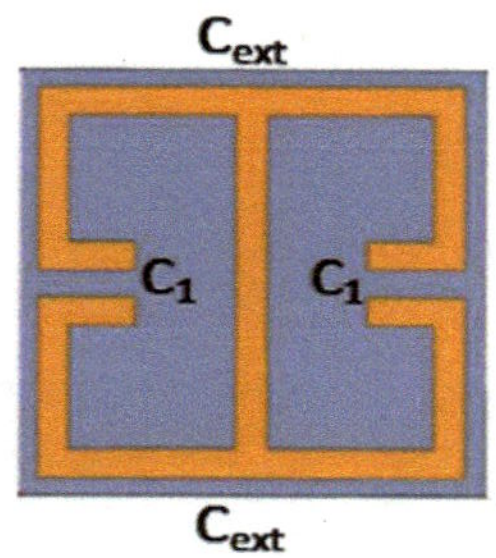	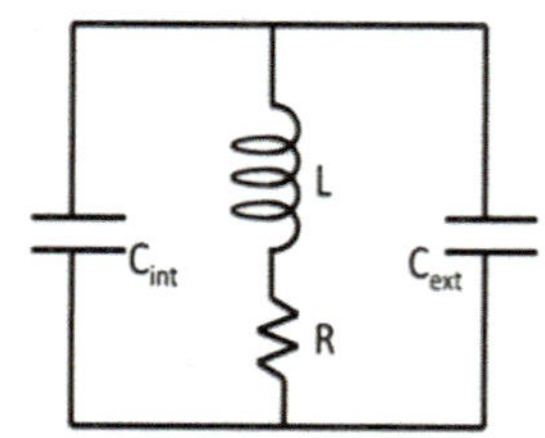	[49]
6	Four Legged	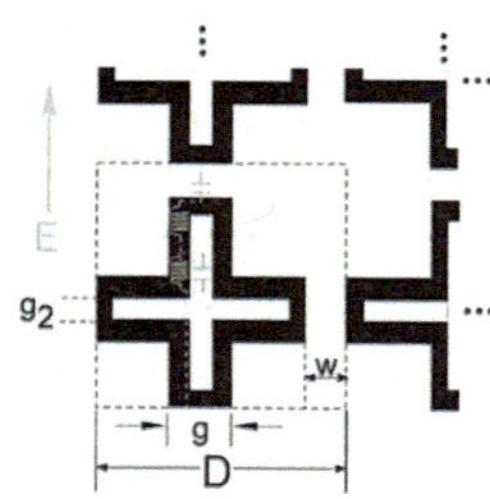	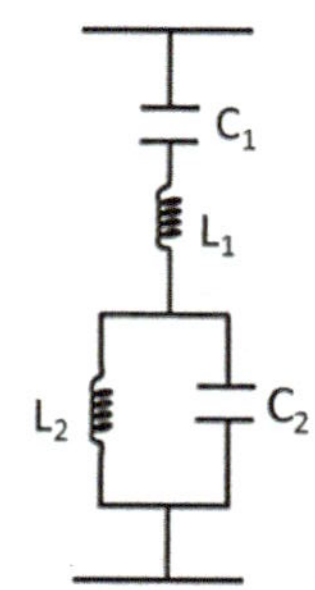	[45]

7	Circular Ring Aperture	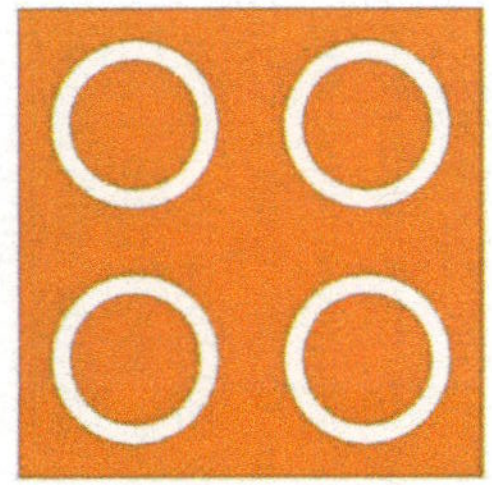	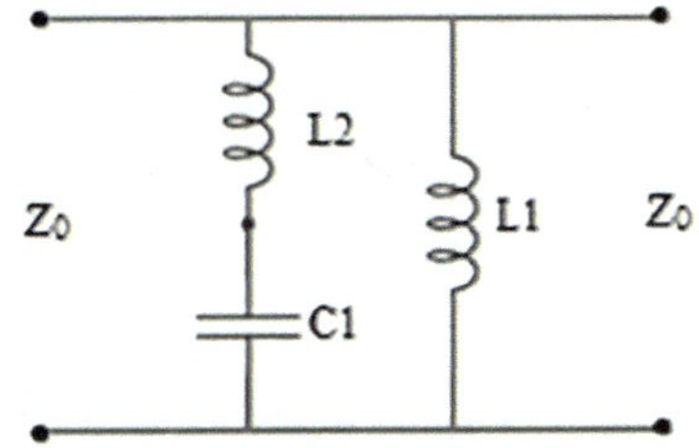	[50]
8	Double Circular Ring Aperture	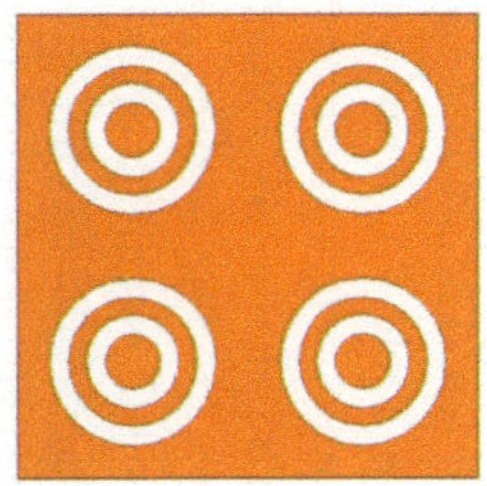	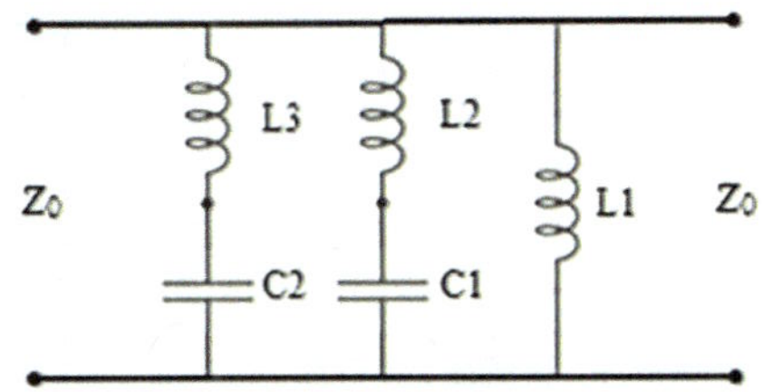	[50]
9	Cross Dipole	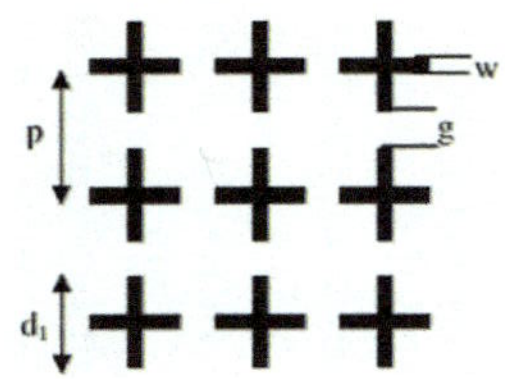	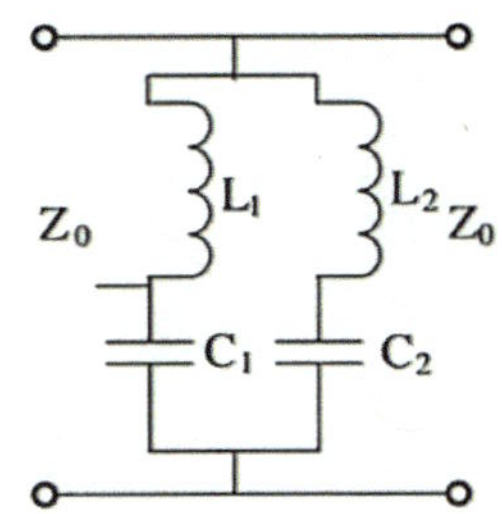	[51]

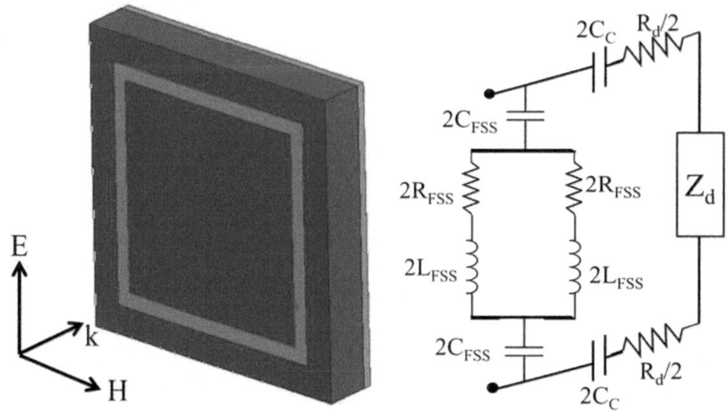

FIGURE 7.5 Equivalent circuit modelling of Square Ring FSS Absorber.

dielectric substrate and is valid limitedly for oblique incidence of wave. An improved and accurate equivalent circuit modelling is proposed by Lee and Langley [52], which is also valid for non-normal incidence. Sometimes using more complex FSS elements in the design of narrow band, multi-band metasurfaces is a necessity. Certain FSS elements can generate a double resonant, or more generally, a multi-resonant behavior in the propagating zone. These types of structures can be modelled by introducing additional lumped elements in the resonant circuit. In general, the simplest form the equivalent circuits of dual band and multi-band metasurfaces consists of repetition of lumped elements in series or shunt. The single band metasurfaces required band responses [48, 50] as shown in Table 7.1, irrespective of consideration of substrate effects. However, multi-resonant elements have the ability to give multiband response such as a common FSS shape like Jerusalem Cross with loading ends [45] as shown in row 4 of Table 7.1. The loading ends enhance the capacitance value thereby additional series LC circuit in shunt with the actual one in the equivalent circuit is required, to model the electric field coupling with the two end-loadings. Further, some multi-resonant elements have the same equivalent circuits but do not resonate due to the same physical behavior [46]. The equivalent circuits of spiral cross array and Jerusalem cross are the same and their resonance response matches with each other. But the cause of resonance is the generation of higher order current modes [47]. Hence, the resonance response entirely depends upon chosen element geometry, current distribution and size irrespective of the equivalent circuit. However, equivalent circuit analysis is an accurate technique to analyze the behavior of the metasurfaces but the geometry and current distribution cannot be ignored.

7.5 EQUIVALENT CIRCUIT MODELLING OF ANISOTROPIC FSSS

In the majority of work the symmetric shape of FSS elements is analyzed using circuital approaches and therefore their response is independent of the azimuth angle [53, 54]. Practically anisotropic FSS elements are generally employed in various applications such as antennas [55], surface wave waveguides [56, 57], polarization converters [58] and metasurfaces [59, 60]. This anticipates the analysis of FSS impedance matrix and extraction of equivalent circuits for the representation of specific element shapes [61]. For the symmetric FSS elements the equivalent circuit can be retrieved independent to polarization of the incident wave. For the anisotropic metasurfaces the reflection coefficient is formulated as [61]:

$$\Gamma = \begin{bmatrix} \Gamma_{TE-TE} & \Gamma_{TE-TM} \\ \Gamma_{TM-TE} & \Gamma_{TM-TM} \end{bmatrix} \tag{7.45}$$

As the metasurface is a passive structure therefore matrix diagonalization is applied by rotating the coordinate system from Cartesian to Crystal plane for a generic FSS element. The angle of rotation, $\phi\, rot$, is evaluated by using initial matrix. After the diagonalization of the matrix, the FSS impedance on crystal axes can be represented in terms of LC equivalent circuit as depicted step by step in [61]. At first, FSS impedance is calculated at normal incidence and then rotation angle is calculated for identification of crystal axis. Finally, impedance on crystal axis is evaluated for the extraction of lumped parameters.

7.6 FSS MODELLING USING COUPLED LINE THEORY

Field coupling is an important phenomenon and has numerous applications in filters, directional couplers, delay transmission line and micro-stripline etc. Various techniques are reported to design and analyze coupled line so far, such as even and odd mode, coupled mode and congruent-transformation techniques [62]. Even and odd mode coupling technique can be utilized to model FSS structures through surface current distribution. The surface currents form a closed loop upon the normal incidence of plane wave on the FSS structure. The distribution of the surface currents is responsible for the generation of even and odd mode couplings among the unit cells of FSS structure. In order to determine the characteristics of coupled lines the self- and mutual inductances and capacitances between the lines are required to be calculated. As per the directions of the surface current in the array structure, each unit cell is coupled with the neighbouring unit cells through even and odd mode field couplings. Equivalent circuit modelling for square ring FSS absorber structure can be done using coupled line theory [63], and even and odd mode coupling is used to accurately determine the lumped parameters and absorption frequency. The square ring FSS on a grounded dielectric substrate is modelled by considering series RLC resonators connected in parallel with coupling capacitance and short-circuited transmission line as represented in Figure 7.5.

The overall impedance is calculated by the parallel combinations of top surface FSS impedance and grounded substrate impedance [63]. Further, the periodic arrangement of metallic square rings introduces extra coupling capacitance that is calculated by using coupled-line geometry of the microstrip line. The effect of vertical coupling capacitance is extracted by even-mode coupling provides only, whereas

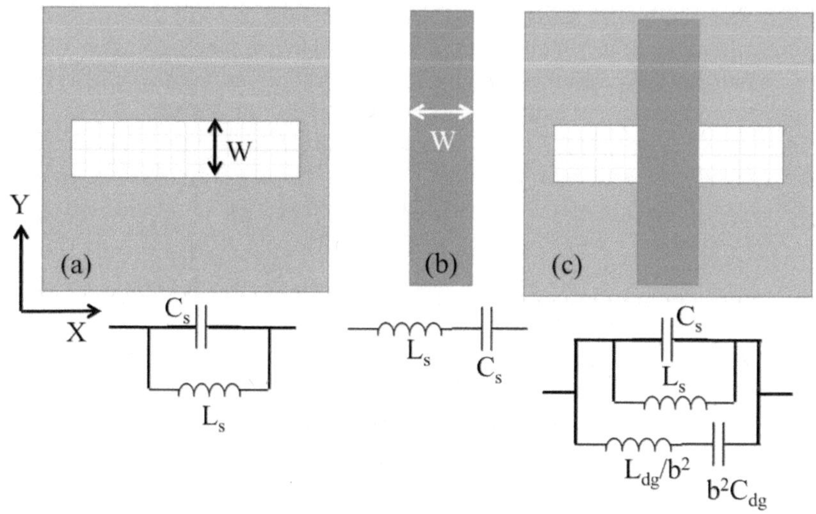

FIGURE 7.6 Equivalent circuit model for metasurfaces (a) Slot Type (b) Dipole Type (c) Coupled complementary [62].

the odd-mode coupling helps to extract the effects of both the vertical and horizontal coupling capacitances. The FSS impedance is a series combination of L_{FSS} (inductance due to length of square ring), R_{FSS}(resistance due to finite conductivity of top metallic pattern) and C_{FSS} (coupling capacitance due to adjacent cells). On the other hand the grounded substrate impedance is calculated as the summation of characteristic impedance (Z_C) for the homogenous dielectric material and dielectric loss impedance (Z_d). This type of equivalent circuit modelling can be considered as a generalized model for analyzing absorber structures of arbitrarily shaped FSSs.

7.7 EQUIVALENT CIRCUIT MODELLING OF COUPLED COMPLEMENTARY METASURFACES

The surfaces inspired from Babinet's Principle are called as coupled complementary metasurfaces. When two complementary surfaces are placed in close proximity of each other, the resulting surface acts as a band pass structure. For these metasurfaces the unit cell size and the distance between the complementary surfaces, are substantially smaller with respect to the free space wavelength at resonant frequency [64]. The two complementary surfaces exhibit band pass and band stop response respectively. When such surfaces are coupled together to form a single metasurface, the resultant response will always be band pass. When subjected to an incident wave, for the analyses of the behavior of the two complementary surfaces such as slot-type (inductive) and the dipole-type (capacitive) suspended in free space surface and considering their metallic parts are lossless MTS transmission line technique is exploited the along the z axis [64]. Around a resonant frequency, the capacitive type surfaces exhibits a band stop response and the inductance type has a band pass response. The

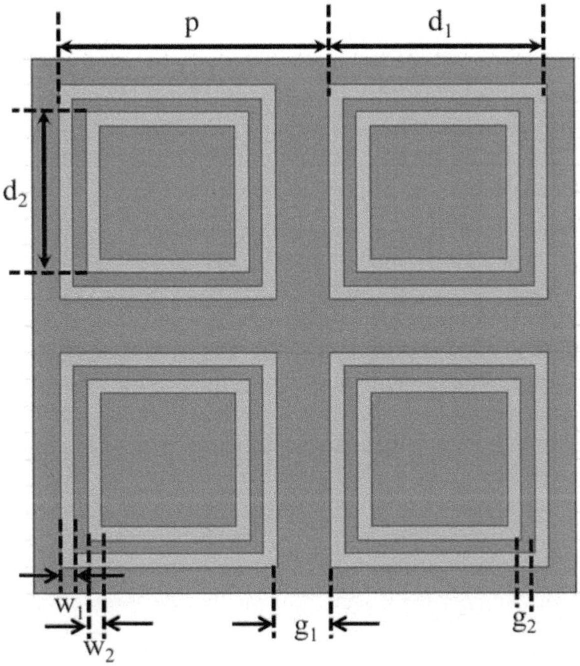

FIGURE 7.7 Equivalent circuit model for metasurfaces (a) slot type, (b) dipole type, and (c) coupled complementary.

slot-type metasurface is represented as a parallel L_s–C_s circuit and the dipole type series L_d–C_d circuit respectively as represented in Figure 7.6.

For the slot type, at low frequencies 'Ls' determines the dominant behavior and capacitance determines the subdominant behavior. On the other hand for the dipoles the C_d is dominant and L_d is subdominant at low frequencies. Using the standard EFIE-MFIE couple of integral equations solved by the Galerkin Method of Moments, equivalent circuit model for closely coupled complementary metasurfaces (CCMTS) was established [64]. The dominant components at low frequency are C_{dg} and L_s [64]. The equivalent impedance of the coupled complementary metasurface (slot/dipole type) consists of parallel combination of the slot inductance and the dipole capacitance. The capacitance C_{dg} is much greater than C_d and the inductance L_{dg} is much smaller than L_d. The term '*b*' represents the reciprocal effects and is a non-dimensional coupling coefficient that takes into account the radiation of the slots on the dipoles. The coupled complementary metasurfaces are potentially used for curved surface and phase gradient radomes. When the gap between the two coupled layers is negligibly small then on the basis of magnetic field integral equation for the complementary structure an analytical equivalent circuit can be developed. This circuit consists of the parallel of the inductive admittance of the slots and the capacitive admittance of the dipoles. As the response of a parallel LC circuit is band pass thereby a complementary metasurface will always have a pass-band response.

7.8 CONCLUSION

Frequency selective surfaces (FSSs) are an assembly of identical elements arranged in a one- or two-dimensional infinite array. In this chapter, a review of various techniques of equivalent circuit modelling of frequency selective surfaces is provided. Mathematical equations to extract the lumped parametric values for equivalent circuits under different techniques of modelling have been discussed in detail. Techniques such as averaged approach, first order retrieving and multimode technique are discussed in detail. Equivalents of generic FSS elements are tabulated to ease the understanding about the development of equivalent circuit modelling. Effects of dielectric and angle of incidence are also detailed. Basic theory to model dual band frequency selective surfaces and phenomenon to achieve multiband response using multi-resonant elements are discussed. Equivalent circuit modelling of anisotropic metasurfaces are discussed in detail. Coupled line theory and modelling of coupled complementary metasurfaces are elaborated to model the FSS based absorbers.

REFERENCES

[1] D. Rittenhouse, "An optical problem, proposed by Mr. Hopkinson, and solved by Mr. Rittenhouse," *Transactions of the American Philosophical Society*, vol. 2, pp. 201–206, 1786.

[2] R. W. Wood, "On a remarkable case of uneven distribution of light in a diffraction grating spectrum," *Philosophical Magazine*, vol. 4, pp. 396–402, 1902.

[3] L. Rayleigh, "On the dynamical theory of gratings," *Proceedings of the Royal Society of London. Series A, Containing Papers of a Mathematical and Physical Character*, vol. 79, no. 532, pp. 399–416, 1907.

[4] A. Hessel and A. A. Oliner, "A new theory of wood's anomalies on optical gratings," *Applied Optics*, vol. 4, no. 10, pp. 1275–1297, 1965.

[5] G. Marconi and C. S. Franklin, "Reflector for use in wireless telegraphy and telephony," US Patent 1,301,473, April 1919.

[6] R. Mitra, C. H. Chan and T. Cwik, "Techniques for analyzing frequency selective surfaces: A review," *Proceedings of the IEEE*, vol. 76, no. 12, December 1988.

[7] G. Von Trentini, "Partially reflecting sheet arrays," *IRE Transactions on Antennas and Propagation*, vol. AP-4, pp. 666–671, 1956.

[8] W. Rotman and A. A. Oliner, "Periodic structures in trough waveguide," *IRE Transactions on Microwave Theory and Techniques*, vol. 7, no. 1, pp. 134, 142, January 1959.

[9] L. Goldstone and A. A. Oliner, "Leaky-wave antennas I: Rectangular waveguides," *IRE Transactions on Antennas and Propagation*, vol. 7, no. 4, pp. 307, 319, October 1959.

[10] D. R. Jackson and A. A. Oliner, "A leaky-wave analysis of the high gain printed antenna configuration," *IEEE Transactions on Antennas and Propagation*, vol. 36, no. 7, pp. 905–910, 1988.

[11] F. O'Nians and J. Matson, "Antenna feed system utilizing polarization independent frequency selective intermediate reflector," U.S. Patent 3-231-892, January 1966.

[12] V. Agrawal and W. Imbriale, "Design of a dichroic cassegrain subreflector," *IEEE Transactions on Antennas and Propagation*, vol. 27, no. 4, pp. 466–473, July 1979.

[13] B. A. Munk, *Frequency selective surfaces-theory and design*, John Wiley & Sons, New York, 2000.

[14] Reuven Shavit, "Frequency selective surfaces (FSS) radomes," *Radome Electromagnetic Theory and Design, IEEE*, pp. 39–88, 2018, doi:10.1002/9781119410850.ch3.

[15] W. Li, Y. Wang, S. Sun and X. Shi, "An FSS-backed reflection/transmission reconfigurable array antenna," *IEEE Access*, vol. 8, pp. 23904–23911, 2020, doi:10.1109/ACCESS.2020.2970611.

[16] Arslan Kiyani, "Design of reflectarray antenna integrated with FSS textured configurations for wireless communication applications," a Thesis submitted at Tun Hussein Onn University of Malaysia, August 2014.

[17] D. Sánchez-Escuderos, M. Cabedo-Fabrés, E. Antonino-Daviu and M. Ferrando-Bataller, "Low-profile planar lens with multilevel FSS for directivity enhancement," *2014 IEEE Antennas and Propagation Society International Symposium (APSURSI), Memphis, TN, USA*, pp. 2076–2077, 2014, doi:10.1109/APS.2014.6905365.

[18] H. Li et al., "Reconfigurable Fresnel lens based on an active second-order bandpass frequency-selective surface," *IEEE Transactions on Antennas and Propagation*, vol. 68, no. 5, pp. 4054–4059, May 2020, doi:10.1109/TAP.2019.2948392.

[19] S. Genovesi, F. Costa and A. Monorchio, "Low-profile array with reduced radar cross section by using hybrid frequency selective surfaces," *IEEE Transactions on Antennas and Propagation*, vol. 60, no. 5, pp. 2327–2335, May 2012, doi:10.1109/TAP.2012.2189701.

[20] F. Costa, C. Amabile, A. Monorchio and E. Prati, "Waveguide dielectric permittivity measurement technique based on resonant FSS filters," *IEEE Microwave and Wireless Components Letters*, vol. 21, no. 5, pp. 273–275, May 2011, doi:10.1109/LMWC.2011.2122303.

[21] M. Bilal, R. Saleem, Q. H. Abbasi, B. Kasi and M. F. Shafique, "Miniaturized and flexible FSS-based EM shields for conformal applications," *IEEE Transactions on Electromagnetic Compatibility*, vol. 62, no. 5, pp. 1703–1710, October 2020, doi:10.1109/TEMC.2019.2961891.

[22] Sunil Jorwal, Ashish Dubey, Rajeev Gupta and Smriti Agarwal, "A review: Advancement in metamaterial based RF and microwave absorbers," *Sensors and Actuators A: Physical*, vol. 354, p. 114283, 2023, ISSN 0924-4247, doi:10.1016/j.sna.2023.114283.

[23] F. Zhang, G.-M. Yang and Y.-Q. Jin, "design and analysis of linear to circular polarization converter with third-order meta-frequency selective surfaces," *IEEE Transactions on Antennas and Propagation*, vol. 68, no. 9, pp. 6646–6655, September 2020, doi:10.1109/TAP.2020.2982503.

[24] Stanislav B. Glybovski, Sergei A. Tretyakov, Pavel A. Belov, Yuri S. Kivshar and Constantin R. Simovski, "Metasurfaces: From microwaves to visible," *Physics Reports*, vol. 634, pp. 1–72, 2016, ISSN 0370-1573, doi:10.1016/j.physrep.2016.04.004.

[25] Chi Ma, Binggang Xiao, Dapeng Zhou and Lihua Xiao, "A novel tunable terahertz wave modulator based on graphene and frequency selective surface (FSS)," *Optics Communications*, vol. 478, p. 126375, 2021, ISSN 0030-4018, doi:10.1016/j.optcom.2020.126375.

[26] S. P. Singh, A. K. Jha and M. J. Akhtar, "Design of broadband superstrate FSS for terahertz imaging and testing applications," *2016 IEEE MTT-S International Microwave and RF Conference (IMaRC), New Delhi, India*, pp. 1–4, 2016, doi:10.1109/IMaRC.2016.7939632.

[27] R. Dickie, R. Cahill, V. Fusco, H. S. Gamble and N. Mitchell, "THz frequency selective surface filters for earth observation remote sensing instruments," *IEEE Transactions on Terahertz Science and Technology*, vol. 1, no. 2, pp. 450–461, November 2011, doi:10.1109/TTHZ.2011.2129470.

[28] Ghobadi, Amir, Turkan Gamze Ulusoy Ghobadi, Ferdi Karadas and Ekmel Ozbay, "Semiconductor thin film based metasurfaces and metamaterials for photovoltaic and photoelectrochemical water splitting applications," *Advanced Optical Materials*, vol. 7, no. 14, p. 1900028, 2019.

[29] Alexander M. Lerer, Pavel E. Timoshenko and Sergei B. Rochal, "Terahertz frequency selective surfaces using heterostructures based on two-dimensional diffraction grating of single-walled carbon nanotubes," *International Journal of Smart and Nano Materials*, vol. 14, 2022, arXiv preprint arXiv:2203.05945.

[30] P. Harms, R. Mittra and Wai Ko, "Implementation of the periodic boundary condition in the finite-difference time-domain algorithm for FSS structures," *IEEE Transactions on Antennas and Propagation*, vol. 42, no. 9, pp. 1317–1324, September 1994, doi:10.1109/8.318653.

[31] R. Mittra, C. H. Chan and T. Cwik, "Techniques for analyzing frequency selective surfaces, a review," *Proceedings of the IEEE*, vol. 76, no. 12, pp. 1593–1615, 1988.

[32] G. G. MacFarlane, "Surface impedance of an infinite wire grid, at oblique angles of incidence," *Journal of the IEE*, vol. 93 (III E), pp. 1523–1527, December 1946.

[33] R. Hornejäger, "Electromagnetic properties of wire grids," *Annalen der Physik*, vol. 4, pp. 25–35, January 1948.

[34] G. Von Trentini, "Gratings as circuit elements of electric waves in space," *Zeitschrift für Angewandte Physics*, vol. 5, pp. 221–231, June 1953.

[35] J. R. Wait, "The impedance of wire grid parallel to a dielectric interface," *IRE Transactions on Microwave Theory and Techniques*, vol. MTT-5, no. 2, pp. 99–102, April 1957.

[36] R. Ulrich, K. F. Renk and L. Genzel, "Tunable submillimeter interferometers of the Fabry-Perot type," *IEEE Transactions on Microwave Theory and Techniques*, vol. 11, no. 5, pp. 363–371, 1963.

[37] N. Marcuvitz (ed.), *Waveguide handbook*, IEE Electromagnetic Waves Series 21, McGraw-Hill, New York, 1951.

[38] J. P. Montgomery, "Scattering by an infinite periodic array of thin conductors on a dielectric sheet," *IEEE Transactions on Antennas and Propagation*, vol. 23, pp. 70–75, 1975.

[39] J. C. Vardaxoglou, and E. A. Parker, "Performance of two tripole arrays as frequency-selective surfaces," *Electronics Letters*, 19, pp. 709–710, 1983.

[40] C. H. Tsao and R. Mittra, "Spectral-domain analysis of frequency selective surfaces comprised of periodic arrays of cross dipoles and Jerusalem crosses," *IEEE Transactions on Antennas and Propagation*, vol. 32, pp. 478–486, 1984.

[41] I. Anderson, "On the theory of self-resonant grids," *Bell System Technical Journal*, vol. 54, pp. 1725–1731, 1975.

[42] O. Luukkonen, C. Simovski, G. Granet, G. Goussetis, D. Lioubtchenko, A. V. Räisänen and S. A. Tretyakov, "Simple and accurate analytical model of planar grids and high-impedance surfaces comprising metal strips or patches," *IEEE Transactions on Antennas and Propagation*, vol. 56, no. 6, pp. 1624–1632, 2008.

[43] V. V. Yatsenko, S. A. Tretyakov, S. I. Maslovski and A. A. Sochava, "Higher order impedance boundary conditions for sparse wire grids," *IEEE Transactions on Antennas and Propagation*, vol. 48, no. 5, pp. 720–973, 2000.

[44] S. Tretyakov, *Analytical modelling in applied electromagnetics*, Artech House, Boston, 2003.

[45] F. Costa, A. Monorchio and G. Manara, "Efficient analysis of frequency selective surfaces by a simple equivalent circuit model," *IEEE Antennas and Propagation Magazine*, vol. 54, no. 4, pp. 35–48, 2012.

[46] F. Costa, A. Monorchio and G. Manara, "An overview of equivalent circuit modeling techniques of frequency selective surfaces and metasurfaces," *ACES Journal*, vol. 29, No. 12, pp. 960–976, December 2014.

[47] M. Garcia-Vigueras, F. Mesa, F. Medina, R. Rodriguez-Berral and J. L. Gomez-Tornero, "Simplified circuit model for arrays of metallic dipoles sandwiched between dielectric slabs under arbitrary incidence," *IEEE Transactions on Antennas and Propagation*, vol. 60, no. 10, pp. 4637, 4649, October 2012.

[48] Asim Egemen Yilmaz and Mustafa Kuzuoglu, "Design of the square loop frequency selective surfaces with particle swarm optimization via the equivalent circuit model," *Radioengineering*, vol. 18, no. 2, pp. 95–101, June, 2009.

[49] B. J. Arritt, D. R. Smith and T. Khraishi, "Equivalent circuit analysis of metamaterial strain-dependent effective medium parameters," *Journal of Applied Physics*, vol. 109, p. 073512, 2011.

[50] B. S. da Silva, A. L. Pereira de Siqueira Campos and A. G. Neto, "Equivalent circuit model for analysis of frequency selective surfaces with ring and double concentric ring apertures," *IET Microwaves, Antennas & Propagation*, vol. 14, pp. 600–607, 2020.

[51] D. Singh, A. Kumar, S. Meena and V. Agarwala, "Analysis of frequency selective surfaces for radar absorbing materials," *Progress in Electromagnetics Research B*, vol. 38, pp. 297–314, 2012.

[52] C. K. Lee and R. J. Langley, "Equivalent-circuit models for frequency selective surfaces at oblique angles of incidence," *IEE Proceedings H—Microwaves Optics and Antennas*, vol. 132, pp. 395–399, 1985.

[53] S. Maci, M. Caiazzo, A. Cucini and M. Casaletti, "A pole-zero matching method for EBG surfaces composed of a dipole fss printed on a grounded dielectric slab," *IEEE Transactions on Antennas and Propagation*, vol. 53, no. 1, pp. 70–81, January 2005.

[54] F. Caminita, E. Martini, G. M. Sardi and S. Maci, "Metadispersion in anisotropic and bianisotropic media," *2011 XXXth URSI General Assembly and Scientific Symposium*, pp. 1–4, August 2011.

[55] A. M. Patel and A. Grbic, "Transformation electromagnetics devices based on printed-circuit tensor impedance surfaces," *IEEE Transactions on Microwave Theory and Techniques*, vol. 62, no. 5, pp. 1102–1111, May 2014.

[56] D. J. Gregoire and A. V. Kabakian, "Surface-wave waveguides," *IEEE Antennas and Wireless Propagation Letters*, vol. 10, pp. 1512–1515, 2011.

[57] M. Mencagli, E. Martini and S. Maci, "Surface wave dispersion for anisotropic metasurfaces constituted by elliptical patches," *IEEE Transactions on Antennas and Propagation*, vol. 63, no. 7, pp. 2992–3003, July 2015.

[58] C. Mao, Y. Yang, X. He, J. Zheng and C. Zhou, "Broadband reflective multi-polarization converter based on single-layer double-l-shaped metasurface," *Applied Physics A*, vol. 123, no. 12, p. 767, November 2017. Online], doi:10.1007/s00339-017-1322-6.

[59] B. H. Fong, J. S. Colburn, J. J. Ottusch, J. L. Visher and D. F. Sievenpiper, "Scalar and tensor holographic artificial impedance surfaces," *IEEE Transactions on Antennas and Propagation*, vol. 58, no. 10, pp. 3212–3221, October 2010.

[60] M. Teniou, H. Roussel, N. Capet, G. P. Piau and M. Casaletti, "Implementation of radiating aperture field distribution using tensorial metasurfaces," *IEEE Transactions on Antennas and Propagation*, vol. 65, no. 11, pp. 5895–5907, November 2017.

[61] M. Borgese and F. Costa, "A simple equivalent circuit approach for anisotropic frequency-selective surfaces and metasurfaces," *IEEE Transactions on Antennas and Propagation*, vol. 68, no. 10, pp. 7088–7098, October 2020, doi: 10.1109/TAP.2020.3008658.

[62] R. Garg and I. J. Bahl, "Characteristics of coupled microstriplines," *IEEE Transactions on Microwave Theory and Techniques*, vol. 27, no. 7, pp. 700–705, July 1979, doi:10.1109/TMTT.1979.1129704.

[63] S. Ghosh, and K. V. Srivastava, "An equivalent circuit model of fss-based metamaterial absorber using coupled line theory," *IEEE Antennas and Wireless Propagation Letters*, vol. 14, 2015.

[64] S. S. Bukhari, W. G. Whittow, J. C. Vardaxoglou and S. Maci, "Equivalent circuit model for coupled complementary metasurfaces," *IEEE Transactions on Antennas and Propagation*, vol. 66, no. 10, pp. 5308–5317, October 2018, doi:10.1109/TAP.2018.2860052.

8 Miniaturized X Band DMTL Phase Shifter Using Space Filling Curves

S. Kanthamani

8.1 INTRODUCTION

The capacity of contemporary microwave and millimeter wave phased array antennas to physically move the antenna elements—a prerequisite for quick beam steering and beam shaping—makes them appealing in space. By electronically adjusting the signal's phase within the antenna system, this might be accomplished. Each antenna element of a standard phased array antenna can receive a phase shifter, which can be used to direct the array beam in various directions, feeding thousands of components in all. In order to construct phased array antennas, phase shifters with low loss, low cost, and minimal weight are crucial [1–6].

MEMS, Ferroelectric, Ferrite, and semiconductor technologies could all be used to create phase shifters [7–10]. At RF and millimeter wave frequencies, MEMS switches and components showed remarkable performance, including low driving power, good isolation, and minimal insertion loss. By alternating between two distinct signal channels that result in switched line phase shifters, the MEMS switch can be set up to produce phase shift. In either case, a MEMS switch can be utilized as a distributed capacitive switch, which modifies the transmission line's effective capacitance [11–16] to create distributed MEMS transmission line phase shifters. MEMS technology combined with an innovative design produces large-phased array antennas that are inexpensive, lightweight, and have a far lower production cost.

Customary Distributed Micro Electromechanical Transmission Line (DMTL) phase shifters are made by applying linear force across a MEMS membrane across the transmission line. The phase shifter's lateral dimension grows as the phase shift intensity does. The device's total size and substrate area are both increased by the increase in lateral dimension [17–18]. The linear phase shifter's minimum substrate area, which measures 8.39mm by 0.49mm, is documented in [17–18]. A small two- and three-bit DMTL phase shifter with a 1.5 × 2.1mm and 3.2 × 2.1mm size reduction is designed, according to [19–20]. It is imperative to make the DMTL phase shifter as tiny as possible without compromising its functionality.

 DOI: 10.1201/9781032656021-8

Utilizing Defected Ground Structures (DGS) is an additional method of accomplishing size reduction in DMTL phase shifters [21]. The actuation voltage, dielectric charging effect, restoring force, and switching speed are all impacted by the MEMS membrane's shrinkage. By employing fractals in the phase shifter architecture at the *Ku*-band, more size reduction is accomplished [22]. The Koch fractal is used in the design of the phase shifter membrane, resulting in a 23% decrease in actuation voltage. The meander structure-designed, five-bit DMTL phase shifter exhibits a reduction in length. When compared to a traditional phase shifter, the length reduction is almost 70% less [23]. The phase shifter's compactness is impacted by the wasted space between the meanders, despite the significantly reduced lateral dimension [24]. For additional size reduction, space filling curves could be included in the formulation of the DMTL phase shifter design.

Fine structures are produced by employing a recursive generating process to construct the space filling curves. In order to create smaller gadgets, the space filling curve is utilized to simulate natural phenomena like clouds, vegetables, and coastlines. By using these designs, more electrical length can be produced in a smaller die area. In [25] presented a fresh concept of a DMTL phase shifter developed employing space filling curve. The central conductor of the CPW transmission line is printed across the substrate while maintaining the shape of the assigned space filling curve. The most compact 180° phase shifter is made with the Moore curve (1.5 × 1.5mm), modified Hilbert curve (0.82 × 0.82mm), and Peano curve (1 × 0.88mm) [26]. This is in contrast to a conventional DMTL phase shifter. As a result, performance is enhanced and size is decreased.

In this chapter the design of the DMTL phase shifters for X band applications is done using fractal geometry. The work presented is on the development of phase shifter structure to achieve the high compactness. Three phase shifter designs are done and compared. In the first design, the phase shifter is designed using conventional CPW in which 11 unit cells with three membranes are placed periodically over the center conductor of the CPW transmission line. In the later, the center conductor of the CPW is modelled using space filling curves up to second and third iterations leading to phase shifters and the dimension gets greatly reduced with acceptable performance in the entire range of frequencies of interest.

8.2 X BAND DISTRIBUTED MICRO ELECTROMECHANICAL TRANSMISSION LINE (DMTL) PHASE SHIFTERS

The CPW-based transmission line and the MEMS bridges are designed as part of the distributed MEMS transmission line phase shifter concept. The phase shifter can be designed as illustrated in Figure 8.1 by spacing several MEMS bridges periodically on a CPW transmission line. The membrane gets actuated when an external electrostatic voltage is applied, producing the desired differential phase shift value. A phase shift at the desired frequency can be produced using the phase shifting unit cell in the previously published designs, which has a small number of miniature-size bridges. Higher phase shift values can be obtained by actually cascading such similar unit cells over a traditional CPW transmission line. Nevertheless, the creation of such a

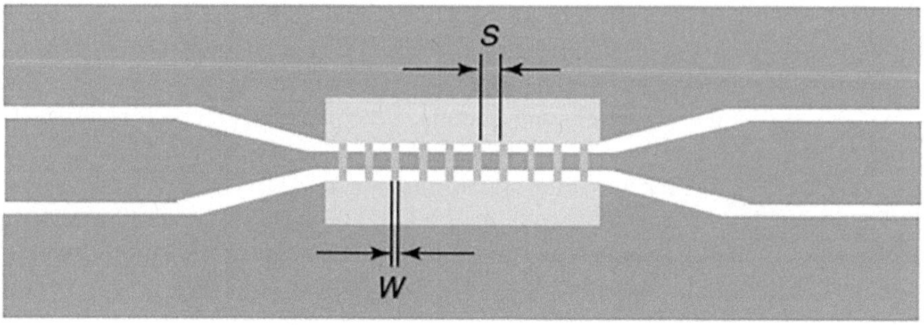

FIGURE 8.1 Distributed MEMS transmission line phase shifter.

structure results from the linear cascade of several unit cells, which occupy a huge amount of lateral dimensions.

To conserve space, the entire phase shifter structure follows a unique space filling curve. In the CPW based transmission line structure, center conductor is designed by employing space filling curve structures and each unit cell is placed over the CPW transmission line at equal spacing 'S'. Thus, the device dimensions are greatly reduced from the conventional one yet keep the return loss, insertion loss, and phase error within the prescribed limits. Three phase shifters are designed in which the first one is designed conventionally, the second structure is designed using the Peano space filling curve and the third structure is designed using the Hilbert space filling curve. Three phase shifters are designed for X band, and their return loss and insertion loss are compared.

8.3 SPACE FILLING CURVE

Because of their special qualities, fractals—non-Euclidean geometries—have more uses in the fields of science and engineering. The majority of RF engineers are drawn to fractal geometries, which has resulted in the development of innovative concepts for RF applications. High actuation voltage is required while developing a MEMS bridge with a rectangular form using DMTL phase shifters. To reduce the actuation voltage, MEMS bridges have recently undergone a variety of material and geometrical changes. Many attempts are made to lower the actuation voltage, including lowering the beam's spring constant, expanding the electrostatic field area, and closing the gap, but all of these measures lead to inadequate isolation loss [27]. By utilizing fractal geometry, the structure can reduce the MEMS bridge's actuation voltage by increasing the membrane's stiffness and bending momentum. Peano proposed the space filling curves for the first time in 1890 [28], followed by the Hilbert curve [29], the Sirpinski space filling curve [30], the Moore curves [31], Lebesgue [32], and Sagan [33] for the other fundamentals of space filling curves. The self-similar structures and substitution rules, which are crucial in the creation of such curves, constitute the basis for the generation of space filling curves. Dekking [34] developed the process for creating and characterizing the space filling curve.

Fractals are geometries that fall into the categories of points, lines, sheets, and volumes; therefore, a line might have the structure of a sheet. The lines can be meandering in a way that satisfies the space filling requirements, which require the lines

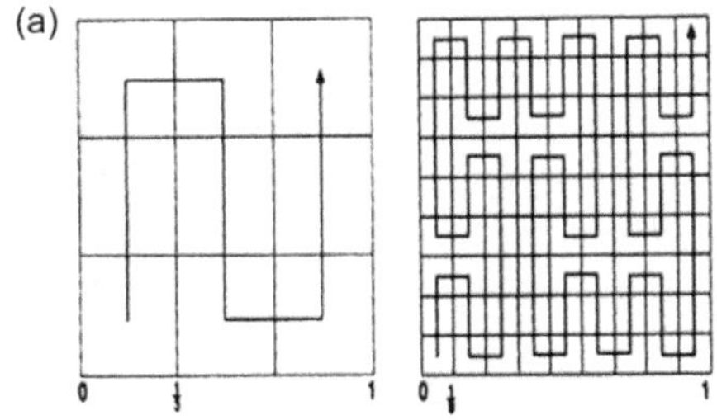

FIGURE 8.2a Peano curves.

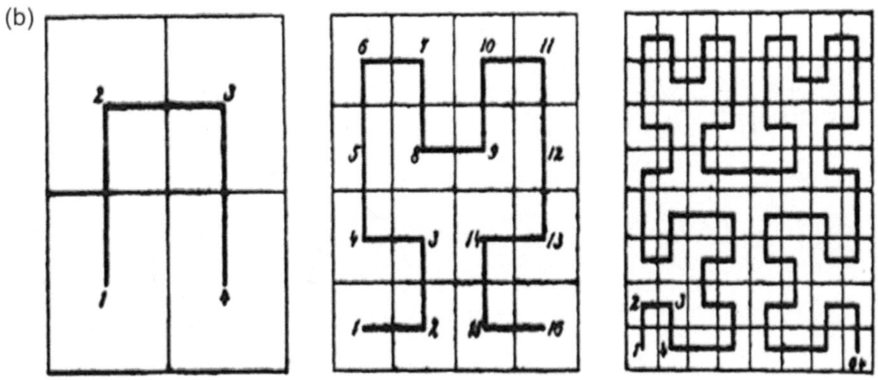

FIGURE 8.2b Hilbert curves.

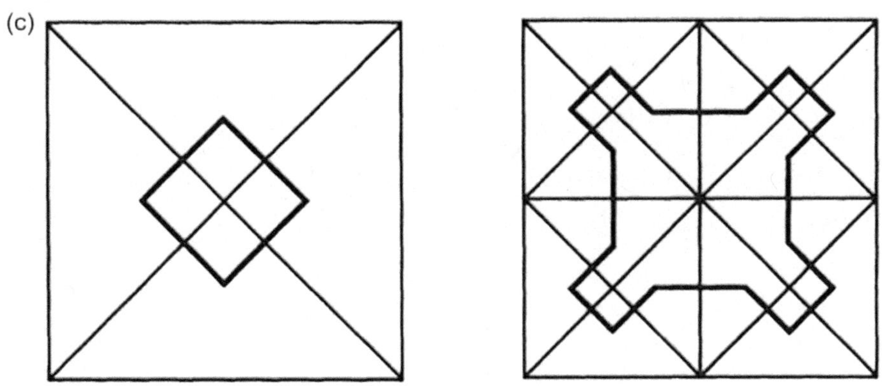

FIGURE 8.2c Approximating polygons for the Sierpinski curve.

to be electrically very long yet fitted in a small physical space, in order to fill the full sheet. Compactly utilizing the available physical space results in the miniaturization of the structure while maintaining the anticipated properties [35]. Figure 8.2 displays a small number of space filling curves.

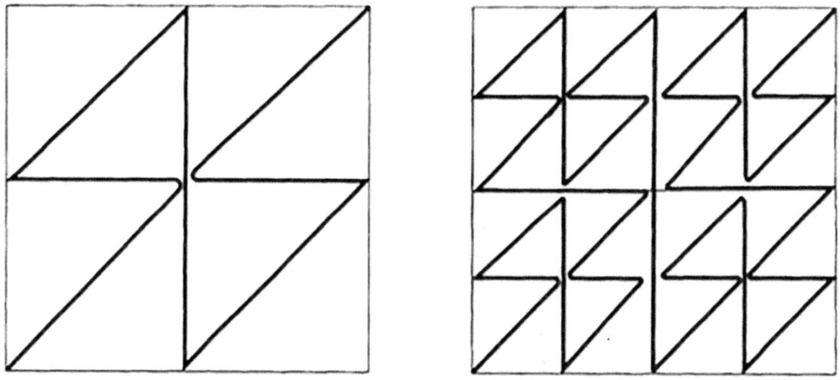

FIGURE 8.2d First and second approximating polygon for Lebesgue's space filling curves $n = 2\ 3\ (1 + 2)\ 2\ n - 1\ 3\ (2 - 2)\ 1\ 2\ n$ {\displaystyle l_{n}={2 \over 3}(1+{\sqrt {2}})2^{n}-{1 \over 3}(2-{\sqrt {2}}){1 \over 2^{n}}}.

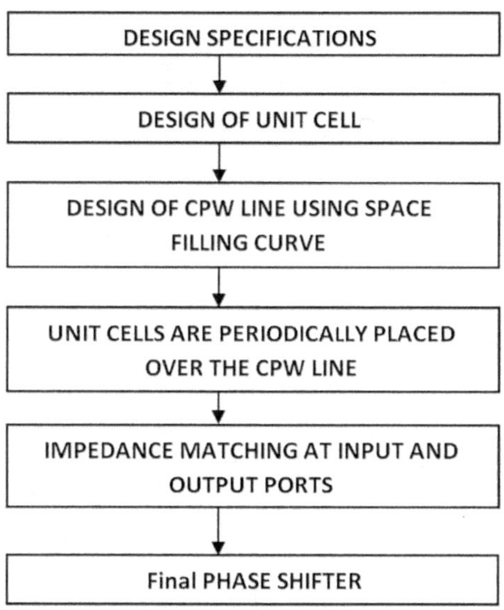

FIGURE 8.3 Flow chart for designing a phase shifter.

8.4 DESIGN AND SIMULATION OF AN X BAND PHASE SHIFTER

Design of a phase shifter is done in the following steps:

Step 1: Design of a unit cell and the unit cell has three membranes. 11 unit cells totaling of 33 membranes are used for this X band design

Step 2: Phase shifter design using conventional CPW

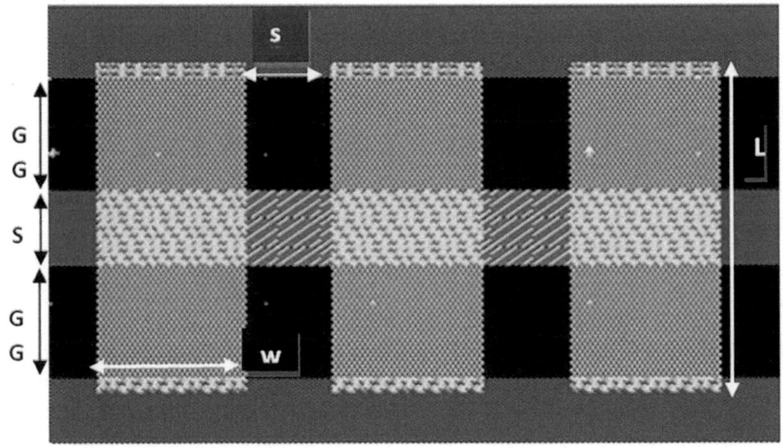

FIGURE 8.4 Unit cell with three bridges.

TABLE 8.1
Parameter Values for the Unit Cell Design

Sl.No	Parameters	Dimensions (mm)
1.	G/W/G	15/10/15
2.	Thickness of dielectric(*td*)	0.15
3.	Air gap (*go*)	0.5
4.	Width of bridge (*w*)	14
5.	Thickness of bridge(*t*)	0.3
6.	Spacing between the bridges(*s*)	8.5
7.	Length of bridge (*L*)	44

Step 3: Phase shifter design using space filling curves:

1. Peano curves

2. Hilbert curves

8.4.1 Design of a Unit Cell

Design of a unit cell involves the selection of a number of membranes, based on the frequency of operation and phase shift desired. For this design of phase shifter a unit cell consists of three membranes (bridges): 11 unit cells totaling of 33 membranes are used to produce a phase shift of 180°. The required dimensions of the membrane and the unit cell are calculated using ADS Line calc. Figure 8.4 shows the unit cell with three membranes. The essential dimensions involved in the design are tabulated in Table 8.1.

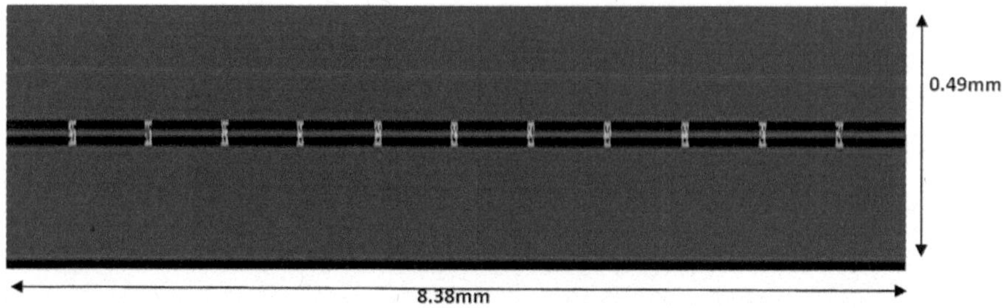

FIGURE 8.5 Phase shifter using the straight CPW line.

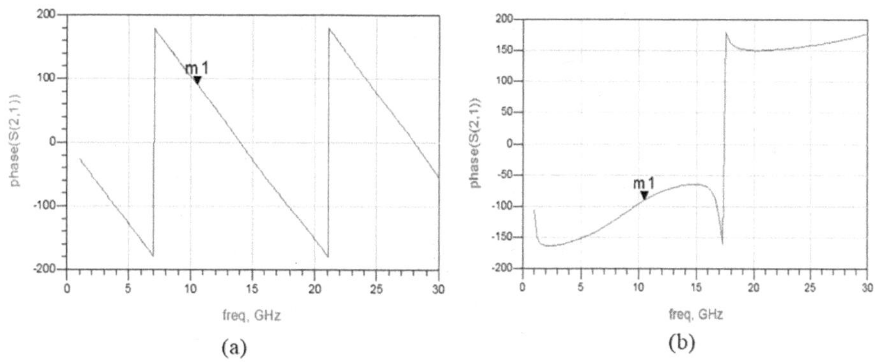

FIGURE 8.6 Phase shift response for (a) Up state and (b) Down state.

8.4.2 Design of a Phase Shifter Using Conventional CPW

The phase shifter is formulated by placing 33 membranes periodically over the CPW line with equal spacing. A differential phase shift of 180° can be produced by the complete structure, as illustrated. Figure 8.5 depicts the phase shifter structure with the straight CPW line. Its overall dimensions are 8.39 × 0.49mm. ADS simulates the structure that is displayed.

The phase variation and average insertion and return loss obtained for the phase shifter at both UP and DOWN states are 179.83°, −1.2dB, and −21.78dB.

From Figure 8.6 it is observed that at 11 GHz, there is a differential phase shift of 179.89°. The return loss and the insertion loss response is shown in Figure 8.7.

8.4.3 Design of a Phase Shifter Using Peano Space Filling Curve

In this design, center conductor of basic CPW line is modeled as Peano space filling curve as shown in Figure 8.8. Peano curves are generated using the procedures available in [26]. To formulate a phase shifter operating at the desired frequency Peano curves up to second iteration are used. The size of the modeled CPW line is 0.88mm × 1mm.

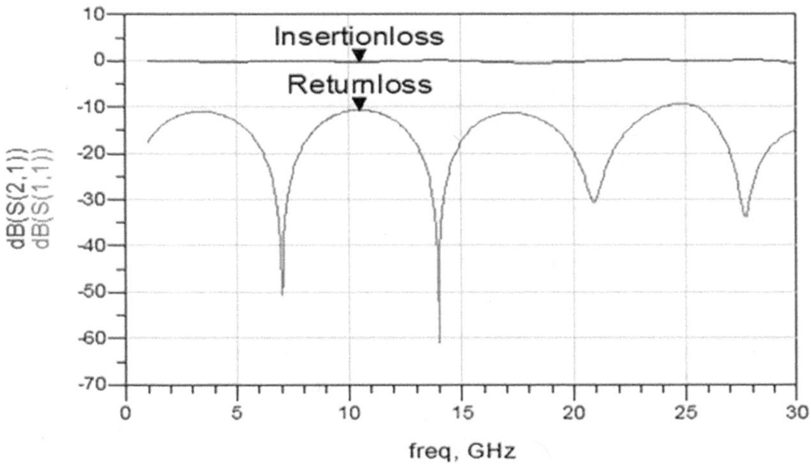

FIGURE 8.7 S-parameter response for the phase shifter using the straight CPW line.

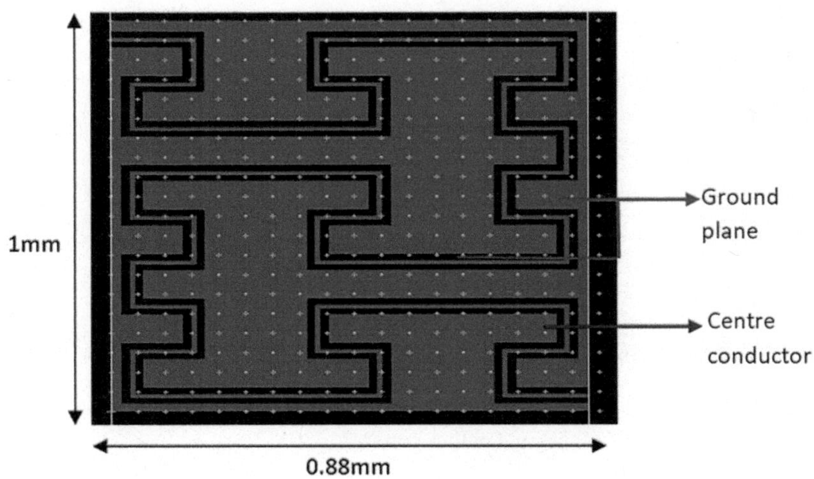

FIGURE 8.8 CPW line modeled as Peano space filling curve.

Figure 8.9 shows the ADS layout of Peano space filling curve along with 11 unit cells and each cell has three membranes.

For the purpose of matching, flaring is done at the input and output sides, to match out the CPW line as shown in Figure 8.10.

The phase shift response of the phase shifter using Peano curve for up and down states is shown in Figure 8.11 (a) and (b). From Figure 8.11 it is observed that at 11 GHz, there is a differential phase shift of 179.6°.

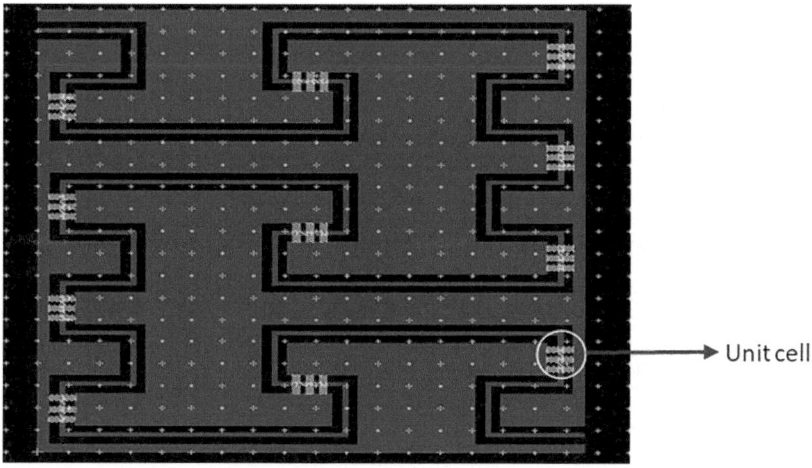

FIGURE 8.9 Phase shifter using the CPW line modeled as Peano space filling curve with.

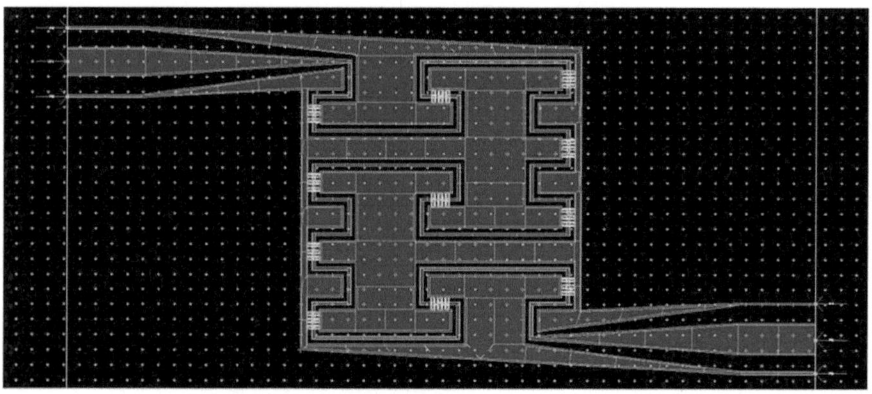

FIGURE 8.10 Phase shifter using the CPW line modeled as Peano space filling curve with impedance matching at both input and output ports.

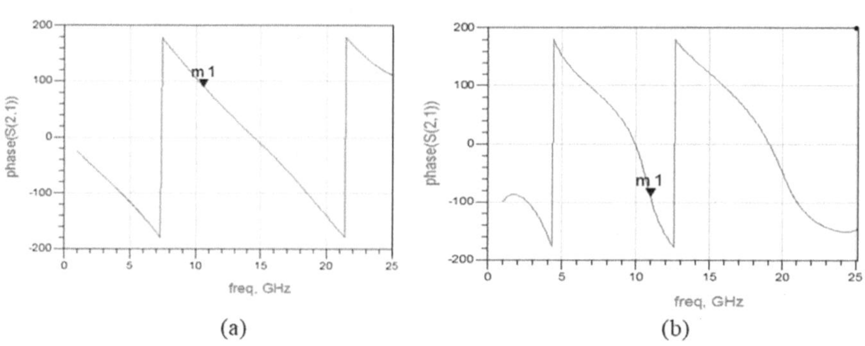

FIGURE 8.11 Phase shift response of the phase shifter using the CPW line modeled as Peano space filling curve (a) for up state and (b) for down state.

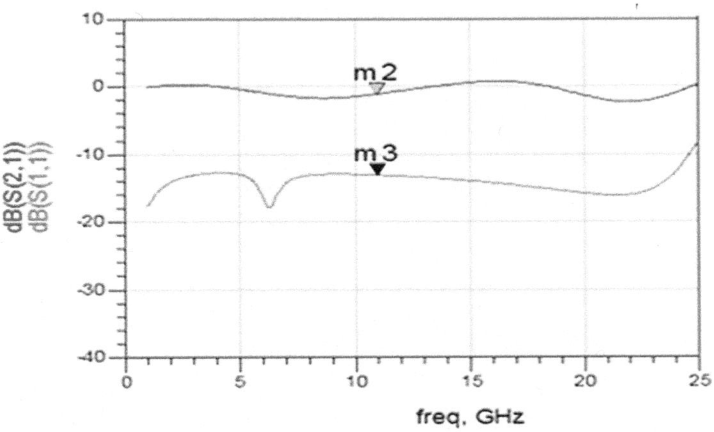

FIGURE 8.12 S-parameter response for the phase shifter using the CPW line modeled as Peano space filling curve.

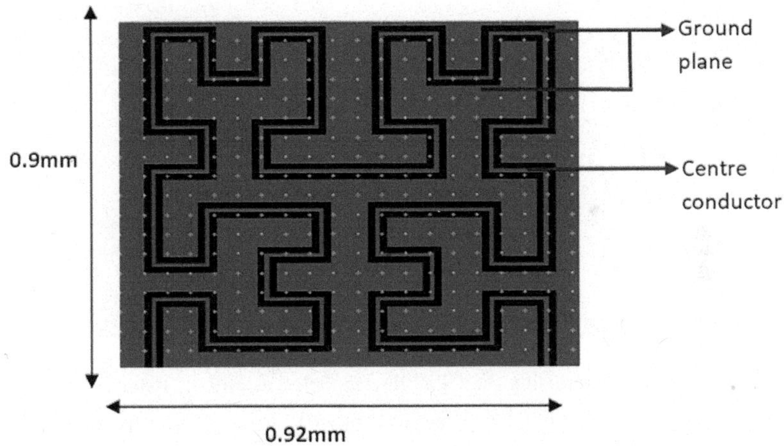

FIGURE 8.13 CPW line modeled as Hilbert space filling curve.

The return loss and the insertion loss response are shown in Figure 8.12. It is observed that insertion loss is obtained as –1.131dB which is greater than –2dB and return loss is obtained as –13.084dB which is less than –10dB.

8.4.4 Design of a Phase Shifter Using Hilbert Space Filling Curve

In this design, center conductor of basic CPW line is modeled as Hilbert space filling curve as shown in Figure 8.13. Hilbert curves are generated using the procedures available in [28]. To formulate a phase shifter operating at the desired frequency Hilbert Peano curves up to third iteration are used. The size of the modeled CPW line is 0.92mm × 0.9mm.

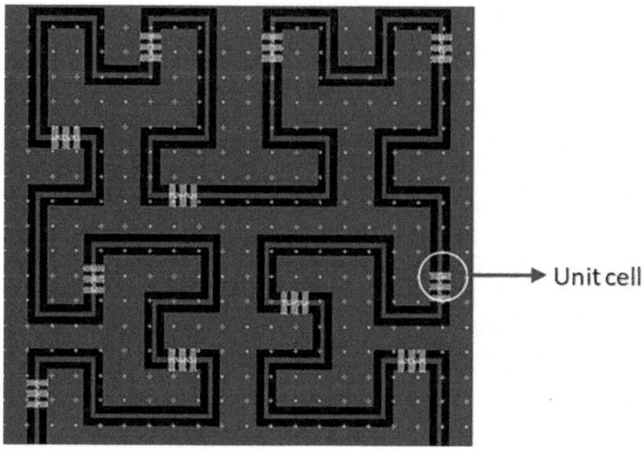

FIGURE 8.14 Phase shifter using the CPW line modeled as Hilbert space filling curve with 11 unit cell bridges.

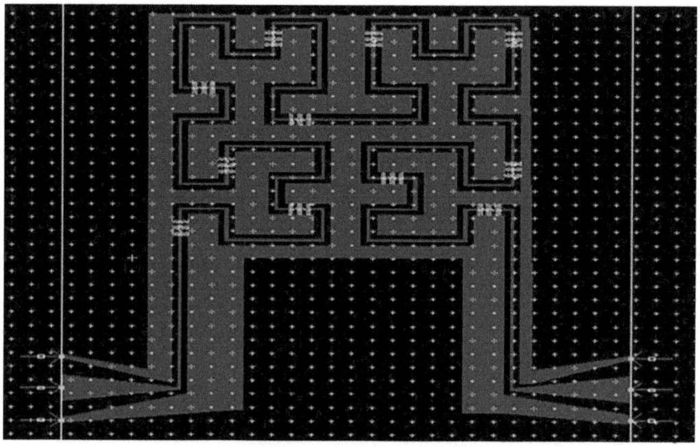

FIGURE 8.15 Phase shifter using the CPW line modeled as Hilbert space filling curve with impedance matching at both input and output ports.

The structural dimensions are chosen to accommodate the 11 unit cells designed in section 8.4.1. The layout of the phase shifter designed using Hilbert space filling curve is shown in Figure 8.14.

As seen in Figure 8.15, the device's termination is intended to generate 50Ω matching. In this, 11 unit cells are periodically placed over the CPW line with equal spacing. It occupies the total area of 0.92mm × 0.9mm.

The return loss and the insertion loss response are shown in Figure 8.16. It is observed that insertion loss is obtained as −1.131dB and return loss is obtained as −13.084dB. The phase shift response of the phase shifter using Hilbert curves for up and down states is shown in Figure 8.17. From the Figure 8.17 it is observed that at 11 GHz, there is a differential phase shift of 180°.

Thus the Table 8.2 implies that the performance of the phase shifter structure designed by using the space filling curves are closer to the conventional one with much reduced size proving as the most compact 180° phase shifters.

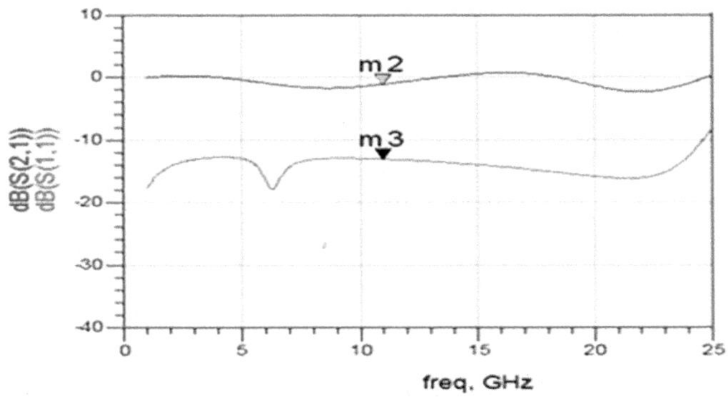

FIGURE 8.16 S-parameter response for the Hilbert space filling curve based phase shifter.

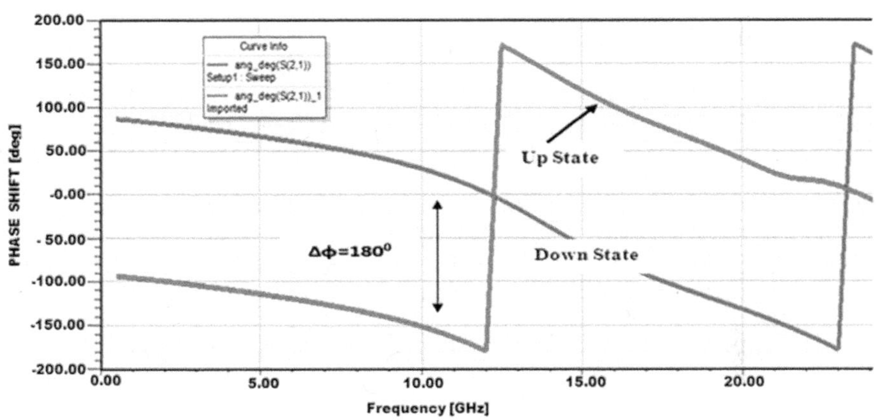

FIGURE 8.17 Phase shift response of the phase shifter using the CPW line modeled as Peano space filling curve (a) for up state and (b) for down state.

TABLE 8.2
Comparison of Results

S.No	Structure chosen	Phase Shift	Space	Insertion Loss	Return Loss
1.	Using straight CPW line	179.89°	8.39mm × 0.49mm	—0.272dB	—10.704dB
2.	Using Peano curve based CPW line	179.68°	0.88mm × 1mm	—1.131dB	—13.084dB
3.	Using Hilbert curve based CPW line	180°	0.92mm × 0.9mm	—1.13dB	—13.07dB

8.5 CONCLUSION

The work presented in this chapter is a trial for the design and development of a phase shifter to achieve the high compactness. Three structures are proposed to verify the compactness. The first phase shifter is designed using the conventional CPW whereas the remaining two phase shifters follow unique space filling curves. In the later phase shifters the dimension gets greatly reduced with acceptable performance in the entire range of frequencies of interest. The phase shifter that follows the Peano curve of second iteration occupies a space of 1mm × 0.88mm with 180° phase shift at 11 GHz with insertion loss of −1.31dB and return loss of −13.084dB and the phase shifter that follows the Hilbert curve of third iteration occupies space of 0.92mm × 0.9mm with 180° phase shift at 11 GHz, whereas the conventional one has the dimension of 8.39mm × 0.49mm with insertion loss of −0.272dB and return loss of −10.704dB. Thus by using the unique space filling curves the phase shifter design becomes more compact with the acceptable performance.

REFERENCES

[1] G. M. Rebeiz, G. L. Tan, and J. S. Hayden, "RF MEMS Phase Shifters: Design and Applications," *IEEE Microwave Magazine*, vol. 3, pp. 72–81, June 2002.

[2] G. M. Rebeiz, *RF MEMS: Theory, Design and Technology*, second ed., Wiley, New York, 2003.

[3] Joseph S. Hayden and Gabriel M. Rebeiz, "2-Bit MEMS Distributed X-Band Phase Shifters," *IEEE Microwave and Guided Wave Letters*, vol. 10, pp. 540–542, 2000.

[4] Joseph S. Hayden and Gabriel M. Rebeiz, "One and Two-Bit Low-Loss Cascadable MEMS Distributed X-Band Phase Shifters," *Proceedings of the IEEE MTT-S International Microwave Symposium Digest*, pp. 161–164, 2001.

[5] J. S. Hayden, A. Malczewski, J. Kleber, C. L. Goldsmith, and G. M. Rebeiz, "2 and 4-Bit DC-18 GHz Microstrip MEMS Distributed Phase Shifters," *IEEE MTT-S International Microwave Symposium Digest*, vol. 1, pp. 219–222, May 2001.

[6] G. L. Tan, R. E. Mihailovich, J. B. Hacker, J. B. DeNatale, and G. M. Rebeiz, "A Very-Low-Loss 2-Bit X-Band RF MEMS Phase Shifter," *IEEE MTT-S International Microwave Symposium Digest*, vol. 1, pp. 333–335, June 2002.

[7] N. S. Barker and G. M. Rebeiz, "Distributed MEMS True-Time Delay Phase Shifters and Wide-Band Switches," *IEEE Transactions on Microwave Theory and Techniques*, vol. 46, pp. 1881–1890, November 1998.

[8] N. Barker and G. Rebeiz, "Optimization of Distributed MEMS Phase Shifters," *IEEE MTT-S International Microwave Symposium Digest*, vol. 1, pp. 299–302, June 1999.

[9] Y. J. Ko, J. Y. Park, and J. U. Bu, "Integrated 3-Bit RF MEMS Phase Shifter with Constant Phase Shift for Active Phased Array Antennas in Satellite Broadcasting Systems," *12th International Conference on Transducers, Solid-State Sensors, Actuators and Microsystems*, vol. 2, pp. 1788–1791, June 2003.

[10] Y. J. Ko, J. Y. Park, H. T. Kim, and J. U. Bu, "Integrated Five-Bit RF MEMS Phase Shifter for Satellite Broadcasting/Communication Systems," *IEEE The Sixteenth Annual International Conference on Micro Electro Mechanical Systems*, pp. 144–148, January 2003.

[11] A. Chakraborty, A. Kundu, S. Dhar, S. Maity, S. Chatterjee, and B. Gupta, "Compact K-Band Distributed RF MEMS Phase Shifter Based on High-Speed Switched Capacitors," *Proceedings of the IEEE MMS*, pp. 25–28, 2011.

[12] D. Mercier, K. V. Caekenberghe, and G. M. Rebeiz, "Miniature RF MEMS Switched Capacitors," *Proceedings of the IEEE MTT-S Digest*, pp. 745–748, 2005.
[13] Joseph S. Hayden and Gabriel M. Rebeiz, "A Low-Loss Ka-Band Distributed MEMS 2-Bit Phase Shifter Using Metal-Air-Metal Capacitors," *Proceedings of the IEEE MTT-S Digest*, pp. 337–340, 2002.
[14] Joseph S. Hayden and Gabriel M. Rebeiz, "Very Low-Loss Distributed X Band And Ka-Band MEMS Phase Shifters Using Metal-Air-Metal Capacitors," *IEEE Transactions on Microwave Theory and Techniques*, vol. 51, pp. 309–314, 2003.
[15] A. Grichener, D. Mercier, and G. M. Rebeiz, "High-Power High-Reliability High-Q Switched RF MEMS Capacitors," *Proceedings of the IEEE MTT-S Digest*, pp. 31–34, 2006.
[16] D. Mercier, K. V. Caekenberghe, and G. M. Rebeiz, "Miniature RF MEMS Switched Capacitors," *IEEE MTT-S International Microwave Symposium Digest*, vol. 3, pp. 1931–1934, June 2005.
[17] B. Lacroix, A. Pothier, A. Crunteanu, and P. Blondy, "Phase Shifter Design Based on Fast RF MEMS Switched Capacitors," *Proceedings of the IEEE 3rd European Microwave ICs Conference*, pp. 478–481, 2008.
[18] Benjamin Lacroix, Arnaud Pothier, Aurelian Crunteanu, Christophe Cibert, Frédéric Dumas-Bouchiat, Corinne Champeaux, Alain Catherinot, and Pierre Blondy, "Sub-Microsecond RF MEMS Switched Capacitors," *IEEE Transactions on Microwave Theory and Techniques*, vol. 55, pp. 1314–1321, 2007.
[19] K. Varian and D. Walton, "A 2-Bit RF MEMS Phase Shifter in Athick-Film BGA Ceramic Package," *IEEE Microwave and Wireless Components Letters*, vol. 12, no. 9, pp. 321–323, 2002.
[20] J. J. Hung, L. Dussopt, and G. M. Rebeiz, "A Low-Loss Distributed 2-Bit W-Band MEMS Phase Shifter," *Proceedings of the 33rd European Microwave Conference, Munich*, 2003, pp. 983–986, 2003.
[21] K. Snehanshu Sengupta and Usha Kiran, "Miniatured Phase Shifter," *Indian Journal of Science and Technology*, vol. 18, no. 19, pp. 1–7, 2015.
[22] M. Jahanbakht, M. N. Moghaddasi, and A. A. Lotfi Neyestanak, "Fractal Beam Ku-Band Mems Phase Shifter," *Progress in Electro magnetics Research Letters*, vol. 5, pp. 73–85, 2008.
[23] Yijia Du, Jingfu Bao, Wenchang Wu, Cheng Tu, and Pengfei Mu, "Design of Novel 5-Bit Distributed MEMS Phase Shifter," International Conference on Communication, Circuits and Systems, Chengdu, China, 2010.
[24] Amrita Chakraborty, Avra Kundu, Sayan Chatterjee, and Bhaskar Gupta, "Design of Compact RF MEMS Phase Shifters Based on Space-Filling," International Conference on Smart Materials Structures and Systems, Bangalore, India, 2012.
[25] Amrita Chakraborty, Avra Kundu, Sayantan Dhar, Santanu Maity, Sayan Chatterjee, and Bhaskar Gupta, "Compact k-Band Distributed RF MEMS Phase Shifter Based on High-Speed Switched Capacitors," Mediterranean Microwave Symposium, Yasmine Hammamet, Tunisia, 2011.
[26] Amrita Chakraborty and Bhaskar Gupta, "Development of Compact 180° Phase Shifters Based on MEMS Technology," Sensors and Actuators A: Physical, vol. 247, pp. 187–198, 2016.
[27] J. M. Huang, K. M. Liew, C. H. Wong, S. Rajendran, M. J. Tan, and A. Q. Liu, "Mechanical Design and Optimization of Capacitive Micro-Machined Switches," *Sensors and Actuators A*, vol. 93, pp. 273–285, 2001.
[28] G. Peano, "Sur une courbe qui remplit toute une aire plane," *Mathematische Annalen*, vol. 36, pp. 157–160, 1890.
[29] D. Hilbert, "Ueber die stetige Abbildung einer Linie auf ein Flächenstück," *Mathematische Annalen*, vol. 38, no. 3, pp. 459–460, 1891.

[30] W. Sierpinski, "Sur une nouvelle courbe continue qui remplit tout rune aire plane," *Bulletin International de l'Académie des Sciences de Cracovie A*, pp. 462–478, 1912.

[31] E. Moore, "On Certain Crinkly Curves," *Transactions of the American Mathematical Society*, vol. 1, pp. 72–90, 1900.

[32] H. Lebesgue, *Lecons Sur L'integration Et La Recherché Des Fonctions Primitives*, Cauthier Villars, Paris, pp. 44–45, 1904.

[33] H. Sagan, *Space Filling Curves*, Springer-Verlag, New York, 1994.

[34] F. M. Dekking, "Recurrent Sets," *Advances in Mathematics*, vol. 44, no. 1, pp. 78–104, 1982.

[35] John P. Gianviffwb and Yahya Rahmat-Samii, "Fractal Antennas: A Novel Antenna Miniaturization Technique, and Applications," *IEEE Antenna's and Propagation Magazine*, vol. 44, no. 1, pp. 20–36, 2002.

9 Microwave Devices and Circuits for Advanced Wireless Communication

Design and Analysis

Diptiranjan Samantaray and Somak Bhattacharyya

9.1 INTRODUCTION

In recent years, the demand for high-frequency communication systems and advanced technologies has surged, necessitating the development of efficient and compact antennas. This chapter introduces the concept of metasurface-based antennas operating in the gigahertz (GHz) frequency range. Metasurfaces, composed of subwavelength resonant structures, offer unprecedented control over the phase, polarization, and amplitude of electromagnetic waves [1–3]. Metasurface-based antennas have emerged as intriguing alternatives for meeting the needs of current GHz frequency applications by leveraging these unique features.

9.1.1 Metasurface

A metasurface is the two-dimensional arrangement of metamaterials or it is simply a surface distribution of electrical field scatters capable of manipulating and controlling electromagnetic waves [1–3]. The geometrical shape of the scatters generated on the dielectric substrate determines the functionality of the metasurfaces [3, 4]. Metasurfaces have the unique properties of having an ultra-low profile and improved gain features with directive radiation pattern and a wider bandwidth [5–8]. Many other advantages of metasurfaces, including their low cost, light weight, excellent flatness, and simplicity of manufacture, have led to the expansion of metasurface research from microwave to optical frequencies over the past decade [4–12]. MS has several potential uses, including antenna performance enhancement, reduction in the size of the resonator, reconfigurable antenna, controllable "smart" surfaces, etc. [4–12].

The metasurface structure depicted in Figure 9.1 comprises squared-metal patches grouped regularly on a substrate [2]. It has a metallic patch size of w_p and a periodicity of *P*. The surface wave resonances of a finite metasurface-based antenna

DOI: 10.1201/9781032656021-9

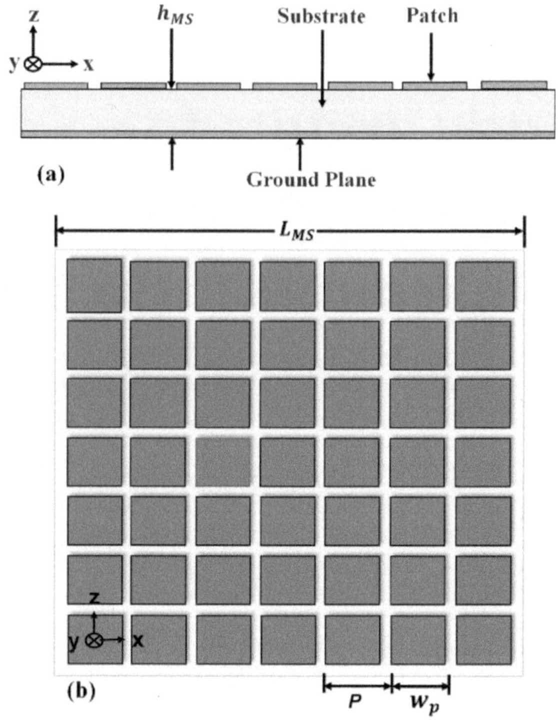

FIGURE 9.1 A generalized metasurface structure (a) side view and (b) top view.

can be estimated qualitatively using the equation (9.1), where β_{SW} is the propagation constant of surface wave resonances, and L_{MS} signifies the overall length of the MS structure as represented in (9.2)

$$\beta_{SW} \times L_{MS} = \pi \tag{9.1}$$

$$L_{MS} = N \times P \tag{9.2}$$

N and P are the numbers of cells and periodicity between the metasurface unit cells, respectively. Putting (9.2) into (9.1) yields (9.3)

$$\beta_{SW} = \frac{\pi}{N \times P} \tag{9.3}$$

The expression in (9.4) relates the decay constant α and the angular frequency ω

$$\beta_{SW} = \sqrt{\eta^2 \omega^2 + \alpha^2} \tag{9.4}$$

For transverse electric (TE) and transverse magnetic (TM) waves, the propagation constant can be determined from (9.5–9.6) respectively,

$$\beta_{TM} = \frac{\omega}{c}\sqrt{1-\frac{Z_s^2}{\eta^2}} \tag{9.5}$$

$$\beta_{TE} = \frac{\omega}{c}\sqrt{1-\frac{\eta^2}{Z_s^2}} \tag{9.6}$$

where β_{TM} and β_{TE} denote the propagation constants of TM- and TE-waves, respectively, c signifies the speed of light in a vacuum, while η and Z_s represent the intrinsic and surface impedances of the metasurface structure respectively [10, 11].

9.1.2 Metasurface Antenna

In recent times, different kinds of convenient radiating antennas are designed using MS. Each design consists of an MS and a simple antenna, such as a patch or slot antenna [13–17]. The characteristics of patch antennas and slot antennas are discussed in this section [13–17]. Numerous low-profile MS antennas with much greater performance than conventional antennas have been built and explored [18–23]. MS can be used to tune the surface impedance, regulate the reflection coefficient phase, manipulate the propagation of surface waves, control the frequency band, control edge diffractions, and control the radiation patterns of antennas [23–29]. One significant difficulty with metasurface antennas is the occurrence of resonances other than the typical resonance induced in the main radiating element due to the occurrence of MS structure with finite size [30, 31]. This chapter uses MS to design various antennas, including miniaturized patch antennas, slot antennas, CPW-fed antennas, and dual-band antennas. At the same time, MS can significantly improve antenna performance in terms of impedance bandwidth, gain, and efficiency [32–39].

9.1.2.1 Design Analysis

An antenna is a device that converts electrical power to broadcast electromagnetic waves (and vice versa). To put it another way, the antenna is a bridge between free space and a directing device. Several metrics have been developed to describe antenna performance, including bandwidth, radiation pattern, beamwidth, directivity, gain, antenna efficiency, polarization, axial ratio, and sidelobe level [40–47]. The primary radiating elements such as microstrip patch antennas and horizontal dipoles generate additional resonances on a finite-sized MS structure [21–23]. Surface waves traveling through the metasurface structure provide this additional resonance, as illustrated in Figures 9.2(a) and 9.2(b). Depending on the application, the driven radiator can be positioned either above or below the MS. During resonances, the overall length of the metasurface is equal to the resonant length of a specific surface wave [21–23]. As a result, additional resonances in an antenna configuration are formed by diverse waves.

The architecture of a standard MSA with their working analysis is shown in Figure 9.3, where the MS is superimposed on a primary element. The radiated electromagnetic wave from the source antenna may be partially reflected and transmitted as a result of the MS. Multiple reflections and transmission waves with diminishing

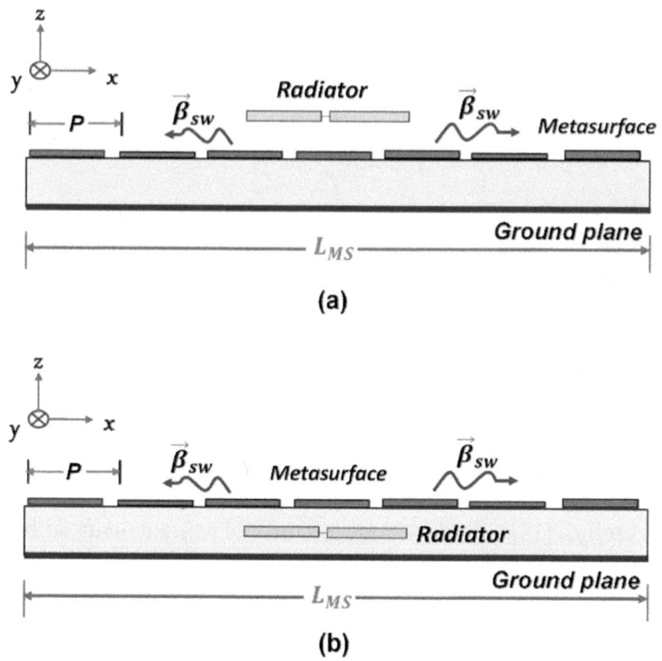

FIGURE 9.2 Radiating element arrangement in an MS antenna (a) radiator above MS and (b) radiator below MS.

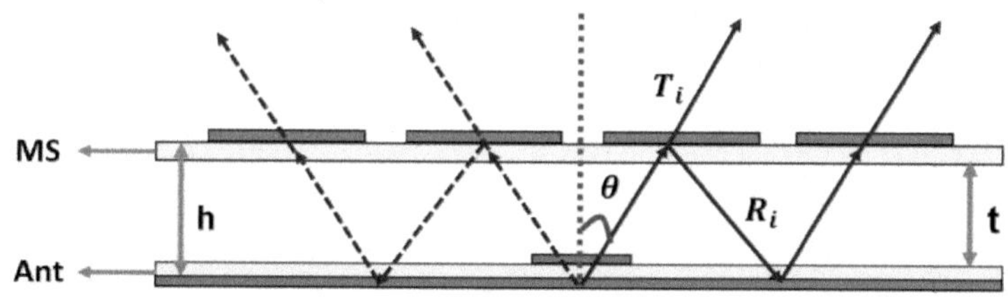

FIGURE 9.3 The architecture of a conventional MS antenna in which electromagnetic waves from the source antenna can be partially transmitted and reflected by the metasurface.

amplitudes are formed between the ground and MS. According to the ray theory presented in [21–23], the electric field in the far-field region is the vector sum of these partially transmitted rays, which can be represented in (9.7)

$$E_T = \sum_{i=-\infty}^{\infty} f(\theta) E_i \tag{9.7}$$

$$E_i = E_0 T e^{j\psi T_i} \tag{9.8}$$

Here, $f(\theta)$ the function of the source's radiated field pattern with respect to the angle of incidence θ and Ei in (9.8) represents the resultant of the successive reflected field vectors during reflections from the ground layer and MS, E_0 denotes the amplitude of the electromagnetic wave from the source, ψT specifies the transmission phase of the MS, and i is the transmitted wave index. If the resonance condition in (9.9) is satisfied, then the power is radiated in the broadside direction,

$$\psi - \pi - \frac{4\pi h}{\lambda_0} = 2N\pi \tag{9.9}$$

Here N specifies as an integer, and λ_0 represents the free-space wavelength. As a result, the height (h) is found to match the requirements provided in (9.10) for the resonance distance:

$$h = \left(\frac{\psi}{360} - 0.5 \right) \frac{\lambda_0}{2} - N \frac{\lambda_0}{2} \tag{9.10}$$

The antenna radiates with high directivity toward the broadside when the MS is made up of homogenous unit cells. The MSA, on the other hand, may radiate at an angle offset from the broadside if the MS is designed with nonhomogeneous unit cells with a gradient phase [40, 41].

9.1.2.2 Analysis of the Unit Cell

The analytical Drude-Lorentz model or the S-parameter retrieval method [48–50] can be used to analyze electromagnetic properties such as complex permeability (μ) and the permittivity (ε). The S-parameter extraction method has been employed. The reference planes of the two ports must adhere to the unit cell, which may be seen in the side view of the simulation setup. The comparable relative permittivity ε_r and permeability μ_r may be calculated using the values of S_{11} and $S_{21.}$ First, using (9.11) and (9.12) from [48–50], the equivalent impedance Z and the reflective index n of the MS are always determined.

The S-parameters of the system can also be indicated in equations (9.11 & 9.12)

$$S_{11} = \frac{R_{01}\left(1 - e^{i2nk_0d}\right)}{1 - R_{01}^2 e^{i2nk_0d}} \tag{9.11}$$

$$S_{21} = \frac{\left(1 - R_{01}^2\right) e^{i2nk_0d}}{1 - R_{01}^2 e^{i2nk_0d}} \tag{9.12}$$

Solving equations (9.11) and (9.12) provide

$$z = \pm \sqrt{\frac{\left(1 + S_{11}\right)^2 - S_{21}^2}{\left(1 - S_{11}\right)^2 - S_{21}^2}} \tag{9.13}$$

$$e^{ink_0d} = \frac{S_{21}}{1 - S_{11} \dfrac{z-1}{z+1}} \tag{9.14}$$

$$n = \frac{1}{k_0 d}\left[\left\{\left[ln\left(e^{ink_0 d}\right)\right]'' + 2m\pi\right\} - i\left[ln\left(e^{ink_0 d}\right)\right]'\right] \tag{9.15}$$

Where $(.)''$ and $(.)'$ denote the real and complex components, respectively; S_{11} and S_{21} are reflection and transmission coefficients, respectively; R_{01} is $\frac{z-1}{z+1}$; z in (9.13) represents the normalized impedance; the refractive index n is calculated using (9.14) and is represented by the equation (9.15); k_0 is the wavenumber; d denotes the maximum length of the unit element, and m signifies the branch caused by the sinusoidal function's periodicity [48, 49]. Additionally, permittivity (ε) and permeability (μ) are proportional to refractive index and normalized impedance as shown by the formulas (9.16 & 9.17)

$$\varepsilon = \frac{n}{z} \tag{9.16}$$

$$\mu = nz \tag{9.17}$$

9.1.3 Different Metasurface Antenna Prototypes

9.1.3.1 Antenna with Radiating Element and Metasurface on the Same Surface

The development of planar magneto-electro-dielectric waveguided metamaterials (MED-WG-MTM) for effective permeability and permittivity extraction is described in Figure 9.4 [51]. The MED-WG-MTM cell has an electric complementary spiral ring resonator (CSR) in the upper metallic layer and a magnetic embedded Hilbert line (EHL) in the ground plane.

The electro-dielectric and magneto-dielectric waveguided substrates are combined to form a magneto-electro-dielectric waveguided MTM element, which is electrically smaller [51]. An upper metallic layer and a lower ground plane make up the planar WG-MTM artificial material in the planar waveguide environment. A two-turn complementary spiral ring resonator (CSR) was etched in the upper metallic plane, while an embedded Hilbert-line (EHL) inspired by the space-filling property of fractal curves was employed in the lower ground plane. Higher refractive index and wave impedance values are achieved by the MTM element, which can be employed to reduce antenna size and boost antenna impedance bandwidth, respectively. The recommended antenna is less sensitive to metal shielding than a monopole antenna, regardless of operating band or BW, allowing for more practical applications.

The second concept in Figure 9.5 employs an aperture coupling to feed a single-layer broadband MS antenna through a coplanar waveguide (CPW) [52]. An MS radiator is fed by a stair-shape aperture on the bottom, which is again fed by a CPW amplification source. The antenna's input impedance can be adjusted for wideband characteristics using a special stair-shape aperture built for this purpose. The benefits of this antenna design include a wide impedance bandwidth, a single-layer structure, and ease of design and manufacture. The use of MS with radiating elements resulted in a wide impedance bandwidth of 67.3% (4.81–9.69 GHz) and a maximum gain of 9.18 dBi.

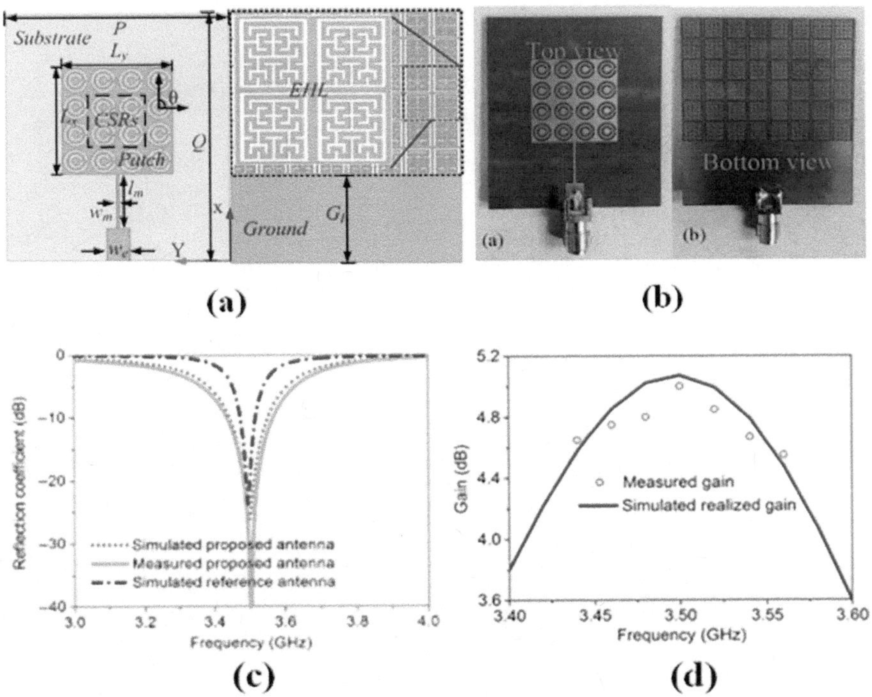

FIGURE 9.4 (a) Antenna prototype. (b) Fabricated structure. (c) Reflection characteristics. (d) Gain characteristics [51].

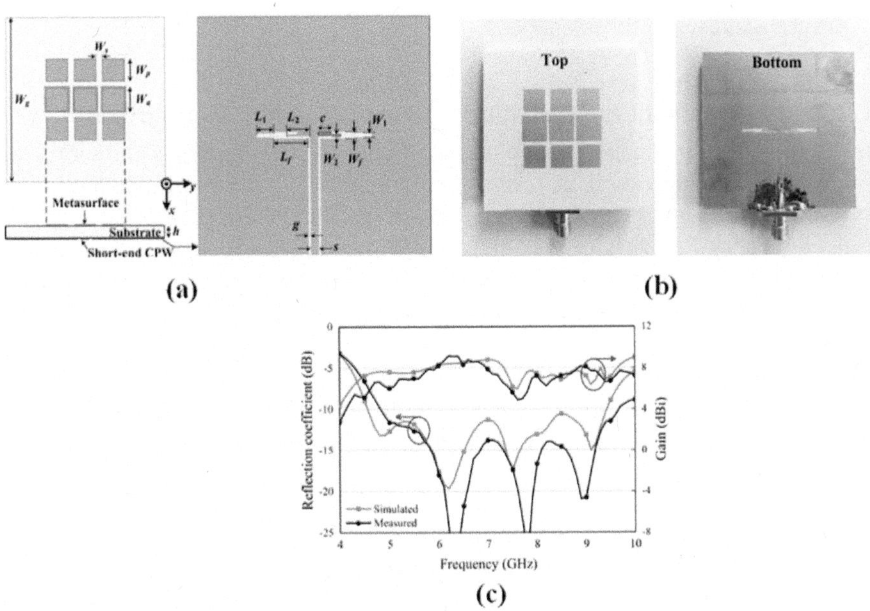

FIGURE 9.5 (a) Antenna prototype. (b) Fabricated structure. (c) Reflection and gain characteristics [52].

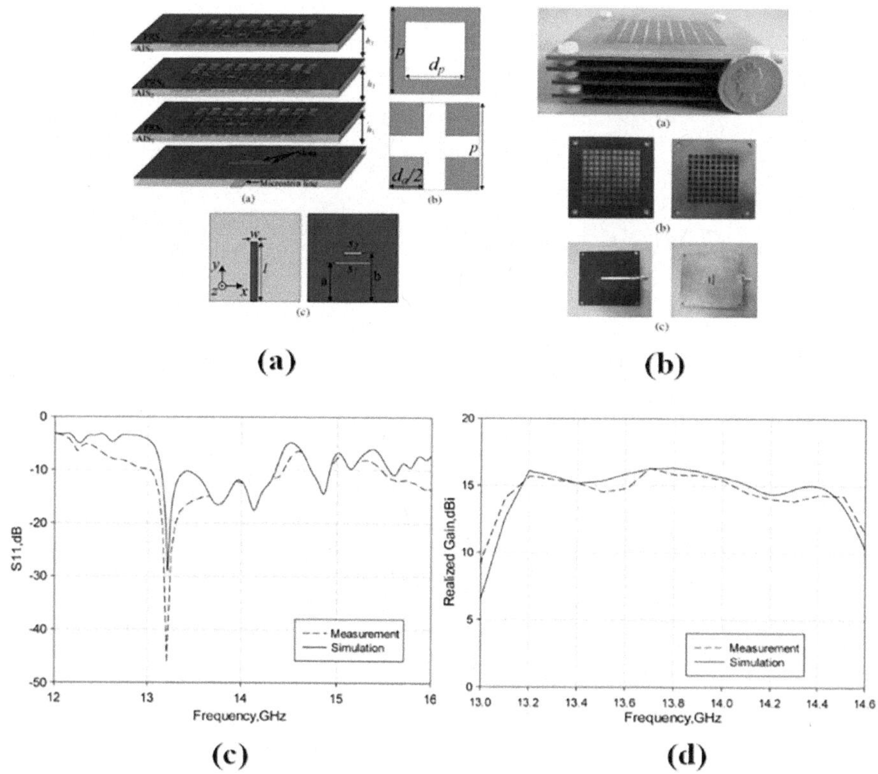

FIGURE 9.6 (a) Antenna prototype. (b) Fabricated structure. (c) Reflection characteristics. (d) Gain characteristics [53].

9.1.3.2 Antenna with the Radiating Element below the Metasurface

A method for developing planar antennas with subwavelength profiles and significant gain over a wide bandwidth is described in Figure 9.6 [53]. An innovative multi-layer periodic array design is developed to increase the bandwidth performance of subwavelength Fabry-Perot cavity-type antennas. Three double-sided periodic arrays are constructed and optimized using an inductive partially reflected surface (PRS) and a capacitive artificial impedance surface (AIS) printed on either side of a dielectric substrate.

The distance between them and the ground plane is approximately one-sixth of a wavelength. A total profile of $\lambda/2$ is accomplished by establishing three air spaces in this manner. The antenna has achieved 16.9 dBi directivity and 10.7% of fractional bandwidth over a range of frequencies.

The next design techniques in Figure 9.7 [54] demonstrates that an FP resonator antenna is capable of simultaneously increasing gain and decreasing the radar cross section (RCS). However, the absorption in the operational band is lowered to maintain the same radiation performance. As a result, RCS reduction inside the band is limited. In addition, the lumped resistance restricts applications that require the superstrate design to be smooth and oscillate. Another innovative superstrate is

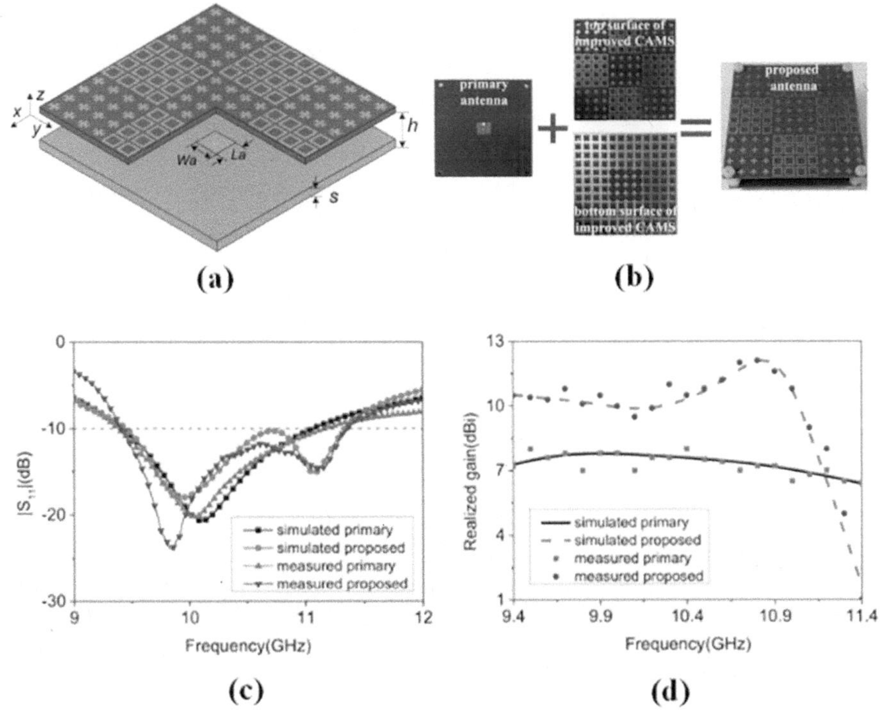

FIGURE 9.7 (a) Antenna prototype. (b) Fabricated structure. (c) Reflection characteristics. (d) Gain characteristics [54].

designed to overcome the foregoing constraints. A chessboard-organized metamaterial superstrate (CAMS) is composed of two distinct types of frequency selective surfaces (FSS) and is based on the RCS-reducing principle of chessboard reflection screens and the gain-enhancing theory of FP resonator antennas. The upper surface of CAMS has a chessboard-style layout that reduces RCS. To increase antenna gain, an FP resonator cavity is formed on the bottom surface of the primary antenna's ground plane. The FP resonator antenna enhances gain by 4.9 dB at 10.8 GHz and has a 3 dB gain bandwidth between 9.4 and 11.1 GHz (16.58%). Additionally, the suggested RCS of the FP resonator antenna is reduced from 8 to 18 GHz, resulting in a peak reduction of 39.4 dB. Additionally, the in-band RCS is significantly reduced as a result of the combined action of CAMS and the FP resonator cavity.

9.1.3.3 Antenna with the Radiating Element above the Metasurface

An antenna with an artificial magnetic conductor (AMC) surface for wireless local area network (WLAN) applications was shown in Figure 9.8 [55]. Bowtie-shaped patches with U-slots and a 4 × 4 AMC surface are used to make this antenna. The bowtie shapes are responsible for wider bandwidth whereas the AMC surface improves antenna radiation characteristics. The radiation pattern of the antenna will improve as the number of AMC cells rises. While a lower reflector size can create edge-effect radiation, it can also cause the front-to-back ratio to be lowered, as well

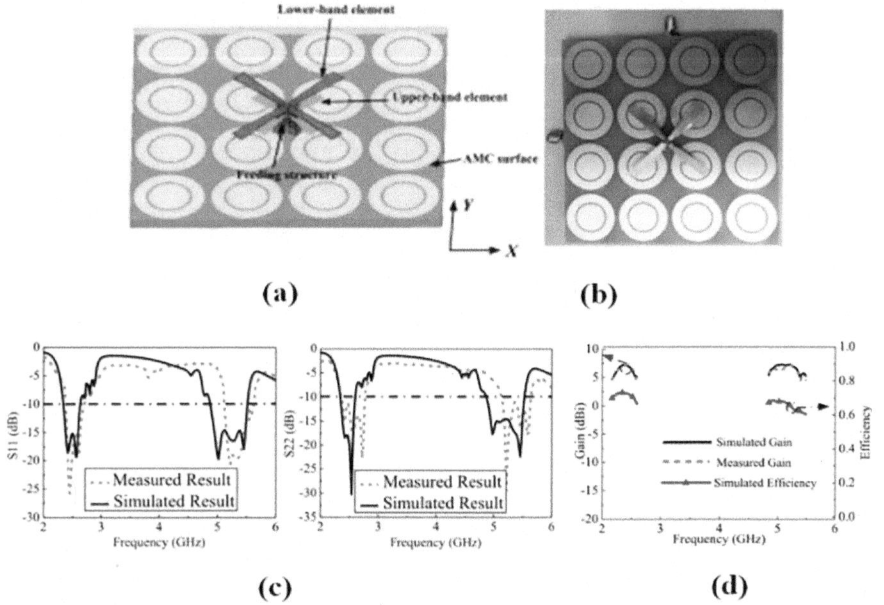

FIGURE 9.8 (a) Antenna prototype. (b) Fabricated structure. (c) Reflection characteristics at port1 and port2. (d) Gain characteristics [55].

as impedance matching to be inadequate. This antenna is capable of functioning in two bands for WLAN applications, a lower band between 2.36 and 2.76 GHz and a higher band between 5.12 and 5.62 GHz. This antenna prototype generates radiation patterns with a peak gain of 7.2 dBi in the lower band and 7.3 dBi in the upper band. The antenna is suitable for usage in multiband base stations for WLAN and other wireless networks.

A dual-polarized, omnidirectional, wideband, and low-profile antenna with an artificial magnetic conductor (AMC) reflector is presented in Figure 9.9 [56]. The antenna construction comprises of a vertically polarized low-profile monopole antenna, horizontally polarized circular loop antenna, and an AMC reflector. Incorporating a circular ring, metallic reflector considered as a monopole, and AMC surface improves the antenna performance and makes the antenna directive at boresight direction. This antenna operates in the frequency range 1.7–2.7 GHz, has a 45% impedance bandwidth, and offers consistent radiation patterns. In order to achieve the desired antenna gain and bandwidth, a wider metallic ground plane with a thicker profile thickness is used in this antenna.

9.2 DESIGN ANALYSIS OF THE SINGLE-LAYERED METASURFACE ANTENNA

9.2.1 Geometry of Antenna

The antenna structure consists of a split ring resonator (SRR) based corner cut rectangular patch antenna with MS printed on a 30×47 mm^2 FR4 substrate of the

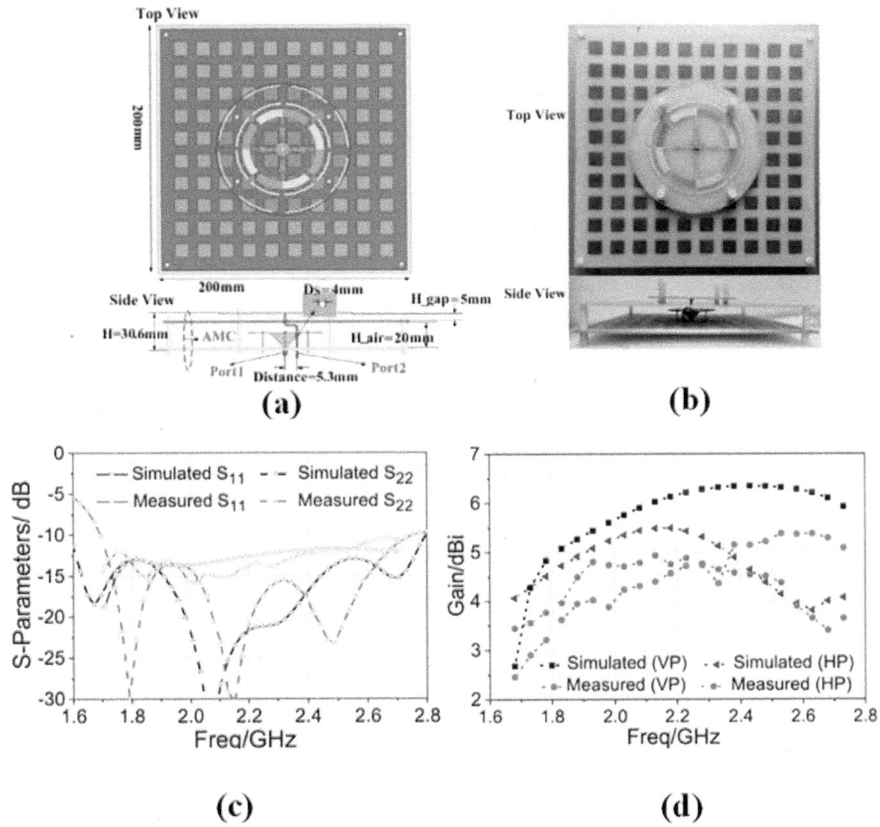

FIGURE 9.9 (a) Antenna prototype. (b) Fabricated structure. (c) Reflection characteristics. (d) Gain characteristics [56].

thickness of 1.6 mm, having a dielectric constant of 4.4 and loss tangent of 0.025. The 4×5 order MS is placed on the upper region of the patch whereas two 1×4 arrays of unit cells are arranged adjacent to the patch surface. The defected ground has been realized by using five symmetric rectangular slots of dimension 1×2 mm^2. The SRR based corner cut rectangular patch antenna and the MS are placed together in the same plane. So it makes the antenna compact and low-profile. The overall structure of the prototype is described in Figure 9.10 [57].

The resonant frequency (*f0*) of the antenna is obtained by using the equivalent circuit analysis method and mentioned in (9.18), where *Leq* and *Ceq* are effective inductance and capacitance of the antenna

$$f_0 = \frac{1}{2\pi\sqrt{L_{eq}C_{eq}}} \tag{9.18}$$

Initially, a profile closer to half wavelength has been selected based on the operating frequency. The patch dimensions *Pw* and *Pl* are tuned for required resonant

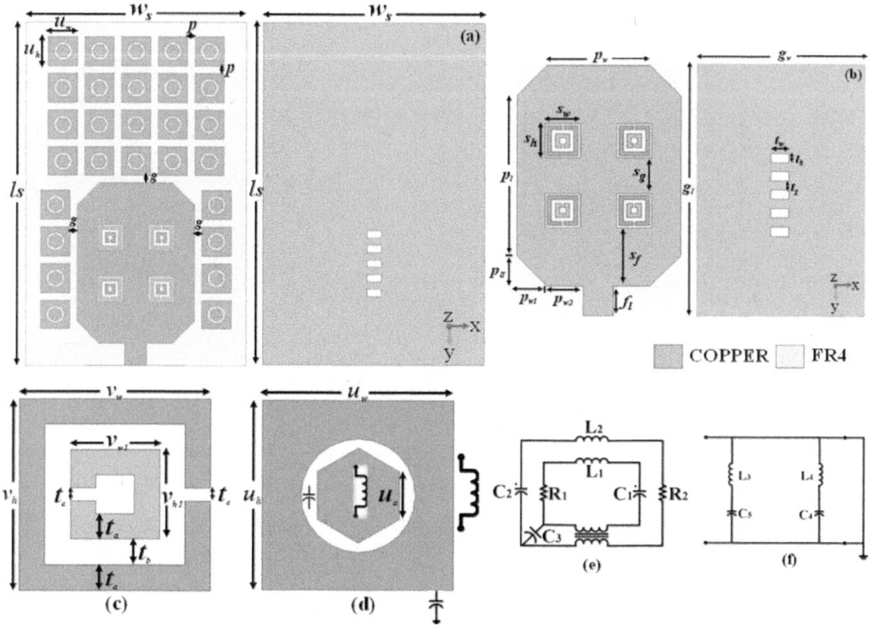

FIGURE 9.10 (a) Antenna structure with top and bottom surfaces. (b) Conventional patch with top and bottom surfaces. (c) SRR. (d) Unit cell. (e) Equivalent circuit model of SRR. (f) Equivalent circuit model of the unit cell. R_1, R_2, L_1, L_2, L_3, L_4, C_1, C_2, C_3, C_4, and C_5 are the resistances, inductances, and capacitances of the SRR and the unit cell [57].

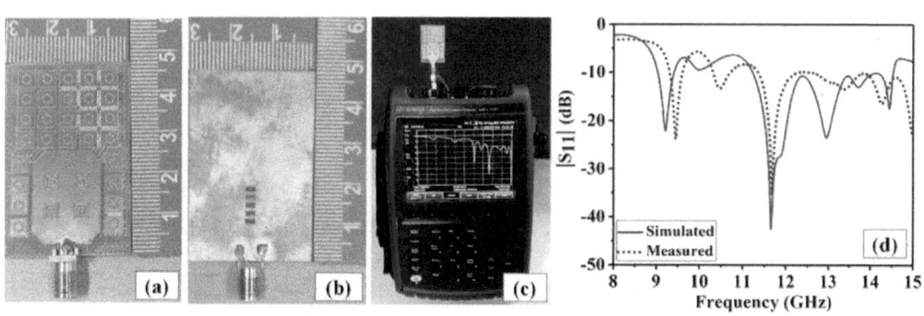

FIGURE 9.11 The fabricated prototype (a) top view, (b) bottom view, (c) S-parameter measurement of using VNA, and (d) Plot of Simulated vs Measured $|S_{11}|$ (dB) with respect to frequency.

frequencies. Next, the positions of the slots in the ground plane along with the slot length dimension can be changed to tune the frequencies. Here, the slots are always placed symmetrically along the feed direction. The individual arm lengths are smaller than quarter-wavelength at the radiating frequencies. The antenna is excited by a 50Ω microstrip feedline applied with respect to the ground plane. The antenna prototype has been shown in Figure 9.11.

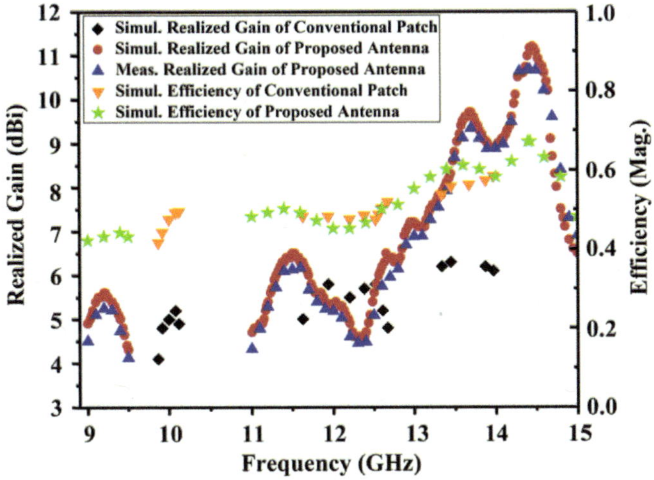

FIGURE 9.12 Responses of realized gain (dBi) and efficiency (Mag.) at different operating frequencies within the conventional patch and the antenna.

The antenna impedance bandwidth has been achieved in the dual-frequency regions ranging from 9.29 GHz–9.60 GHz and 11.30 GHz–15 GHz. Moreover, the maximum return loss of 35 dB has been achieved at 11.67 GHz, resulting in good matching with the simulated response. Some discrepancies have been noticed between the simulated and measured responses in the range of 11–15 GHz which may be attributed to the dispersive nature of the FR4 substrate.

The performance responses of the realized gain (dBi) and efficiency of the conventional patch and the MS antenna at different operating frequencies are illustrated in Figure 9.12. As revealed from the plots, the maximum realized gain of 10.66 dBi has been achieved at 14.44 GHz which is nearly 4.5 dBi higher than the conventional antenna. The gain of the antenna has been enhanced to a significant level after the incorporation of 4×5 order MS at the upper surface and two 1×4 array of unit cells placed adjacent to the radiating patch. Besides, it has been further observed from Figure 9.12 that the maximum radiation efficiency of the antenna is 67% at 14.44 GHz.

The radiation pattern characteristics of the antenna have been depicted in Figure 9.13 at the operating frequencies of 11.65 GHz and 14.44 GHz. Unidirectional radiation pattern has been achieved both in E-plane and H-plane at the operating frequencies as evident from the co-polarized radiation patterns. The cross-polarization levels are at least 20 dB below that of the co-polarization levels both in E and H-planes. In the far-field measurement, the cross-polarization levels are suppressed at different frequencies due to the incorporation of the metasurface placed just above the patch.

In the last design, the MS antenna is single-layered, dual, and it operates in the X-band region only. The performance of the antenna has been improved by the MS with the combination of two different unit cells. In the next antenna design, the efficiency and the antenna features have been improved using multi-layered

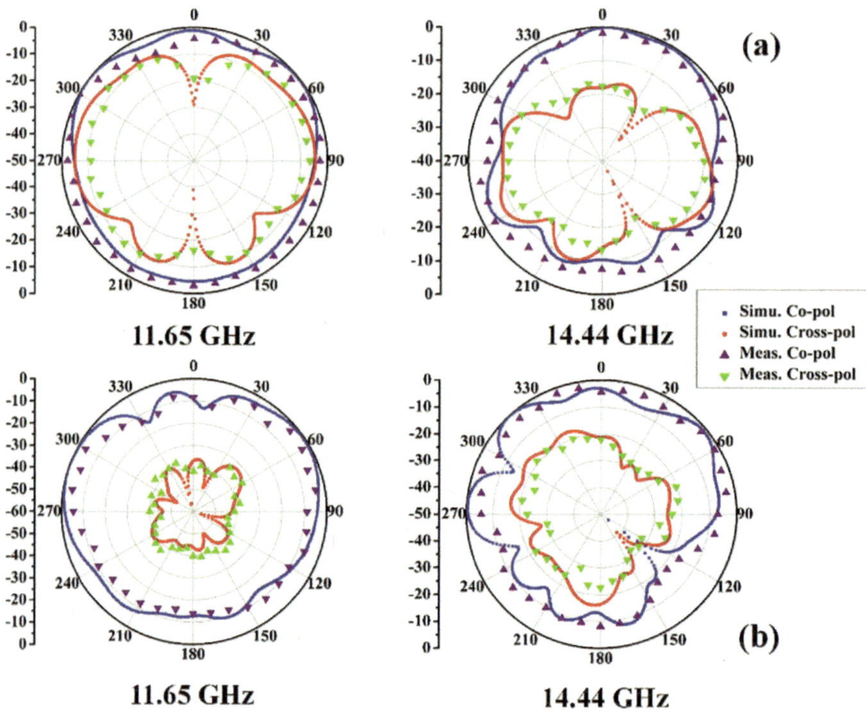

FIGURE 9.13 Co-polarized and cross-polarized (a) E-plane and (b) H-plane at different operating frequencies within the structure.

configuration. Additionally, the next section covers the design, analysis of the circuit model, and implementation of an innovative low-profile slotted patch antenna based on metasurfaces, which is intended for dual-band applications in the C and X bands [58–73].

9.3 DESIGN ANALYSIS OF THE MULTI-LAYERED METASURFACE ANTENNA

9.3.1 Antenna Design with a Radiating Element above the Metasurface

The antenna design comprises a modified slotted patch and a 7×7 array of periodic metasurface (MS) as superstrate configuration. The MSA comprises a slotted patch and a 7×7 order MS structure in which the patch is formed by square annular-shaped slots at the four corners. The MS layer is integrated with unit cells where each meta-atom composes of a centred patch surrounded by four rectangular strips. Both the radiating patch and the MS layer have been printed on a 1.6 mm thick FR4 dielectric layer (loss tangent = 0.025 and relative permittivity = 4.4) and separated from each other by the 2.2 mm thick air spacer. The three-dimensional (3D) geometry of the structure with the conventional patch and 7×7 order MS superstrate layer have been represented in Figure 9.14 [74].

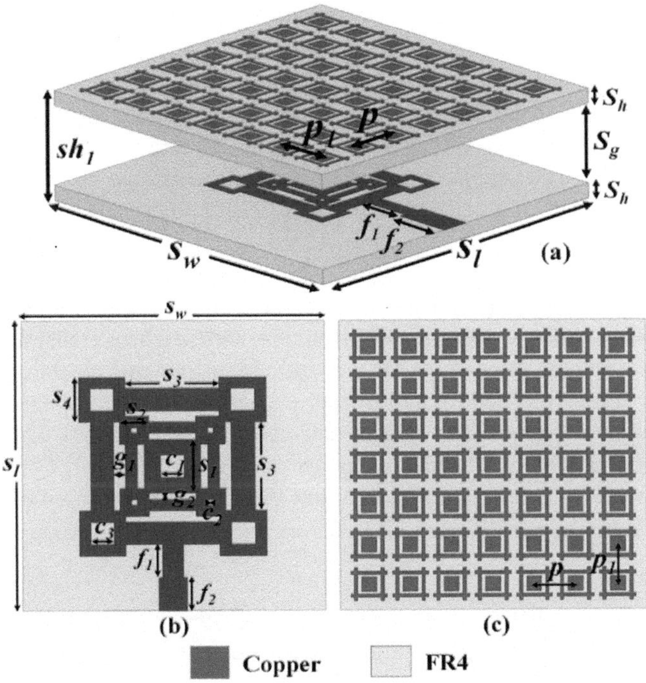

FIGURE 9.14 (a) 3D geometry of the antenna prototype, (b) conventional patch, and (c) 7×7 order MS structure [74].

The resultant input impedance of the equivalent circuit of the antenna can be expressed in (9.14) [72, 73]

$$Z_n = \sum_{k=1}^{n} \frac{j\omega R_k L_k}{R_k\left(1-\omega^2 C_k L_k\right)+j\omega L_k} \tag{9.19}$$

Here, R_k represents the radiation resistance. L_k and C_k indicate the respective inductor and capacitor values corresponding to the slot resonator. The antenna elements are realized by the Foster canonical expression in which R_k takes care of both radiation and ohmic losses [72, 73]. RLC parallel circuits in a series combination of the Foster canonical form is suitable to model the "electric antennas" which behave as an open circuit under DC excitation, whereas the magnetic antennas are represented by parallel connection of series RLC circuits and behave as short circuits under DC excitation [72, 73].

The circuit representation for the slotted patch antenna with quarter-wave transformer feed has been realized by incorporating parallel combinations of RLC circuits in series [73, 75]. The lumped parameters in Figure 4.20(a) have been determined from the circuit model simulation studies as R_1 = 39.6 Ω, R_2 = 74.46 Ω, R_3 = 94.5 Ω, L1 = 0.48 nH, L_2 = 0.35 nH L_3 = 0.18 nH, C_1 = 3.38 pF, C_2 = 1.11 pF, C_3 = 1.31 pF and Z_0 = 50 Ω.

9.3.1.1 Design of the Metasurface Antenna

The design of the MS layer has been formed with a periodical arrangement of 7×7 order unit cells. The MS layer has been placed above the conventional patch and separated by an air spacer of a thickness of 2.2 mm. The unit cell of the MS has been designed according to the operating wavelength and comprises a centre patch surrounded by four rectangular strips of identical dimensions.

The patterns of the unit cell are based on different practical applications [76–83]. Floquet port is used in the simulator to invoke the periodic boundary conditions with various Floquet modes. Fundamentally, Floquet modes are plane waves with propagation direction set by the frequency and geometry of the periodic structure. Floquet ports are used exclusively with periodic structures defined by primary and secondary boundaries (Coupled Lattice Pair). They contain plane waves whose frequency, phasing, and the geometry of the periodic structure determine the propagation direction viz., planar phased arrays and frequency selective surfaces [81–83]. The periodically arranged unit cell exhibits 0° reflection phase at 10 GHz and 17.65 GHz as observed from Figure 9.15(b). The variation in the reflection phase is bounded between ±180° over the frequency region 1 GHz to 20 GHz as seen from Figure 9.15(b).

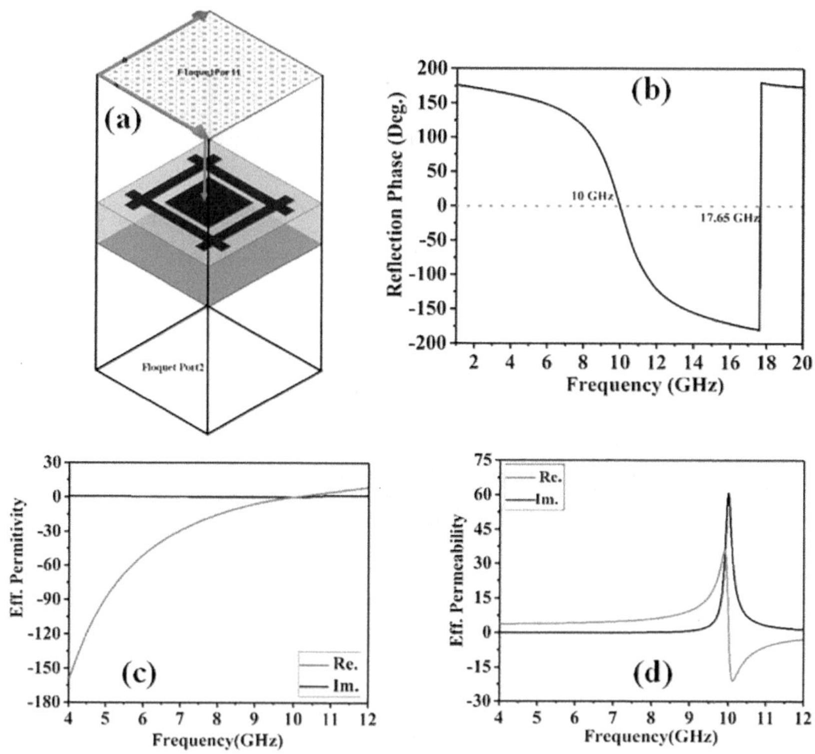

FIGURE 9.15 (a) The three-dimensional view of the unit cell, (b) reflection phase characteristics with respect to frequency, (c) effective permittivity of the unit cell, and (d) effective permeability of the unit cell.

The real and imaginary parts of the effective permittivity and the effective permeability have been extracted by the S-parameter retrieval method [48–50], and the retrieved parameters of the same have been shown in Figure 9.15. The MS layer behaves as an ENG material ($\varepsilon < 0$ and $\mu > 0$) below 10 GHz while the MNG behaviour ($\varepsilon > 0$ and $\mu < 0$) has been observed above 10 GHz. Hence, 10 GHz is considered as the plasma frequency since the effective permittivity is zero at 10 GHz; thereby yielding a zero-index metamaterial. The zero-phase reflection bandwidth of the MS significantly influences the bandwidth of the array of unit elements. In MS structures, the expansion of bandwidth can be attributed to the increase in equivalent inductance [76, 77].

The equivalent input impedance of an MS unit element can be expressed as the parallel combination of the impedances of the periodic pattern and the grounded dielectric layer impedance. This results in the equality between the MS reflection bandwidth and the bandwidth of a parallel LC circuit with net inductance and capacitance of *Lm* and *Cm*, respectively as seen from (9.20)

$$BW = \frac{\omega_0^2 L_m}{Z_0} = \frac{1}{C_m Z_0} \tag{9.20}$$

where $\omega_0 L_m$ represents the impedance, ω_0 which has a dimension of rad/sec represents the bandwidth, Z_0 is the free space impedance. This implies that by lowering C_m, the MS zero-phase bandwidth can also be enhanced which can be accomplished by forming a wide slot gap between the centred patch and the rectangular strips on each arm of the MS unit cell. The etched slot gap between the centred patch and the rectangular strips of the unit cell generates a capacitive effect and the series combination of these capacitances reduces the total equivalent capacitance of the MS structure; thereby improving the zero phase reflection bandwidth.

In this MS antenna design, a 7×7 array of periodic unit cells have been placed above the modified patch antenna in a superstrate configuration. Here, the superstrate and the ground layers have been separated by a height of sh_1. The resonant conditions of this superstrate based MS antenna can be expressed in (9.21 & 9.22) [78, 79]

$$\varphi_s + \varphi_g - \frac{4\pi h}{\lambda} = 2N\pi, \quad N = 0, \pm 1, \pm 2\ldots \tag{9.21}$$

$$f = \frac{c}{4\pi h}\left(\varphi_s + \varphi_g - 2N\pi\right) \tag{9.22}$$

where h is the spacing between the superstrate layer and the ground, φ_s and φ_g are the reflection phases of the superstrate layer and the metal ground plane respectively, and λ is the operating wavelength in the substrate, differing from the free space wavelength λ_0 in the conventional air-filled antenna. Since λ is shorter than λ_0, the height h of the MS antenna is smaller compared to the conventional air-filled one.

The equivalent circuit model of the antenna loaded with the MS superstrate has been subsequently developed as shown in Figure 9.16(a). The air gap layer between the patch and the MS has been represented with the parameters R_a and C_a connected

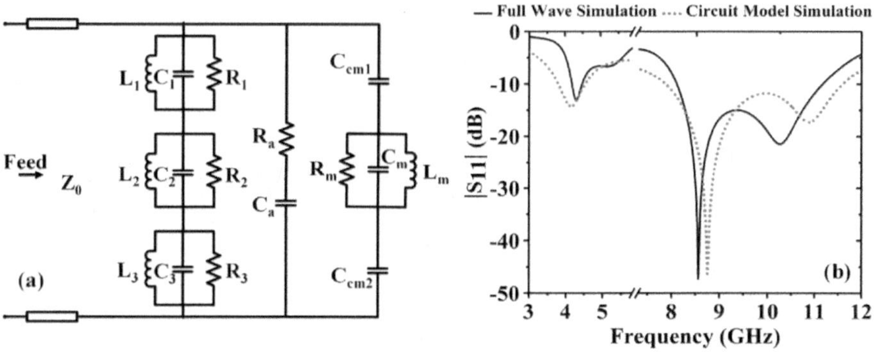

FIGURE 9.16 (a) The equivalent electrical circuit model of the patch antenna with the metasurface as a superstrate. (b) Plot of $|S_{11}|$ (dB) with respect to frequency using full wave simulation vs circuit model simulation characteristics comparison.

in parallel with the antenna circuit while R_m, C_m, and L_m are the equivalent lumped elements used for unit cells of the MS. Further, R_m represents the radiation resistance of the metasurface (MS) resonator. L_m and C_m are the respective equivalent inductance and capacitance values of the MS slot resonator. R_m in the Foster canonical form representation for the MS takes care of both radiation and ohmic losses as mentioned earlier [72, 73]. C_{cm1} and C_{cm2} are the equivalent coupling capacitances between the neighbouring unit cells along with vertical and horizontal directions of the MS layer.

The lumped circuit parameters in Figure 9.16(a) have been calculated from Keysight ADS simulation analysis and these are: $R_a = 377\ \Omega$, $C_a = 220.65$ pF, $R_m = 94.5\ \Omega$, $C_m = 1.186$ pF, $L_m = 0.273$ nH, $C_{cm1} = 0.35$ pF, $C_{cm2} = 0.35$ pF and $Z_0 = 50\ \Omega$. The additional circuit component values for the MS as a superstrate have been derived by fixing the same circuit component values of the radiating patch a. The comparison of the reflection coefficient responses of the numerical simulation and the circuit model simulation is shown in Figure 9.16(b) which offers a good resemblance between the two. The antenna loaded with MS superstrate layer operates at dual frequency regions with the fractional bandwidths of 6.1% and 35.15% at 4.26 GHz and 8.62 GHz, respectively. This antenna design has been considered as the antenna configuration due to the significantly improved characteristics.

9.3.1.2 Analysis of the Multi-Layered Metasurface Antenna

The reflection coefficient characteristics of the MS antenna along with the radiating patch antenna have been shown in Figure 9.17. The radiating patch operates over two frequency regions ranging from 8.3 GHz to 8.51 GHz and 10.19 GHz to 10.87 GHz with fractional bandwidths of 2.4% and 6.5% respectively. This patch antenna exhibits the narrowband behaviour of impedance bandwidth with low return loss. The performance of this radiating patch has been significantly improved by incorporating a 7×7 order MS layer with air spacer in the superstrate configuration. The loading of the MS superstrate layer results in the dual-band response of the antenna operating

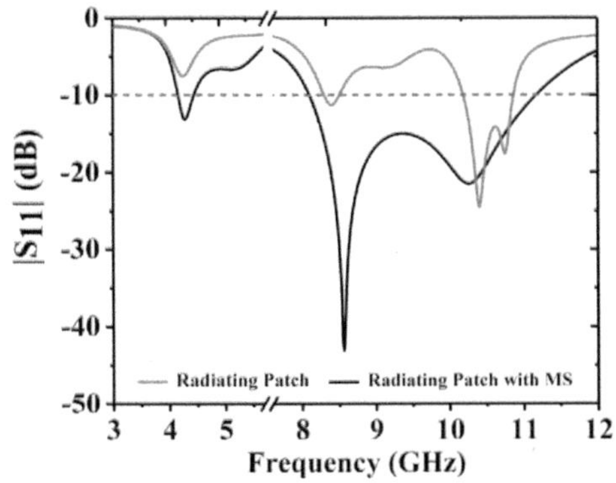

FIGURE 9.17 The plot of $|S_{11}|$ (dB) with respect to frequency (GHz) of the antenna.

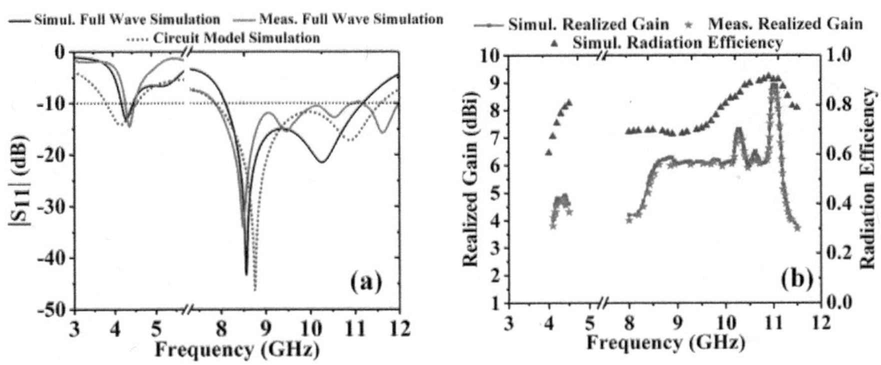

FIGURE 9.18 (a) The plot of the simulated and measured $|S_{11}|$ (dB) along with circuit model simulation with respect to frequency. (b) Responses of realized gain (dBi) and radiation efficiency with respect to frequency.

from 4.18 GHz to 4.44 GHz and 8.11 GHz to 11.14 GHz with the respective fractional bandwidths of 6.1% and 35.15% at 4.26 GHz and 8.62 GHz. The maximum return loss of 42 dB has been achieved at 8.62 GHz.

The comparison of the simulated and the measured reflection coefficient characteristics of the prototype along with the reflection coefficient response achieved by the ADS circuit model have been illustrated in Figure 9.18(a). The measured 10 dB impedance bandwidth has been realized in the frequency ranges 4.24 GHz–4.48 GHz and 7.87 GHz–11.01 GHz. Hence, the experimental reflection coefficient response has followed the simulated characteristics as well as the circuit model response.

The far-field antenna gains and radiation efficiencies of the prototype with respect to frequencies have been shown in Figure 9.18(b). In the prototype, the maximum simulated realized gain is 8.91 dBi whereas the experimentally measured realized gain of 8.72 dBi has been achieved. There is a variation in the radiation patterns across the operational bands that may be due to leakage of out of phase fields from

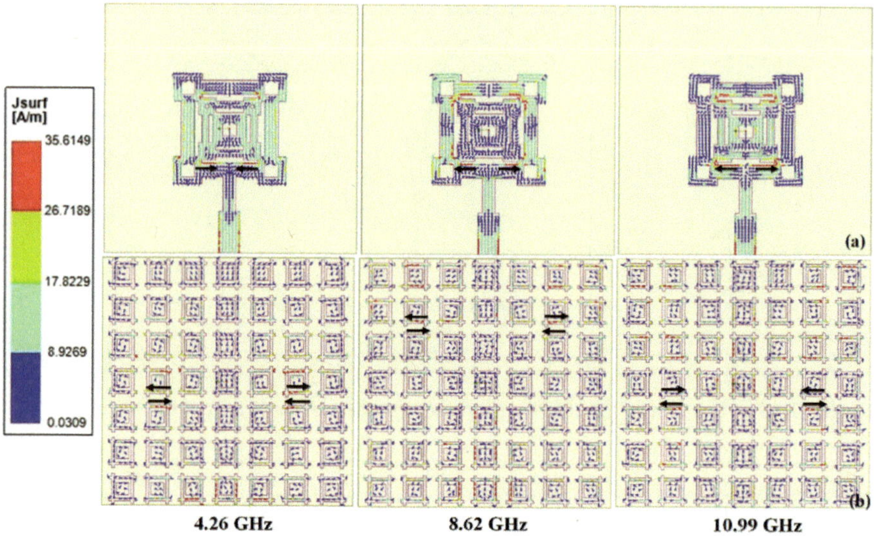

FIGURE 9.19 Surface current distributions at (a) radiating patch and (b) 7×7 order metasurface at different operating frequencies.

few slots at patch, ground slots, or the unit cells of the metasurface. The presence of coupling currents in the device may be due to less interspacing of the unit cells and the patch. However, such distortion at high frequency is still acceptable for wireless applications under multipath environment as the antenna's cross-polarized levels remain lower than the co-polarized ones. Besides, it has been further observed that the maximum radiation efficiency of the antenna is 91% at 10.99 GHz. The radiation efficiency improvement is due to the combined effect of the slotted patch and the 7×7 order metasurface. The planar MS structure is usually a high impedance surface that can modulate the electromagnetic fields excited by a patch resonator. The antenna efficiency value is influenced by a number of factors, including antenna geometry, beam pattern, and frequency reuse strategy. In some cases, the highest antenna efficiency does not lead to more effectiveness in some practical applications; hence, an intermediate efficiency value is preferable. Here the efficiency value is increased to 91% as it includes the conductor loss, dielectric loss, and ohmic loss. The actual radiation efficiency will be lower in practise.

The surface current distributions in the radiating patch and the MS layer at 4.26 GHz, 8.62 GHz, and 10.99 GHz have been demonstrated in Figure 9.19. The surface currents are more intense at the slot in the patch and also in the aperture of metasurface unit cells. It is further observed that the concentration of surface current at 10.99 GHz is more as compared to other operating frequencies due to maximum field distribution among unit cell slots on the top surface.

The simulated and measured radiation patterns of the antenna along E-plane ($\varphi = 0^0$) and H-plane ($\varphi = 90^0$) at 4.26 GHz, 8.62 GHz, and 10.99 GHz have been illustrated in Figure 9.20. The MS antenna radiates along the boresight direction both in

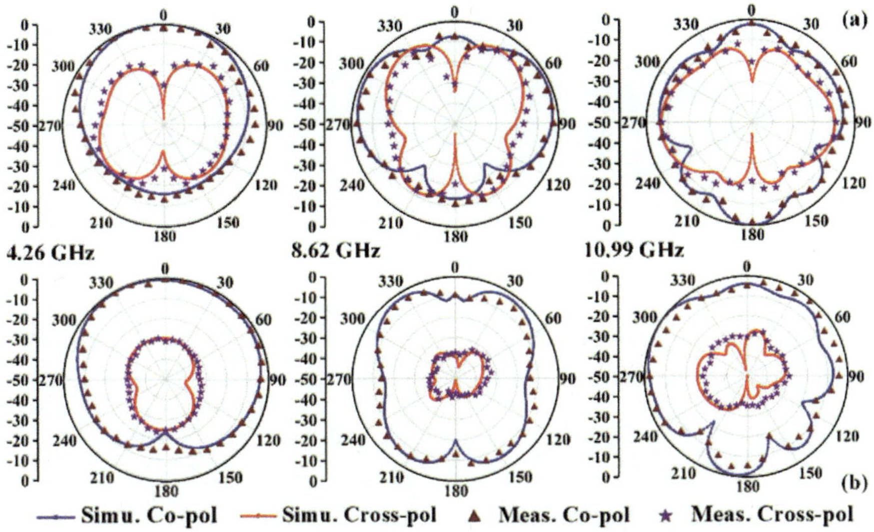

FIGURE 9.20 Radiation pattern: (a) E-plane co-polarized and cross-polarized and (b) H-plane co-polarized and cross-polarized at different operating frequencies within the structure.

E- and H-planes at all the operating frequencies; however, along H-plane at 8.62 GHz the antenna pattern deviates toward $\pm 30^0$. The cross-polarized levels of the antenna remain below 10 dB than that of the co-polarized ones in the E-plane while in the H-plane, the cross-polarized levels remain below 25 dB from the co-polarized levels at all the operating frequencies. Hence, the antenna's radiation patterns are stable at the operating frequencies centred at 4.26 GHz, 8.62 GHz, and 10.99 GHz.

In this dual-band antenna prototype, the design complexity is low as compared to the previously reported ones. In addition, the efficiency of this prototype has been enhanced to 91% due to the improved realized gain of 8.72 dBi. Thus, the antenna can be used for various C- and X-band applications as the antenna performances are significantly enhanced.

9.3.2 Antenna Design with a Radiating Element above the Metasurface

In the second multi-layered MS antenna, the configuration consists of a CPW-fed slotted patch antenna and a 7×7 order AMC surface separated by a layer of Teflon. Teflon has good electrical properties and provides mechanical robustness to the antenna, as well as resistant to temperature variation [4]. In this configuration, the radiating patch is affixed onto the initial dielectric, whereas the MS layer is positioned onto the subsequent dielectric.

9.3.2.1 Design of the Metasurface Antenna

The antenna consists of a slotted patch fed by CPW and a 7×7 order MS in which the patch and MS reflector are constructed on two distinct FR4 dielectric substrates

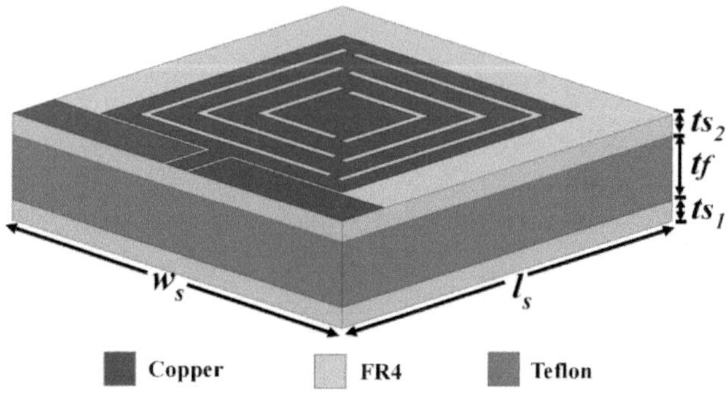

FIGURE 9.21 The 3D view of the metasurface structure [84].

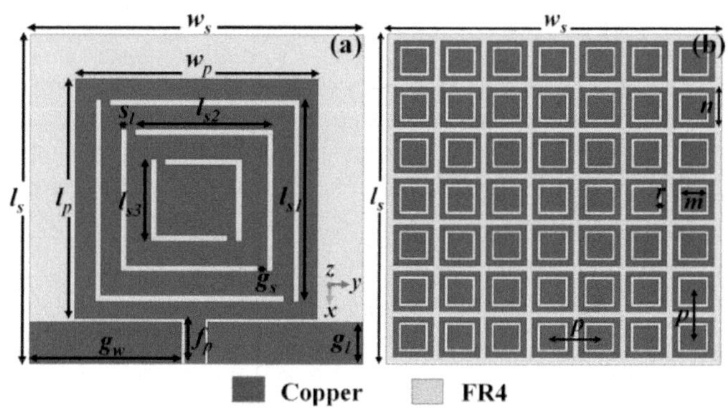

FIGURE 9.22 AMC based MS antenna with (a) radiating patch and (b) 7×7 order AMC surface.

(relative permittivity of 4.4 and loss tangent of 0.025) of 1.6 mm thickness, separated by a 5 mm thick Teflon layer (loss tangent = 0.001 and relative permittivity = 2.1). The MS unit cell comprises a centre patch surrounded by a square ring-shaped patch. A three-dimensional schematic of the MS structure is shown in Figure 9.21. Moreover, Figure 9.22 shows the radiating patch and 7×7 order MS surface of the MSA [84].

The MS surface is a copper-based conducting surface that is placed above the second FR4 substrate. The equivalent LC parallel resonance of the MS unit cell has been achieved by adjusting the space between the centre patch and the square ring-shaped patch surrounding it. The equivalent input impedance of an MS unit element can be expressed as the parallel combination of the impedances of the periodic pattern and the grounded dielectric layer impedance [85–90]. There is an improvement in bandwidth due to the combined effect of slot impedance in the patch and the MS.

The reflection coefficient characteristics of the antenna prototype have been shown in Figure 9.23 and Figure 9.24. The CPW-fed slotted patch antenna exhibits improved bandwidths from 2.31 GHz to 2.41 GHz, and 7.54 GHz to 10.37 GHz, with a maximum return loss of 31 dB at 7.73 GHz.

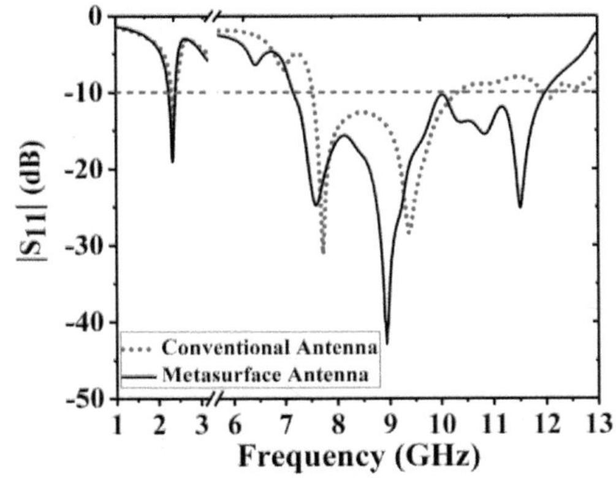

FIGURE 9.23 The comparison of the reflection characteristics of the conventional antenna and the metasurface antenna with respect to frequency.

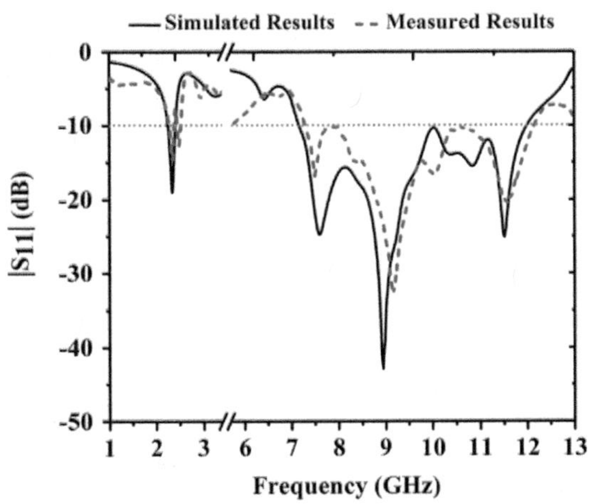

FIGURE 9.24 The plot of the simulated and measured $|S_{11}|$ (dB) with respect to frequency.

The antenna offers an impedance bandwidth of −10 dB in the frequency ranges 2.27 GHz–2.41 GHz and 7.26 GHz–12.04 GHz with a maximum return loss of 33 dB at 9.1 GHz. The analysis of the surface current distributions of the CPW-fed patch and the AMC surface have been carried out at 2.34 GHz, 11.52 GHz, and 11.98 GHz as illustrated in Figure 9.25. The area between the patch surface contact and the CPW feed has a higher density of surface currents. Nevertheless, the field distribution is highest in the vicinity of slots, as well as the patch and unit cell surface interfaces of the MS. The directions of surface currents at the unit cell interface are parallel to each other, establishing a magnetic coupling between the adjacent unit cells [91–96].

The far-field gain characteristics of the antenna and radiation efficiencies are also shown in Figure 9.26, where the measured parameters have been compared to the simulated ones and found to be in close agreement. The maximum simulated realized gain is 9.2 dBi, while the experimentally measured realized gain has been

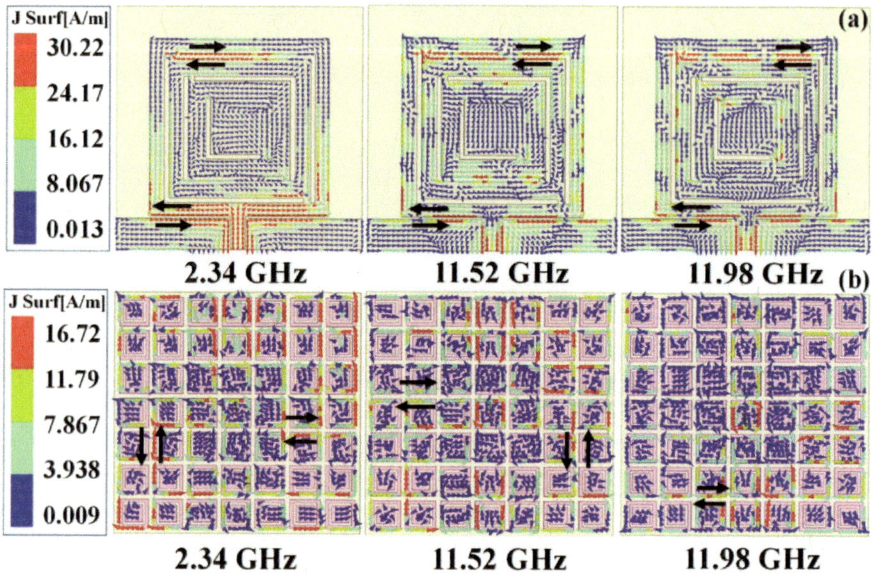

FIGURE 9.25 Surface current distributions at (a) radiating patch and (b) 7×7 order MS at different operating frequencies.

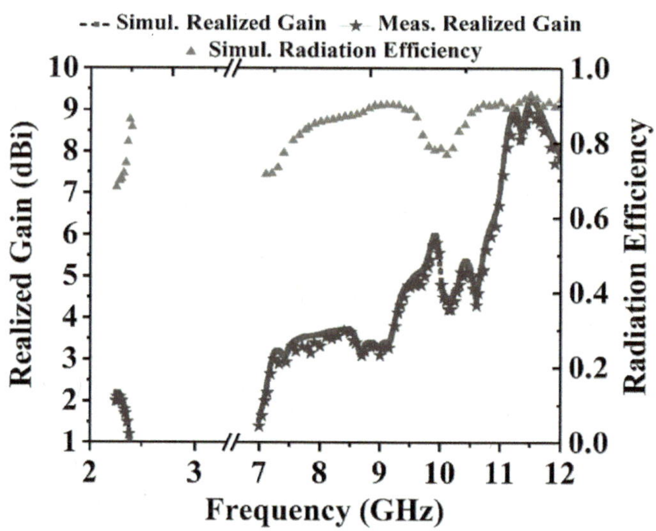

FIGURE 9.26 Responses of realized gain (dBi) and radiation efficiency with respect to frequency.

achieved as 9.1 dBi at 11.98 GHz. The AMC surface provides a gain improvement of nearly 5 dB over the standard patch, which has a maximum realized gain of 4.3 dBi. Moreover, the antenna offers the maximum radiation efficiency of 91.2% at 11.98 GHz. Antenna losses like dielectric loss, conductor loss, and other losses are

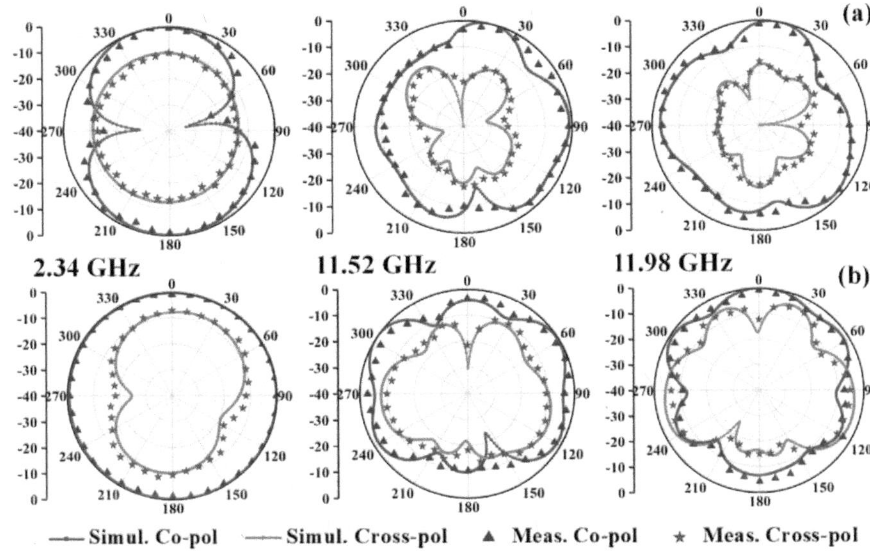

FIGURE 9.27 Radiation patterns in the (a) E-plane and (b) H-plane at frequencies of 2.34 GHz, 11.52 GHz, and 11.98 GHz.

not included in this efficiency value, which rises to 91.2%. The actual radiation efficiency will be lower in practise. The radiation patterns are unstable over the operating bands maybe due to leakage out of phase fields from few slots at patch or the unit cells of the metasurface. The vector surface current distributions of the conventional patch and the metasurface are shown in Figure 9.27 at 2.34 GHz, 11.52 GHz, and 11.98 GHz, respectively. It can be observed that there is a strong coupling current at both the radiating patch and the metasurface at 2.34 GHz while at 11.52 GHz and 11.98 GHz, the coupling currents are less in the slots of the patch and the unit cells of the metasurface. The presence of strong coupling currents in the device may be due to less interspacing of the unit cells. However, such distortion at high frequency is still acceptable for wireless handheld devices working at multipath environment as the antenna's cross-polarized levels remain lower than the co-polarized ones in the boresight direction.

Figure 9.27 depicts the simulated and experimental radiation patterns of the antenna along with the E- and H-planes at 2.34 GHz, 11.52 GHz, and 11.98 GHz, respectively. This antenna has directional radiation characteristics that radiate along the boresight direction in both planes at all the operating frequencies. Moreover, at all operational frequencies, the cross-polarized levels of the antenna are at least 10 dB lower than the co-polarized ones. This antenna can be used in a variety of IoT applications, including WMAX/WLAN, civil and military surveillance, and more.

9.4 CONCLUSION

This chapter highlights the significant capabilities of metasurface-based antennas in transforming GHz frequency applications. Metasurface antennas possess advantageous characteristics such as their compact dimensions, beam-steering capabilities, and versatility in integration with diverse systems. Consequently, these antennas are poised to play a pivotal role in shaping the future of communication and radar technologies. This chapter serves as a comprehensive guide to this exciting field, offering insights into metasurface fundamentals, design strategies, practical applications, and future prospects.

REFERENCES

[1] C. Holloway, A. Dienstfrey, E. F. Kuester, J. F. O'Hara, A. K. Azad, and A. J. Taylor, "A discussion on the interpretation and characterization of metafilms/metasurfaces: The two-dimensional equivalent of metamaterials," *Metamaterials*, vol. 3, no. 2, pp. 100–112, Oct. 2009.

[2] C. L. Holloway, E. F. Kuester, J. A. Gordon, J. O'Hara, J. Booth, and D. R. Smith, "An overview of the theory and applications of metasurfaces: The two-dimensional equivalents of metamaterials," *Antennas and Propagation Magazine, IEEE*, vol. 54, pp. 10–35, 2012.

[3] R. Nakamura and T. Fukusako, "Broadband design of circularly polarized microstrip patch antenna using artificial ground structure with rectangular unit cells," *IEEE Transactions on Antennas and Propagation*, vol. 59, p. 2103, 2011.

[4] D. Samantaray and S. Bhattacharyya, "A gain-enhanced slotted patch antenna using metasurface as superstrate configuration," *IEEE Transactions on Antennas and Propagation*, vol. 68, no. 9, pp. 6548–6556, Sept. 2020.

[5] M. Chandra, Nilotpal, D. Samantaray, M. Thottappan, and S. Bhattacharyya, "A broadband transmissive type metasurface cross-polarization converter for EMC application," *IEEE Transactions on Electromagnetic Compatibility*, vol. 65, no. 1, pp. 186–194, Feb. 2023.

[6] F. Costa, O. Luukkonen, C. R. Simovski, A. Monorchio, S. A. Tretyakov, and P. M. de Maagt, "TE surface wave resonances on high-impedance surface-based antennas: Analysis and modeling," *IEEE Transactions on Antennas and Propagation*, vol. 59, no. 10, pp. 3588–3596, Oct. 2011.

[7] I. Park, "Application of metasurfaces in the design of performance-enhanced low-profile antennas," *EPJ Applied Metamaterials*, vol. 5, p. 11, 2018.

[8] S. X. Ta and I. Park, "Artificial magnetic conductor based circularly polarized crossed-dipole antennas _ 1. AMC structure with grounding pins," *Radio Science*, vol. 52, p. 630, 2017.

[9] S. X. Ta and I. Park, "Artificial magnetic conductor based circularly polarized crossed-dipole antennas _ 2. AMC structure without grounding pins," *Radio Science*, vol. 52, p. 642, 2017.

[10] C. A. Balanis, *Antenna Theory Analysis and Design*, Third Edition, John Willy and Sons, Inc. Hoboken, New Jersey, Wiley Reprint, India, 2010.

[11] D. M. Pozar and D. H. Schaubert, *Microstrip Antennas*, New York: IEEE Press, 1995.

[12] C. Saha, J. Y. Siddiqui, and Y. Antar, *Multifunctional Ultra Wideband Antennas: Trends, Techniques and Applications*, Boca Raton, FL: Taylor and Francis, 2019.

[13] P. Jin and R. W. Ziolkowski, "Multi-frequency, linear and circular polarized, metamaterial-inspired, near-field resonant parasitic antennas," *IEEE Transactions on Antennas and Propagation*, vol. 59, no. 5, pp. 1446–1459, May 2011.

[14] Y. Dong and T. Itoh, "Metamaterial-based antennas," *Proceedings of the IEEE*, vol. 100, no. 7, pp. 2271–2285, July 2012.

[15] R. O. Ouedraogo, E. J. Rothwell, A. R. Diaz, K. Fuchi, and A. Temme, "Miniaturization of patch antennas using a metamaterial-inspired technique," *IEEE Transactions on Antennas and Propagation*, vol. 60, no. 5, pp. 2175–2182, May 2012.

[16] N. Zhu and R. W. Ziolkowski, "Broad-bandwidth, electrically small antenna augmented with an internal non-foster element," *IEEE Antennas and Wireless Propagation Letters*, vol. 11, pp. 1116–1120, 2012.

[17] L. -M. Si, W. Zhu, and H. -J. Sun, "A compact, planar, and CPW-fed metamaterial-inspired dual-band antenna," *IEEE Antennas and Wireless Propagation Letters*, vol. 12, pp. 305–308, 2013.

[18] A. K. Bhattacharyya and G. Goyette, "A novel horn radiator with high aperture efficiency and low cross-polarization and applications in arrays and multibeam reflector antennas," *IEEE Transactions on Antennas and Propagation*, vol. 52, no. 11, pp. 2850–2859, Nov. 2004.

[19] H. Huang, Y. Liu, S. Zhang, and S. Gong, "Multiband metamaterial-loaded monopole antenna for WLAN/WiMAX applications," *IEEE Antennas and Wireless Propagation Letters*, vol. 14, pp. 662–665, 2015.

[20] J. C. Myers, P. Chahal, E. Rothwell, and L. Kempel, "A multilayered metamaterial-inspired miniaturized dynamically tunable antenna," *IEEE Transactions on Antennas and Propagation*, vol. 63, no. 4, pp. 1546–1553, Apr. 2015.

[21] A. H. Naqvi and S. Lim, "A beam-steering antenna with a fluidically programmable metasurface," *IEEE Transactions on Antennas and Propagation*, vol. 67, no. 6, pp. 3704–3711, June 2019.

[22] G. Von Trentini, "Partially reflecting sheet arrays," *IRE Transactions on Antennas and Propagation*, vol. 4, no. 4, pp. 666–671, Oct. 1956.

[23] F. Costa, A. Monorchio, and G. Manara, "An overview of equivalent circuit modeling techniques of frequency selective surfaces and metasurfaces," *Applied Computational Electromagnetics Society Journal*, vol. 29, no. 12, 2014.

[24] F. Costa, A. Monorchio, and G. Manara, "Efficient analysis of frequency-selective surfaces by a simple equivalent circuit model," *IEEE Antennas and Propagation Magazine*, vol. 54, no. 4, pp. 35–48, Aug. 2012.

[25] M. Borgese and F. Costa, "A simple equivalent circuit approach for anisotropic frequency-selective surfaces and metasurfaces," *IEEE Transactions on Antennas and Propagation*, vol. 68, no. 10, pp. 7088–7098, 2020.

[26] D. -H. Kwon, "Lossless scalar metasurfaces for anomalous reflection based on efficient surface field optimization," *IEEE Antennas and Wireless Propagation Letters*, vol. 17, no. 7, pp. 1149–1152, July 2018.

[27] O. Quevedo-Teruel, M. Ebrahimpouri, and F. Ghasemifard, "Lens antennas for 5G communications systems," *IEEE Communications Magazine*, vol. 56, no. 7, pp. 36–41, July 2018.

[28] D. Samantaray, S. Bhattacharyya, and K. V. Srinivas, "A modified fractal-shaped slotted patch antenna with defected ground using metasurface for dual band applications," *International Journal of RF and Microwave Computer-Aided Engineering*, vol. 29, no. 12, pp. 1–9, Dec. 2019.

[29] M. D. Gregory *et al.*, "A low cost and highly efficient metamaterial reflector antenna," *IEEE Transactions on Antennas and Propagation*, vol. 66, no. 3, pp. 1545–1548, Mar. 2018.

[30] M. Li, Y. Zhang, and M. Tang, "Design of a compact, wideband, bidirectional antenna using index-gradient patches," *IEEE Antennas and Wireless Propagation Letters*, vol. 17, no. 7, pp. 1218–1222, July 2018.

[31] S. Yu, H. Liu, and L. Li, "Design of near-field focused metasurface for high-efficient wireless power transfer with multifocus characteristics," *IEEE Transactions on Industrial Electronics*, vol. 66, no. 5, pp. 3993–4002, May 2019.

[32] S. Zarbakhsh, M. Akbari, F. Samadi, and A. Sebak, "Broadband and high-gain circularly-polarized antenna with low RCS," *IEEE Transactions on Antennas and Propagation*, vol. 67, no. 1, pp. 16–23, Jan. 2019.

[33] W. Yang, S. Chen, Q. Xue, W. Che, G. Shen, and W. Feng, "Novel filtering method based on metasurface antenna and its application for wideband high-gain filtering antenna with low profile," *IEEE Transactions on Antennas and Propagation*, vol. 67, no. 3, pp. 1535–1544, Mar. 2019.

[34] K. Liu, Y. Ge, and C. Lin, "A compact wideband high-gain metasurface-lens-corrected conical horn antenna," *IEEE Antennas and Wireless Propagation Letters*, vol. 18, no. 3, pp. 457–461, Mar. 2019.

[35] T. Yue, Z. H. Jiang, and D. H. Werner, "A compact metasurface-enabled dual-band dual-circularly polarized antenna loaded with complementary split ring resonators," *IEEE Transactions on Antennas and Propagation*, vol. 67, no. 2, pp. 794–803, Feb. 2019.

[36] Y. Juan, W. Yang, and W. Che, "Miniaturized low-profile circularly polarized metasurface antenna using capacitive loading," *IEEE Transactions on Antennas and Propagation*, vol. 67, no. 5, pp. 3527–3532, May 2019.

[37] Y. M. Pan, P. F. Hu, X. Y. Zhang, and S. Y. Zheng, "A low-profile high-gain and wideband filtering antenna with metasurface," *IEEE Transactions on Antennas and Propagation*, vol. 64, no. 5, pp. 2010–2016, May 2016

[38] M. J. Al-Hasan, T. A. Denidni, and A. R. Sebak, "Millimeter-wave EBG-based aperture-coupled dielectric resonator antenna," *IEEE Transactions on Antennas and Propagation*, vol. 61, no. 8, pp. 4354–4357, Aug. 2013.

[39] Z. Guo, H. Tian, X. Wang, Q. Luo, and Y. Ji, "Bandwidth enhancement of monopole UWB antenna with new slots and EBG structures," *IEEE Antennas and Wireless Propagation Letters*, vol. 12, pp. 1550–1553, 2013.

[40] T. Vu, S. Collardey, A. Tarot, and K. Mahdjoubi, "Input impedance of Fabry-Perot, EBG, and leaky-wave antennas excited by a line source," *IEEE Antennas and Wireless Propagation Letters*, vol. 8, pp. 248–252, 2009.

[41] J. d. D. Ntawangaheza, L. Sun, C. Yang, Y. Pang, and G. Rushingabigwi, "Thin-profile wideband and high-gain microstrip patch antenna on a modified AMC," *IEEE Antennas and Wireless Propagation Letters*, vol. 18, no. 12, pp. 2518–2522, Dec. 2019.

[42] S. Shi *et al.*, "Wideband planar phased array antenna based on artificial magnetic conductor surface," *IEEE Transactions on Circuits and Systems II: Express Briefs*, vol. 67, no. 10, pp. 1909–1913, Oct. 2020.

[43] H. Wu *et al.*, "A low-profile wideband dual-polarized antenna based on an improved HIS and its broad-angle beam-scanning array," *IEEE Antennas and Wireless Propagation Letters*, vol. 19, no. 3, pp. 383–387, Mar. 2020.

[44] N. M. Mohamed-Hicho, E. Antonino-Daviu, M. Cabedo-Fabrés, and M. Ferrando-Bataller, "A novel low-profile high-gain UHF antenna using high-impedance surfaces," *IEEE Antennas and Wireless Propagation Letters*, vol. 14, pp. 1014–1017, 2015.

[45] J. d. D. Ntawangaheza, L. Sun, Y. Li, D. Biao, Z. Xie, and G. Rushingabigwi, "A single-layer planar low-profile wideband microstrip line-fed metasurface antenna," *IEEE Antennas and Wireless Propagation Letters*, vol. 20, no. 9, pp. 1641–1645, Sept. 2021.

[46] J. Zhu, Y. Yang, S. Li, S. Liao, and Q. Xue, "Dual-band dual circularly polarized antenna array using FSS-integrated polarization rotation AMC ground for vehicle satellite communications," *IEEE Transactions on Vehicular Technology*, vol. 68, no. 11, pp. 10742–10751, Nov. 2019.

[47] H. Fernández Álvarez, M. E. de Cos, and F. Las-Heras, "AMC's angular stability improvement through the introduction of lumped components," *IEEE Antennas and Wireless Propagation Letters*, vol. 17, no. 5, pp. 813–816, May 2018.

[48] D. R. Smith, S. Schultz, P. Markoš, and C. M. Soukoulis, "Determination of effective permittivity and permeability of metamaterials from reflection and transmission coefficients," *Physical Review B*, vol. 65, p. 195104, April 19, 2002.

[49] C. L. Holloway, E. F. Kuester, J. Baker-Jarvis, and P. Kabos, "A double negative (DNG) composite medium composed of magnetodielectric spherical particles embedded in a matrix," *IEEE Transactions on Antennas and Propagation*, vol. 51, pp. 2596–2603, 2003.

[50] X. Chen, T. M. Grzegorczyk, B. I. Wu, J. Pacheco, Jr., and J. A. Kong, "Robust method to retrieve the constitutive effective parameters of metamaterials," *Physical Review E*, vol. 70, p. 016608, July 26, 2004.

[51] T. Cai, G. -M. Wang, X. -F. Zhang, Y. -W. Wang, B. -F. Zong, and H. -X. Xu, "Compact microstrip antenna with enhanced bandwidth by loading magneto-electro-dielectric planar waveguided metamaterials," *IEEE Transactions on Antennas and Propagation*, vol. 63, no. 5, pp. 2306–2311, May 2015.

[52] J. Wang, H. Wong, Z. Ji, and Y. Wu, "Broadband CPW-fed aperture coupled metasurface antenna," *IEEE Antennas and Wireless Propagation Letters*, vol. 18, no. 3, pp. 517–520, Mar. 2019.

[53] K. Konstantinidis, A. P. Feresidis, and P. S. Hall, "Broadband sub-wavelength profile high-gain antennas based on multi-layer metasurfaces," *IEEE Transactions on Antennas and Propagation*, vol. 63, no. 1, pp. 423–427, Jan. 2015,

[54] Y. Zheng *et al.*, "Wideband gain enhancement and RCS reduction of Fabry–Perot resonator antenna with chessboard arranged metamaterial superstrate," *IEEE Transactions on Antennas and Propagation*, vol. 66, no. 2, pp. 590–599, Feb. 2018.

[55] H. Zhai, K. Zhang, S. Yang, and D. Feng, "A low-profile dual-band dual-polarized antenna with an AMC surface for WLAN applications," *IEEE Antennas and Wireless Propagation Letters*, vol. 16, pp. 2692–2695, 2017.

[56] J. Wu, S. Yang, Y. Chen, S. Qu, and Z. Nie, "A low profile dual-polarized wideband omnidirectional antenna based on AMC reflector," *IEEE Transactions on Antennas and Propagation*, vol. 65, no. 1, pp. 368–374, Jan. 2017.

[57] D. Samantaray and S. Bhattacharyya, "A metasurface based gain enhanced dual band patch antenna using SRRs with defected ground structure," *Radio Science*, vol. 56, no. 2, pp. 1–13, article no. 2020RS007192, Feb. 2021.

[58] Y. F. Cao and X. Y. Zhang, "A wideband beam-steerable slot antenna using artificial magnetic conductors with simple structure," *IEEE Transactions on Antennas and Propagation*, vol. 66, no. 4, pp. 1685–1694, Apr. 2018.

[59] J. H. Kim, C. Ahn, and J. Bang, "Antenna gain enhancement using a holey superstrate," *IEEE Transactions on Antennas and Propagation*, vol. 64, no. 3, pp. 1164–1167, Mar. 2016.

[60] Y. Liu and X. Zhao, "Perfect absorber metamaterial for designing low-RCS patch antenna," *IEEE Antennas and Wireless Propagation Letters*, vol. 13, pp. 1473–1476, 2014.

[61] A. Sharma, D. Gangwar, B. Kumar Kanaujia, S. Dwari, and S. Kumar, "Design of a wideband polarisation conversion metasurface and its application for RCS reduction and gain enhancement of a circularly polarised antenna," *IET Microwaves, Antennas & Propagation*, vol. 13, no. 9, pp. 1427–1437, 2019.

[62] Y. Zheng *et al.*, "Wideband gain enhancement and RCS reduction of Fabry–Perot resonator antenna with chessboard arranged metamaterial superstrate," *IEEE Transactions on Antennas and Propagation*, vol. 66, no. 2, pp. 590–599, Feb. 2018.

[63] A. Baba, R. M. Hashmi, K. P. Esselle, J. G. Marin, and J. Hesselbarth, "Broadband partially reflecting superstrate-based antenna for 60 GHz applications," *IEEE Transactions on Antennas and Propagation*, vol. 67, no. 7, pp. 4854–4859, July 2019.

[64] M. A. Al-Tarifi, D. E. Anagnostou, A. K. Amert, and K. W. Whites, "The puck antenna: A compact design with wideband, high-gain operation," *IEEE Transactions on Antennas and Propagation*, vol. 63, no. 4, pp. 1868–1873, Apr. 2015.

[65] R. M. Hashmi, B. A. Zeb, and K. P. Esselle, "Wideband high-gain EBG resonator antennas with small footprints and all-dielectric superstructures," *IEEE Transactions on Antennas and Propagation*, vol. 62, no. 6, pp. 2970–2977, June 2014.
[66] Y. Zheng, J. Gao, Y. Zhou, X. Cao, H. Yang, S. Li, and T. Li, "Wideband gain enhancement and RCS reduction of Fabry–Perot resonator antenna with chessboard arranged metamaterial superstrate," *IEEE Transactions on Antennas and Propagation*, vol. 66, no. 2, pp. 590–599, 2017.
[67] M. Long, W. Jiang, and S. Gong, "Wideband RCS reduction using polarization conversion metasurface and partially reflecting surface," *IEEE Antennas and Wireless Propagation Letters*, vol. 16, pp. 2534–2537, 2017.
[68] K. Li, Y. Liu, Y. Jia, and Y. J. Guo, "A circularly polarized high-gain antenna with low RCS over a wideband using chessboard polarization conversion metasurfaces," *IEEE Transactions on Antennas and Propagation*, vol. 65, no. 8, pp. 4288–4292, 2017.
[69] T. Sabapathy, M. Jusoh, R. B. Ahmad, M. R. Kamarudin, and P. J. Soh, "A ground-plane-truncated, broadly steerable Yagi–Uda patch array antenna," *IEEE Antennas and Wireless Propagation Letters*, vol. 15, pp. 1069–1072, 2016.
[70] T. Debogovic and J. Perruisseau-Carrier, "Array-fed partially reflective surface antenna with independent scanning and beamwidth dynamic control," *IEEE Transactions on Antennas and Propagation*, vol. 62, no. 1, pp. 446–449, 2014.
[71] X. Ding, Y.-F. Cheng, W. Shao, H. Li, B.-Z. Wang, and D. E. Anagnostou, "A wide-angle scanning planar phased array with pattern reconfigurable magnetic current element," *IEEE Transactions on Antennas and Propagation*, vol. 65, no. 3, pp. 1434–1439, 2017.
[72] S. B. T. Wang, A. M. Niknejad, and R. W. Brodersen, "Circuit modeling methodology for UWB omnidirectional small antennas," *IEEE Journal on Selected Areas in Communications*, vol. 24, no. 4, pp. 871–877, Apr. 2006.
[73] K. H. Sayidmarie and L. S. Yahya, "Modeling of dual-band crescent-shape monopole antenna for WLAN applications," *International Journal of Electromagnetics and Applications*, vol. 4, no. 2, pp. 31–39, 2014.
[74] D. Samantaray, S. K. Ghosh, and S. Bhattacharyya, "Modified slotted patch antenna with metasurface as superstrate for dual-band applications," *IEEE Antennas and Wireless Propagation Letters*, vol. 22, no. 1, pp. 109–113, Jan. 2023.
[75] F. Yu and C. Wang, "Design of a CPW-fed dual band-notched planar wideband antenna for UWB applications," *Ultra Wideband Communications Mohammad Abdul Matin,* IntechOpen, London, UK, pp. 239–254, Aug. 2011.
[76] F. Costa, S. Genovesi, and A. Monorchio, "On the bandwidth of high impedance frequency selective surfaces," *IEEE Antennas and Wireless Propagation Letters*, vol. 8, pp. 1341–1344, 2009.
[77] Y. -F. Cheng, J. Feng, C. Liao, and X. Ding, "Analysis and design of wideband low-RCS wide-scan phased array with AMC ground," *IEEE Antennas and Wireless Propagation Letters*, vol. 20, no. 2, pp. 209–213, Feb. 2021.
[78] E. B. Lima, J. R. Costa, and C. A. Fernandes, "Multiple-beam focal-plane dual-band Fabry–Pérot cavity antenna with reduced beam degradation," *IEEE Transactions on Antennas and Propagation*, vol. 67, no. 7, pp. 4348–4356, July 2019.
[79] P. Xie, G. Wang, H. Li, Y. Wang, and X. Gao, "Reconfigurable Fabry-Pérot cavity antenna basing on phase controllable metasurfaces," *Advanced Radio Frequency Antennas for Modern Communication and Medical Systems*, IntechOpen, London, UK, chapter 72936, Sept. 2020.
[80] J. Zhang, S. Yan, and G. A. E. Vandenbosch, "Radial CRLH-TL-based dual-band antenna with frequency agility," *IEEE Transactions on Antennas and Propagation*, vol. 68, no. 7, pp. 5664–5669, July 2020.
[81] H. Yang and X. Liu, "Wearable dual-band and dual-polarized textile antenna for on- and off-body communications," *IEEE Antennas and Wireless Propagation Letters*, vol. 19, no. 12, pp. 2324–2328, Dec. 2020.

[82] M. Mattsson, C. I. Kolitsidas, and B. L. G. Jonsson, "Dual-band dual-polarized full-wave rectenna based on differential field sampling," *IEEE Antennas and Wireless Propagation Letters*, vol. 17, no. 6, pp. 956–959, June 2018.
[83] F. Liu, J. Guo, L. Zhao, G. Huang, Y. Li, and Y. Yin, "Dual-band metasurface-based decoupling method for two closely packed dual-band antennas," *IEEE Transactions on Antennas and Propagation*, vol. 68, no. 1, pp. 552–557, Jan. 2020.
[84] D. Samantaray and S. Bhattacharyya, "A metasurface-backed planar low-profile dual-band monopole antenna," *Journal of Electromagnetic Waves and Applications*, vol. 37, no. 7–9, pp. 884–897, May 2023.
[85] H. Zhai, K. Zhang, S. Yang, and D. Feng, "A low-profile dual-band dual-polarized antenna with an AMC surface for WLAN applications," *IEEE Antennas and Wireless Propagation Letters*, vol. 16, pp. 2692–2695, 2017.
[86] J. Wu, S. Yang, Y. Chen, S. Qu, and Z. Nie, "A low profile dual-polarized wideband omnidirectional antenna based on AMC reflector," *IEEE Transactions on Antennas and Propagation*, vol. 65, no. 1, pp. 368–374, Jan. 2017.
[87] A. Youse, K. Nader, K. Majid, and A. Iman, "Demonstration of a self-polarizing dual-band single-feed circularly polarized Fabry-Pérot cavity antenna with a broadband axial ratio," *AEU - International Journal of Electronics and Communications*, vol. 111, 2019, Art. no. 152909.
[88] C. Chen, Z. Liu, H. Wang, and Y. Guo, "Metamaterial-inspired self-polarizing dual-band dual-orthogonal circularly polarized Fabry–Pérot resonator antennas," *IEEE Transactions on Antennas and Propagation*, vol. 67, no. 2, pp. 1329–1334, Feb. 2019.
[89] A. Ahmed, M. R. Robel, and W. S. T. Rowe, "Dual-band two-sided beam generation utilizing an EBG-based periodically modulated metasurface," *IEEE Transactions on Antennas and Propagation*, vol. 68, no. 4, pp. 3307–3312, Apr. 2020.
[90] L. Gu, W. Yang, Q. Xue, and W. Che, "A dual-band steerable dual-beam metasurface antenna based on common feeding network," *IEEE Transactions on Antennas and Propagation*, vol. 69, no. 10, pp. 6340–6350, Oct. 2021.
[91] F. Sun, F. Zhang, H. Zhang, H. Zhang, C. Li, and C. Feng, "A frequency diversity printed-Yagi antenna element for apertures selectivity wideband array application," *IEEE Transactions on Antennas and Propagation*, vol. 66, no. 10, pp. 5634–5638, Oct. 2018.
[92] H. Malekpoor and M. Hamidkhani, "Performance enhancement of low-profile wideband multi-element MIMO arrays backed by AMC surface for vehicular wireless communications," *IEEE Access*, vol. 9, pp. 166206–166222, 2021.
[93] R. Joshi, E. Faridah, N. M. Hussin, P. J. Soh, M. F. Jamlos, H. Lago, A. A. Al-Hadi, and S. K. Podilchak, "Dual-band, dual-sense textile antenna with AMC backing for localization using GPS and WBAN/WLAN," *IEEE Access*, vol. 8, pp. 89468–89478, 2020.
[94] E. B. Lima, J. R. Costa, and C. A. Fernandes, "Multiple-beam focal-plane dual-band Fabry–Pérot cavity antenna with reduced beam degradation," *IEEE Transactions on Antennas and Propagation*, vol. 67, no. 7, pp. 4348–4356, July 2019.
[95] H. Zhai, K. Zhang, S. Yang, and D. Feng, "A low-profile dual-band dual-polarized antenna with an AMC Surface for WLAN applications," *IEEE Antennas and Wireless Propagation Letters*, vol. 16, pp. 2692–2695, 2017.
[96] Q. Liu, H. Liu, W. He, and S. He, "A low-profile dual-band dual-polarized antenna with an AMC reflector for 5G communications," *IEEE Access*, vol. 8, pp. 24072–24080, Jan. 2020.

10 Design of a mmWave Reconfigurable Intelligent Surface for Futuristic Wireless Communications

B Anil Babu, BTP Madhav, P Ravi Kumar, and T Anil Kumar

10.1 INTRODUCTION

In recent years, the Reconfigurable Intelligent Surface (RIS) has gained much research interest due to its ability to control electromagnetic (EM) waves dynamically. These RIS surfaces have the potential to provide a smart radio environment with exciting applications for wireless communications. It offers energy and spectrum-efficient solutions by providing extended radio coverage that provides a Line of Sight (LoS) between transmitter and receiver in wireless infrastructure. These devices have the capability of manipulating EM wave propagation by adjusting the phase of the individual unit cells. The RIS consists of a periodic arrangement of Metasurface unit cells backed with a control circuit for getting the reconfigurable states. These RIS have the advantages of low profile, ease of fabrication, low complexity, and high-gain beam scanning with reconfigurable characteristics.

Current wireless communications face the problem of EM signal blockage, low signal coverage, and severe path losses at mmWave and terahertz applications [1]. These RISs become the solution to the mentioned problems with enhanced communication. The RIS can be easy to deploy as it consists of a planar array of Metasurface unit cells that can be aligned with installation platforms. It can be able to provide spectrum efficiency with LoS between the transmitter and receiver. Moreover, the hardware design of RIS is simple and scalable with several unit cells. Each unit cell can adjust the phase of the reflected wave. The existing RIS literature is broadly classified into two types: 1. EM-based RIS and 2. communication-based RIS design [2] as shown in Figure 10.1. This chapter is focused on EM-based RIS literature along with an exploration of RIS fundamentals, applications, challenges, and directions with an example of RIS unit cell design and analysis using numerical simulation software tools.

 DOI: 10.1201/9781032656021-10

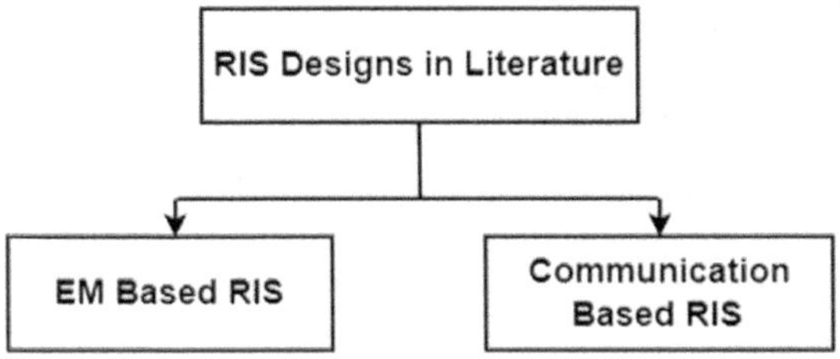

FIGURE 10.1 RIS design classification from literature.

The rest of the chapter is organized into six sections as follows.

1. In section 2, the fundamentals of RIS surface are introduced which include working principle, reconfiguration mechanism, programmability, and multifunctionality with performance indicators of RIS.
2. In section 3, the applications of the RIS are discussed with a focus on IoT, mmWave, and wireless communications.
3. The challenges and future directions in the EM-based RIS research are discussed in section 4.
4. In section 5, the design procedure of the RIS is given as a case study using numerical simulation software.
5. The simulation results and analysis of the realized structure are given in section 6.
6. Finally in the last section, the conclusion is provided.

10.2 RIS FUNDAMENTALS

The RIS consists of subwavelength Metasurface unit cells with reconfigurable elements that provide phase shift in reflection/refraction mode to the incident waves. The incident wave parameters such as frequency, polarization, amplitude, and phase that provide manipulated reflected waves are controlled [3] as shown in Figure 10.2. This section discusses the RIS working principle with its features as follows.

a. **Working Principle**
   The RIS surface consists of an array of subwavelength Metasurface unit cells integrated with reconfigurable elements that can manipulate the reflected wave characteristics. The RIS surface works in two modes of operation such as reflection and refraction mode. Most of the existing literature on RIS is based on reflection mode, which is suitable for most real-time wireless platforms. Each unit cell with subwavelength size provides the reflection phase adjustment based on reconfigurable elements. The scattering properties of the reflected wave generate beamforming, beam steering, and polarization conversion for the incident wave [4]. The conceptual illustration of beamforming using RIS is displayed in Figure 10.3.

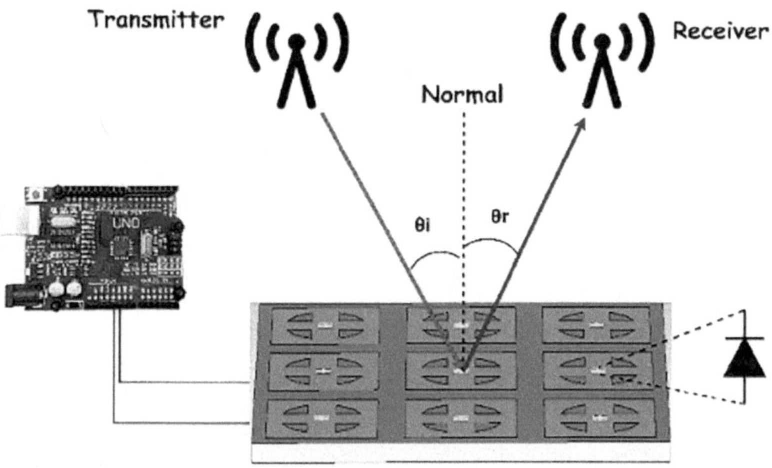

FIGURE 10.2 Illustration of RIS working principle.

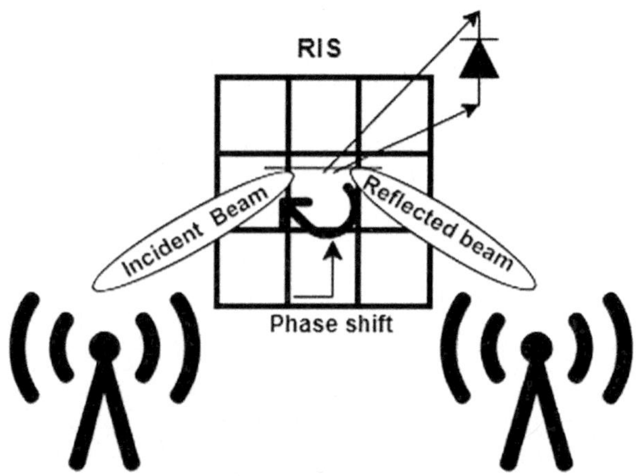

FIGURE 10.3 Beamforming operation using RIS.

The functional operation of the RIS surface is limited to short-range communication that can provide extended radiation coverage in complex environments. It acts as a bridge between transmitter and receiver by providing an economical solution by comparing with existing techniques such as multi antennas and relays. Initially, the research was carried out in reflective RIS and subsequently turned towards full space realization in transmissive and reflective mode RIS surfaces.

b. **Reconfiguration Mechanism**

The control mechanism of the RIS surface using reconfigurable elements is discussed in this section. The RIS surface is realized using lumped elements operating with external stimuli dependent on voltage, force, temperature, etc. Based on the operating states of the components the physical structure of the RIS is altered which in turn provides variable reflection characteristics. These reconfigurable elements in the literature are based on electronic, mechanical, optical, Phase Change Materials (PCM), and microfluid-controlled switches.

Varactor diode, pin diode, liquid crystals, etc. are examples of electronically controlled switches. MEMS and microfluidics are examples of mechanically controlled switches. Optical switches are controlled using infrared or ultrafast optical pulses. Examples of optical switches are Indium Tin Oxide (ITO), graphene, and transparent conductive light-sensitive materials. The choice of the reconfigurable component is specific to robustness at specific frequency bands and adaptability to the hardware design.

c. **Programmability in RIS**

The real-time control of RIS can be done using programmable devices such as FPGA, Arduino, etc. This programmability can be adapted to the RIS by controlling the reconfigurable element. This feature gives the platform to integrate and apply smart and programming aspects to the RIS using dynamic control over the incident EM waves via programming control of reconfigurable elements. Hence this programmable feature gives a platform to the competitive embedded technologies for getting dynamic control over EM waves. The EM scattering properties of the smart programmable surface give an extended platform to integrate programmable control over EM wave properties.

d. **Multifunctionality in RIS**

The multifunctional RIS surfaces are gaining more attention in wireless communication due to their low profile, low cost, and high gain. These multifunctional behaviors can be obtained in either reflection or refraction mode with simultaneous control. The single functionality was realized using most of the existing reconfigurable Metasurfaces for the incident waves. To counter this, a limited Multifunctional Metasurface-based literature exists that can be obtained by controlling the reconfigurable elements.

The half-space and full-space realization of these multifunctional characteristics is another research interest among the researchers. The multifunctional behavior depends on RIS elementary operation on incident waves such as reflection, refraction, focusing, absorption, polarization, and splitting operations. Hence the multifunctionality in RIS can further expand the potential of RIS surface in wireless communications.

e. **RIS Performance Parameters**

The EM performance characteristics of the RIS surface can be assessed using scattering parameter behavior to the angular stability, polarization insensitivity, and reduced losses in the reconfigurable elements. Angular stability is defined as the incident angle range for which the performance characteristics are stable. The parameters such as unit cell size, control board complexity, and cost are also significant for the EM-based RIS design.

The number of unit cells and the number of reconfigurable elements per unit cell also affect the scattering performance for incident waves. Macroscopically it provides energy and spectrum efficiency to wireless communications. Hence the operating bandwidth of the RIS surface is also a significant performance parameter. These parameters are to be considered while designing RIS hardware for the targeted application frequencies.

10.3 RIS APPLICATIONS

RIS-aided wireless communication changes the scenario by providing extended signal coverage in complex environments, spectrum, and energy efficiency. These features enable the applications of RIS to different upgraded wireless platforms. Another significant merit of RIS is low-power wireless communication. The following are some of the potential applications for getting smart radio environments.

a. **RIS-Aided Wireless Communications**
   Over the last few years, different channel optimization and advanced design techniques have been proposed in the literature to improve the robustness of wireless communications. The techniques include advanced modulation, encoding/decoding methods, and design protocols like multiple input and multi-output (MIMO) based transmitter systems, high power transmission, and retransmission. According to this design paradigm, communication engineers usually design the transmitters, the receivers, and the transmission protocols based on the specific properties of the wireless channels to achieve the desired performance. For example, transmitters equipped with multiple radiating elements may be configured differently as a function of the specific characteristics of the wireless channel where they operate, to achieve the desired trade-off in terms of spatial multiplexing, spatial diversity, and beamforming gains.

   Alternatively, RIS-aided wireless communication gives the potential to expand applications in different real-time scenarios. The quality of service (QoS), channel capacity, and increased data rate are achieved for the current demand of wireless communications. Smart radio environments are obtained by using beamforming and interference null operations. This in turn can generate a signal-to-noise ratio for the transmitted signal.
b. **RIS for Futuristic 5G/6G Communications**
   In this section, RIS-aided communication and its integration into futuristic 5G/6G communications are discussed. The deployment of RISs is also flexible for wireless channels to support multi-channel transmissions, especially for the high bandwidth channels in millimeter-wave (mmWave) and terahertz (THz) communications. Benefiting from these advantages, RISs are expected to bring significant capacity gains for future 6G wireless communications.

   Intelligent and autonomous transportation systems, RF localization and sensing, and Non-orthogonal Multiple Access (NOMA) are some of the realizable solutions for the futuristic 5G/6G wireless communications using RIS.

c. **RIS for IOT**

The RIS surface provides a potential solution to the IOT network applications such as smart buildings, smart hospitals, smart cars, smart clothing, smart factories, etc. along with its integration with the embedded platforms. These applications need short-range communication with a high degree of freedom in infrastructure platforms that provide direct communication links with transceiver devices.

The RIS's deployment and integration capability feature gives IOT system designers flexibility for good performance. Moreover, RIS can improve the signal strength using localization and sensing approaches.

10.4 CHALLENGES AND RESEARCH DIRECTIONS

a. **RIS Design for EM Control**

The major research focus in the EM-based RIS design is to achieve beamforming with multifunctional characteristics. The optimized RIS has to be realized for fundamental operations on EM waves. The RIS surface size is another design constraint as RIS functionality is limited to near-field regions only. The mutual coupling between the RIS subwavelength unit cells is another design challenge, which is to be minimal. This gives the challenge to the RIS designers to address mutual coupling between RIS elements [5].

The selection of materials and reconfiguration methods are also significant in the RIS design. The design of near-field-focused RIS surface is another research direction that has applications in interference reduction and EM concentration in desired directions. Finally, The RIS installation platform and investigation of platform-dependent materials for RIS is another important requirement and research direction for wireless communications.

b. **RIS Design for Communications**

The deployment strategy of the RIS is focused on getting direct links between transmitter and receiver in wireless communication. These RIS surfaces are used for reduced energy consumption and spectral efficiency by controlling EM wave propagation. The channel state information (CSI) is one of the main problems in RIS-aided wireless communications. Channel estimation and channel acquisition are other challenges in RIS-aided wireless communication due to near-field operation. The fundamental challenges are not only limited to channel state information estimation but also passive information transfer of RIS, and low-complexity-based robust system design using RIS [6].

c. **RIS Design Using AI Approach**

Nowadays, Artificial Intelligence (AI)-based wireless communication is getting research interest. The Deep Learning and Machine Learning Techniques have been applied to RIS-aided wireless communication in the optimization scenarios of design and communication. The AI-based RIS design solves channel estimation and RF sensing problems with hardware optimization. AI approaches can deal with the maximization of performance, signal coverage, and noise reduction. Most existing design methods rely on model-based optimization to realize phase shift operations which require multiple simulations. Hence the AI-based RIS communication gives open challenges to the researcher by targeting the aforementioned requirements for wireless communications [7].

The AI-based approach to the RIS gives new directions to the researchers by solving the research challenges of getting an optimized design, improving signal coverage, channel estimation, etc.

d. **RIS Design Using Reconfigurable Technique**
The performance of the RIS surface depends on reconfigurable materials used in the design. Hence the exploration of novel reconfigurable materials is required. The material compatibility, response time, and application of external stimuli are the major concerns and challenges in robust RIS design. Most of the reconfigurable elements are electrically controlled in which the control mechanism is used. Nowadays, due to developments in the integration of embedded RF technologies, various development boards are used to get control over the EM waves [8]. The programmable RIS design is gaining more research focus that provides the programmability feature for EM waves. The research trend is continuing in the programmable RIS surfaces using embedded boards such as FPGA, Arduino, and Raspberry Pi to get dynamic control over the reflected EM waves [9–15]. However, the challenge is independent control of EM parameters for multifunctionality. Hence the EM wave propagation can meet the computing capabilities of embedded devices.

10.5 RIS UNIT CELL DESIGN: A CASE STUDY

In this section, the design of the RIS unit cell using the CST numerical simulation tool is demonstrated using simulation results. The RIS unit cell is realized using an FR4 substrate (dielectric constant = 4.4 and loss tangent of 0.025) with the dimensions of $10 \times 10 \times 1.6$ mm^3. The unit cell structure consists of a rectangular patch with circle sectors. The patch structure is split into two parts using a horizontal slot as shown in Figure 10.4. In the center, the AlGaAs pin diode is connected and is responsible for getting reconfigurable states.

The commercially available MADP-000907–14020P pin diode from the company Macom is used in the design as it gives a suitable performance at mmWave frequencies [15]. The RLC-equivalent circuit of the switchable states is presented in Figure 10.5, which is considered in the simulation tool. These RLC components are integrated into the passive Metasurface using lumped elements in the CST tool. The methodology for the RIS unit cell is presented using the flowchart presented in Figure 10.6.

10.6 SIMULATION RESULTS AND BIASING TECHNIQUE

Generally, The RIS surface EM characteristics are analyzed by designing a single unit cell, as these characteristics are replicated by controlling active components. The resonant frequency, reflection amplitude, phase, and polarization of the reflected wave can be manipulated by controlling the characteristics of each unit cell. In the

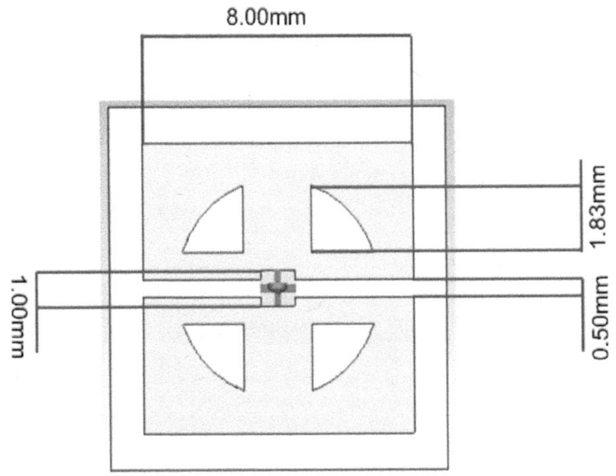

FIGURE 10.4 Unit cell geometry.

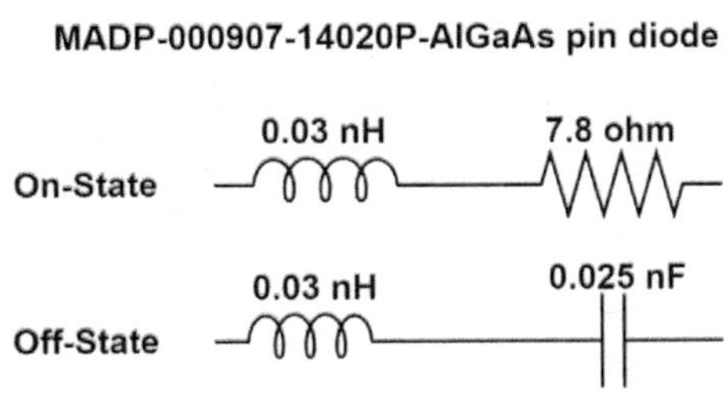

FIGURE 10.5 LC equivalent circuit of AlGaAs pin diode.

design of the RIS unit cell, the EM wave characteristics depend on reconfigurable elements and surface current distribution. The simulated reflection characteristics in On-state are presented in Figures 10.7 and 10.8, in which it provides a resonance peak at 33.72 GHz. In Off-state, it provides dual resonance at 32 and 33.46 GHz with a phase difference of $180 \pm 20°$.

Angular stability is one of the significant parameters defining the RIS surface's performance. The angular stability analysis of the RIS unit cell with various incident (theta) and polarization (Phi) angles is presented in Figures 10.9 and 10.10. It provides angular stability up to 90° incidents and polarization angles at the resonant frequencies. The figure shows the reflection coefficient magnitude distribution in both the switchable states. It is observed that in the Off-state, the reflection bandwidth is more which offers angular stability. In real-time, due to arbitrary incident EM wave direction, the RIS design has to provide high angular stability that can sustain EM performance.

The design of the RIS design is indifferent to the reflective array as the operation of RIS depends on surface current distribution. The surface current distribution of the surface at both the operating states is depicted in Figure 10.11. It is observed that

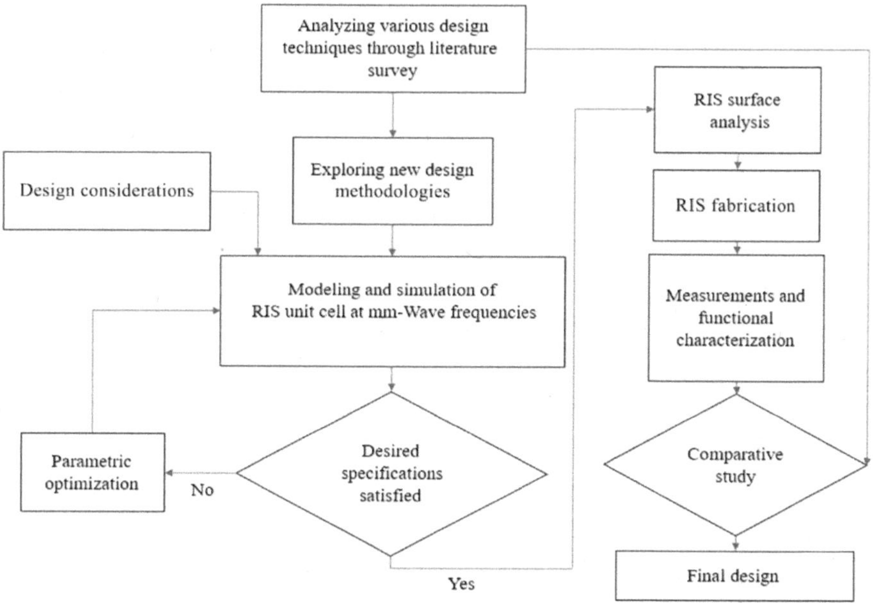

FIGURE 10.6 Design methodology flow chart.

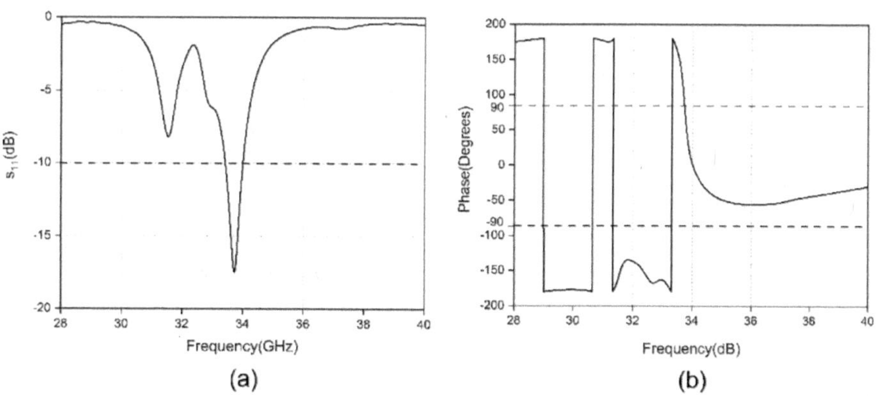

FIGURE 10.7 Reflection characteristics in ON-state (a) amplitude and (b) phase.

the surface charge density is concentrated along the horizontal slit in the Off-state that makes resonance at 33.46 GHz.

Beamforming is one of the applications of the RIS surface, which can be able to reduce the limitation of radiation coverage. The beamforming capability of the RIS unit cell is demonstrated using simulated radiation patterns as shown in Figure 10.12. A peak gain of 11.7 and 13.2 dB is obtained at the normal incidence of the plane wave in bistate operation.

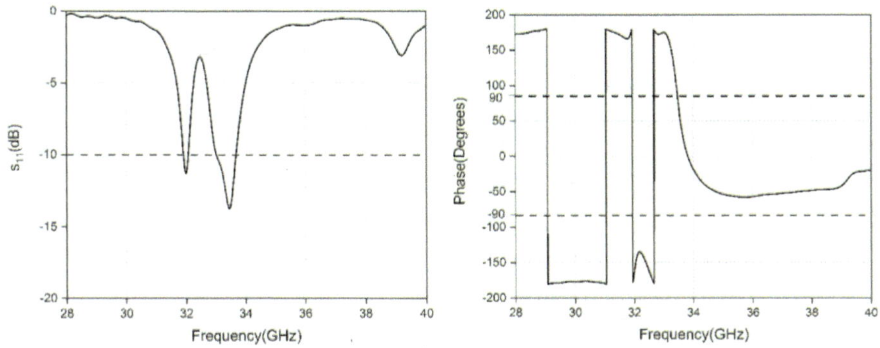

FIGURE 10.8 Reflection characteristics in Off-state (a) amplitude and (b) phase.

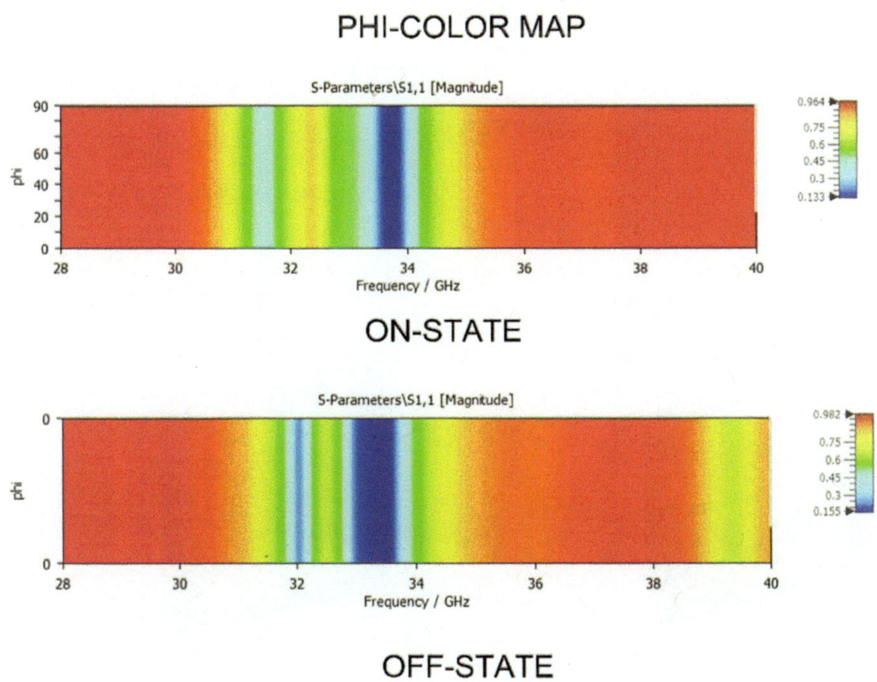

FIGURE 10.9 Reflection magnitude distribution under Phi-Variation.

10.6.1 Biasing Model Using Arduino

The integration of embedded technology in controlling the EM waves is a trending topic in smart radio communications. The control circuit for the AlGaAs pin diode is realized using an Arduino board as shown in Figure 10.13. The biasing voltage can be obtained at the end terminals and connected to the pin diode. In the biasing setup model using Arduino a control voltage of 1.8 V is required for making the diode

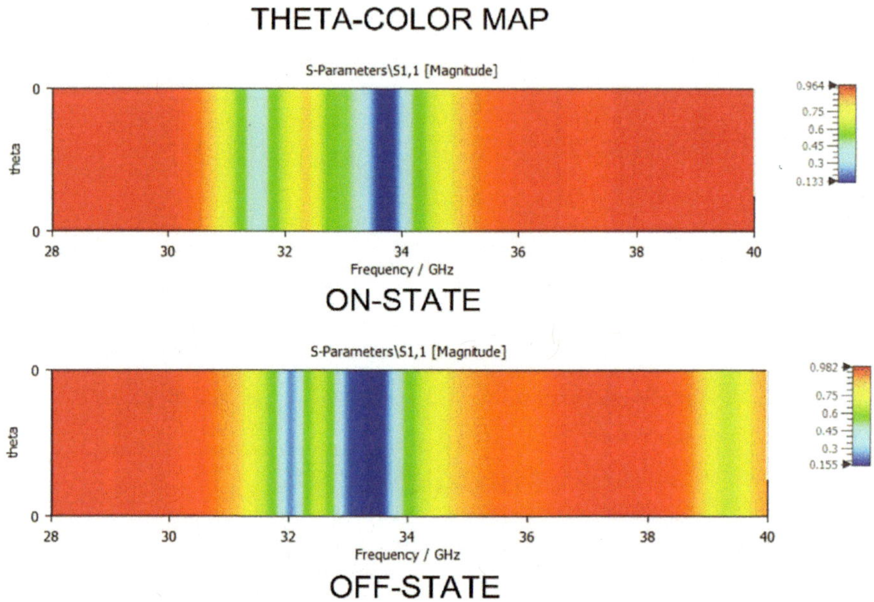

FIGURE 10.10 Reflection magnitude distribution under Theta-Variation.

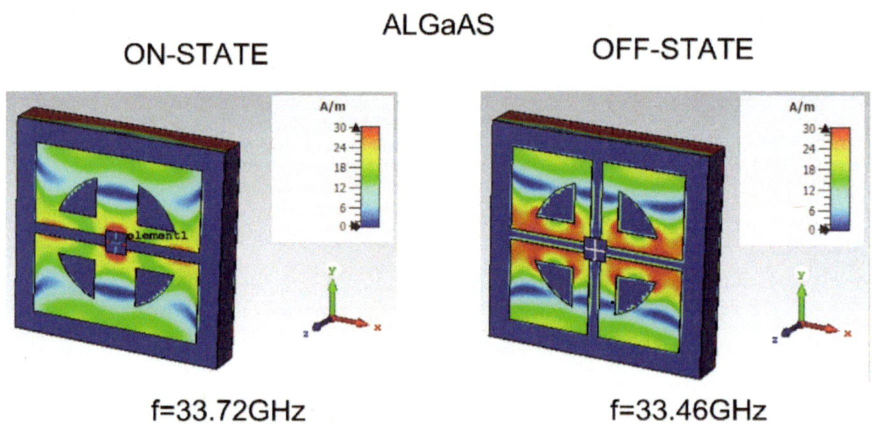

FIGURE 10.11 Surface current distribution in bistate.

switch-on and 0V can be configured for Off-state. This bias configuration can be controlled using a programmable feature that provides suitable bias voltage.

The preceding biasing setup model is realized using Tinker CAD software. The control of reconfigurable elements using embedded development boards enables a programmable feature with an application of embedded technologies to the EM waves.

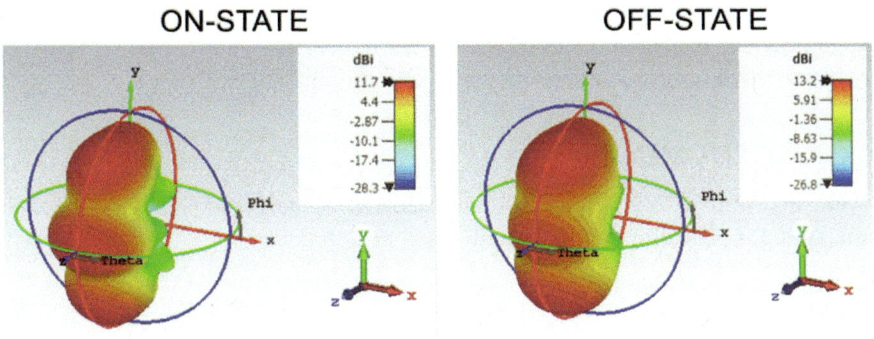

FIGURE 10.12 Simulated beamforming capability of RIS unit cell.

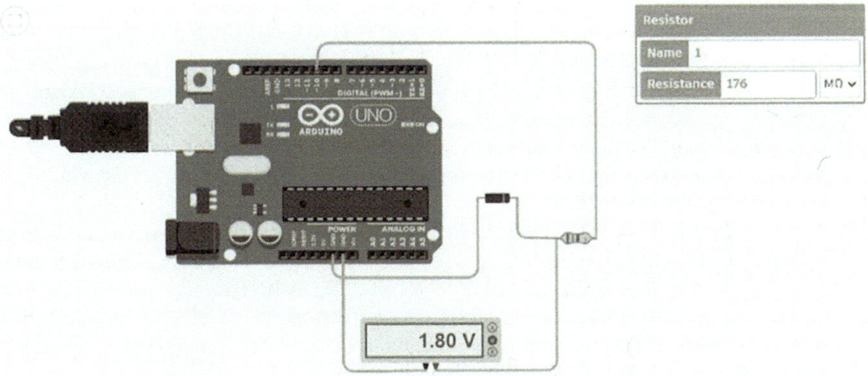

FIGURE 10.13 Biasing setup model using Arduino.

10.7 CONCLUSION

In this chapter, a comprehensive view of RIS fundamentals, challenges, applications, and research directions is presented. Energy efficiency, extended radiation coverage, and spectrum efficiency can be achieved in wireless communication using scalable and low-cost RIS. The focus is on the EM design of the RIS surface using reconfigurable materials, applications, and control techniques. The current trends in RIS design are discussed using RIS hardware design perspectives. The applications of RIS are extending to all wireless trends that can be adapted to computing and programmable aspects using smart EM waves.

The design and analysis of the RIS unit cell for mmWave applications is presented as a case study. This study provides an understanding of RIS design using the CST numerical simulation tool. The design is realized using an FR4 substrate and the AlGaAs pin diode is used as a reconfigurable element. The reflection characteristics are explained using reflection magnitude and phase. The beamforming capability is explained using simulated radiation patterns and achieved a gain >11 dBi in both states. An Arduino-based bias model is exemplified as a control module. Finally, this chapter provides insight into RIS basics, research directions, and trends that pave the way for research in futuristic 5G/6G wireless communications.

ACKNOWLEDGMENTS

We are thankful to the ALRC Research Center, Koneru Lakshmaiah Education Foundation for providing research facilities.

REFERENCES

[1] C. Pan et al., "Reconfigurable intelligent surfaces for 6G systems: Principles, applications, and research directions," *IEEE Communications Magazine*, vol. 59, no. 6, pp. 14-20, June 2021.
[2] Y. Saifullah, et al., "Recent progress in reconfigurable and intelligent metasurfaces: A comprehensive review of tuning mechanisms, hardware designs, and applications," *Advanced Science*, vol. 9, no. 33, 2022, p. 2203747. doi: 10.1002/advs.202203747.
[3] S. Basharat, M. Khan, M. Iqbal, U. S. Hashmi, S. A. R. Zaidi, and I. Robertson. "Exploring reconfigurable intelligent surfaces for 6G: State-of-the-art and the road ahead," *IET Communications*, vol. 16, no. 13, 2022, pp. 1458–1474. doi: 10.1049/cmu2.12364.
[4] E. Basar, "Reconfigurable intelligent surface-based index modulation: A new beyond MIMO Paradigm for 6G," *IEEE Transactions on Wireless Communications*, vol. 68, no. 5, pp. 3187–3196, May 2020.
[5] Z. Zhang and L. Dai, "Reconfigurable intelligent surfaces for 6G: Nine fundamental issues and one critical problem," *Tsinghua Science and Technology*, vol. 28, no. 5, pp. 929–939, October 2023. doi: 10.26599/TST.2023.9010001.
[6] L. Dai et al., "Reconfigurable intelligent surface-based wireless communications: Antenna design, prototyping, and experimental results," *IEEE Access*, vol. 8, pp. 45913–45923, 2020. doi: 10.1109/ACCESS.2020.2977772.
[7] Z. Zhang et al., "Macro modeling of reconfigurable intelligent surface based on microwave network theory," *IEEE Transactions on Antennas and Propagation*, vol. 70, no. 10, pp. 8707–8717, October 2022. doi: 10.1109/TAP.2022.3187645.
[8] F. Wu, R. Lu, J. Wang, Z. H. Jiang, W. Hong, and K. -M. Luk, "A circularly polarized 1 bit electronically reconfigurable reflectarray based on electromagnetic element rotation," *IEEE Transactions on Antennas and Propagation*, vol. 69, no. 9, pp. 5585–5595, September 2021. doi: 10.1109/TAP.2021.3069551.
[9] J. -B. Gros, V. Popov, M. A. Odit, V. Lenets, and G. Lerosey, "A reconfigurable intelligent surface at mmWave based on a binary phase tunable metasurface," *IEEE Open Journal of the Communications Society*, vol. 2, pp. 1055–1064, 2021. doi: 10.1109/OJCOMS.2021.307627
[10] L. Wu et al., "A wideband amplifying reconfigurable intelligent surface," *IEEE Transactions on Antennas and Propagation*, vol. 70, no. 11, pp. 10623–10631, November 2022. doi: 10.1109/TAP.2022.3187137.
[11] J. C. Liang et al., "An angle-insensitive 3-bit reconfigurable intelligent surface," *IEEE Transactions on Antennas and Propagation*, vol. 70, no. 10, pp. 8798–8808, October 2022. doi: 10.1109/TAP.2021.3130108.
[12] A. Sayanskiy et al., "A 2D-programmable and scalable reconfigurable intelligent surface remotely controlled via digital infrared code," *IEEE Transactions on Antennas and Propagation*, vol. 71, no. 1, pp. 570–580, January 2023. doi: 10.1109/TAP.2022.3217327.

[13] X. Cao, Q. Chen, T. Tanaka, M. Kozai, and H. Minami, "A 1-bit time-modulated reflectarray for reconfigurable-intelligent-surface applications," *IEEE Transactions on Antennas and Propagation*, vol. 71, no. 3, pp. 2396–2408, March 2023. doi: 10.1109/TAP.2022.3233659.

[14] H. Luyen, Z. Zhang, J. H. Booske, and N. Behdad, "Wideband, beam-steerable reflectarray antennas exploiting electronically reconfigurable polarization-rotating phase shifters," *IEEE Transactions on Antennas and Propagation*, vol. 70, no. 6, pp. 4414–4425, June 2022. doi: 10.1109/TAP.2022.3141060.

[15] R. Wang, Y. Yang, B. Makki, and A. Shamim, "A wideband reconfigurable intelligent surface for 5G millimeter-wave applications," *arXiv preprint arXiv:2304.11572*, 2023.

11 Design of Four Element Multi-Band MIMO Antenna for 6G mmWave Wireless Applications

Painam Surendra Kumar, Medikonda Karuna, and T Krishna Chaitanya

11.1 INTRODUCTION

Our daily activities depend heavily on mobile communication. We must provide incredibly high data rates, extremely broad coverage, extremely low latency, extremely high dependability, and massive connection and sensing as the number of mobile users increases daily. Future standards can only be met with 6G technology. It is anticipated to take effect in 2030 [1]. Connecting the real as well as virtual environments through machine-to-machine (MTM) communication is among the goals of 6G technology.

Antenna for MIMO is a good option for streaming huge video material and offering services such as higher number of mobile connections, higher data rates, more health care sector service, large coverage, virtual reality, augmented reality, wireless cognition, sensing, imaging, communication, positioning, Internet of Things (IoT), Bio-Nano-Internet of Things, Artificial Intelligence (AI), space travel, machine learning (ML), smart cities, driverless cars, and machine to machine communication [2].

6G technology provides several advantages like individual data rate of 100 Gbps, mobility more than 1000 km/hr, peak data rate of 1 Tbps, latency of 0.1ms, maximum spectral efficiency of 100bps/Hz, connection density of more than 10 million devices/Km2, coverage percentage of more than 99%, end to end reliability of 99.9999%, fully integrated satellite, fully developed AI, fully developed autonomous vehicles, and extended reality [3]. The spectrum in millimetre wave (mmWave) frequency band over 26 GHz to 450 GHz makes it possible to attain exceptionally high capacity and extremely low latency [4]. As a result, there are multiple antenna designs introduced at 6G mmWave. However, many academicians have started to offer the foundation of the norm. One disadvantage of this technology is, it uses more energy as speed and reliability rise, making the construction and upkeep of 6G networks expensive [5].

DOI: 10.1201/9781032656021-11

The features of 6G technology include THz communications, blockchain, AI and ML-based computing, quantum communications, and spectrum sharing. High-speed data transfer is necessary for the 6G technologies of the future. Multiple input, multiple output—referred to as MIMO—systems are crucial in today's modern wireless communication for the purpose of fulfilling the demands of fast data speeds [6]. The demand for high-speed data transfer with multi-band operations has significantly expanded because of the quick development of wireless mobile communication [7]. The antennas' compact designs allowed them to resonate at higher frequencies, allowing the devices to achieve data speeds faster than 5G technology [8].

In multi-element antenna systems like MIMO systems, where antennas are near together, the impact of surface waves increases mutual interaction between the adjacent antennas. The effectiveness of the system is lowered as a result of the electromagnetic waves that interact between the adjacent antennas in the antenna array because of the reciprocal coupling [9]. In order to lower mutual coupling, a variety of decoupling techniques have been developed, including stubs at the ground plane; defected ground structures (DGS); tree-like designs, which are used to reduce surface waves; neutralization lines, which cancel out the initial coupling; and electronic bandgap (EBG) structures in the form of mushrooms [10]. However, we are bringing out a novel technique to reduce mutual coupling for 6GmmWave antenna and enhance antenna performance. MIMO is a suggested technology for 6G communications. While requiring no more MIMO technology for bandwidth or transmission power can boost efficiency of the spectrum and data transfer rate.

In a MIMO setup, several antennas are used at both ends of the communication channel. In these systems, mutual interaction is possible because antenna elements are spaced closely together. Thus, it is challenging to increase isolation or decrease coupling between adjacent antennas that are near together [11]. The previous art antennas have adequate isolation; the reported antenna designs are not compact enough and have complicated designs with more than one layer. Therefore, for 6G mmWave applications, a straightforward, isolation-enhanced, compact, and low cost MIMO antenna is required. MIMO systems are important for raising the data rate in limited bandwidth environments [12]. The design aspects that are most important for a MIMO setup are high isolation and poor correlation. The reciprocal coupling has a detrimental impact on the effectiveness of radiation from the antenna. By placing flawed ground conducting planes beneath the radiating components and introducing parasitic patches among the conducting components, mutual interaction between the radiators is reduced. Since the deterioration of impedance matching, the disadvantage of these approaches is that it is challenging to supply minimal correlation and strong isolation in addition to a wide bandwidth [13]. Consequently, it is crucial to create a new millimetre-wave antenna via minimal inter-element interaction. Predominantly, the two primary high tech issues—mmWave technology and MIMO antennas—are having an effect on the performance and functionality of mobile communication networks, in addition on 5G and upcoming 6G technologies, high link connection, and improved link reliability [14]. In this regard, the researchers want to investigate designing MIMO antennas for mmWave bands as future 6G technologies advance.

The analysis Friis transmission equation shows that as the frequency is increased, the signal wavelength decreases and, as a result, the free space path loss increases. Therefore, the implementation of higher gain (dBi) MIMO antenna is preferred to reduce the losses. The construction of mmWave antennas frequencies presents various challenges to the antenna community [15].

There are several major difficulties for developing mmWave MIMO antennas, including the size of antenna, mutual coupling, losses from high-frequency radiation, and bandwidth and diversity [16]. Consequently, high-gain (dBi) MIMO antennas must be developed in the band of frequency spans of the mmWave, keeping above all these issues in mind. For several mmWave frequencies, MIMO antennas have recently become accessible to provide assistance to 6G cell phones and wireless communications [17]. Therefore, the most important aspects are privacy, confidentiality, and security. To realize this fundamental objective, the systematic framework, essential technologies, and anticipated difficulties were established. Following that, a survey of several machine learning (ML) technologies was carried out, with communications, networking, and security [18]. In conjunction with societal requirements and technical changes that strengthen those requirements will be the key drivers of the generational hop beyond current mobile communication in order to fulfil every one of these needs and realize the future prospects. When all of these elements are taken into account, a compelling case may be made for the next wireless network borders. The benefits of 6th generation wireless technology have advanced beyond ground mobile networks' knowledge, dependability, flexibility, and safety [19]. Antennas with high gain can make up for route loss, while array antennas are utilized to lower the multipath fading and disruption [20].

A MIMO with two linear arrays was presented, with each array having four cylindrical resonator antennas made of dielectric operating at 30 GHz. Beam tilting is achieved using a passive microstrip feed network [21]. A recent triple pairs of metamaterial arrays in an end-fire bowtie antenna was published by Haixin Jiang et al. to get a two-beam radiation pattern, increased gain (dBi), and bandwidth [22]. Mutual coupling among the radiating parts when using MIMO technology causes significant problems, which lowers its effectiveness. For mmWave MIMO applications, a tapered metallic slot antenna array is proposed that offers superior beam-forming performance, better gain, and a reduction in the influence of mutual coupling [23]. Using the idea of linear linked arrays, structure of a MIMO antenna for future smart phone applications, achieving wideband coverage. Making slots and using a power splitter to get the slots excited to improves their separation among the elements [24]. It is suggested to use a small, in a linear polarization, wideband antipodal Vivaldi type antenna using a network of power separator feeds. By reducing interdependence among the radiating elements, a Vivaldi antenna improves gain (12.7 dBi) and exhibits good radiation characteristics [25]. Using a dielectric resonator in rectangle shape driven via a microstrip feed utilizing an aperture coupling, the MIMO antenna's DRA elements are more isolated from one another [26]. Dipole array filled with metamaterial framework, capacitive with loops, an effective MIMO antenna for upcoming cellular communications is also possible. The interdependence among the elements is reduced by reducing the metamaterial region's boundaries [27]. The development of patch antennas for microstrip technology 6G mmWave bands such as

(7–24 GHz), 32GHz Band (31.80–33.40GHz), 40GHz Upper Band (40.5–43.30GHz), and 50GHz Band (50.40–52.60GHz) is the main goal of this work.

Throughout this chapter, an updated solution is suggested without the application of any further separation components or decoupling methods. Even though the antenna elements are positioned fairly close to each other, there is a significant port isolation along with poor correlation by placing the radiating patch at the upper and lower side of the substrate. The suggested design is an easy-to-implement small millimetre-wave broadband MIMO antenna. This antenna performs well in relation to gain (dBi), fractional impedance bandwidth in percentage (FBW), correlation coefficient value, and mutual coupling. It also has good radiation properties. Sections 2 and 3, respectively, explain in detail the antenna configuration and performance.

11.2 PROPOSED ANTENNA DESIGN

The suggested MIMO antenna illustrated in Figure 11.1 is made of RT Duriod 5880™ dielectric material and has dimensions of 25 mm × 28 mm × 0.508 mm. This material has a loss tangent of dielectric 0.0009 and $\varepsilon r = 2.2$. Figure 11.4 illustrates the designed MIMO antenna with optimized parameter values. All the values are measured in mm. The designed antenna comprises of four Hx-HpDs (Hx Hexagon-Hp

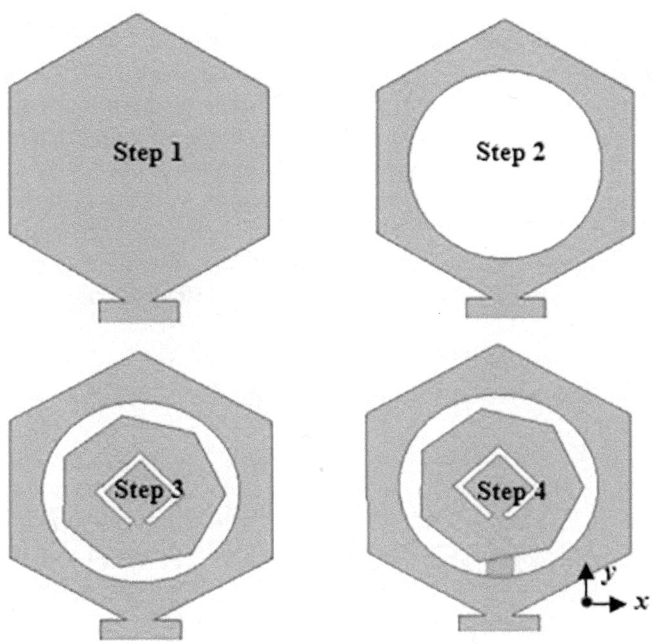

FIGURE 11.1 Single band antenna design process. To design a four element MIMO, one element is created at first. There are four simple steps required to design a single element in MIMO. Steps 1 to 4 indicate the design idea of a single element. To form a four element MIMO this single element is arranged as shown in Figure 11.2.

Heptagon with Ds Diamond slot) radiating elements or patches (denoted by R1, R2, R3, and R4) on the upper layer of the substrate material the radiating elements R1 and R2 are loaded. R3 and R4 are loaded near the base of the substrate material. The radiating elements (R1, R2, R3, and R4) are fed by 4.5 mm × 1.0 mm sized 50-ohm microstrip line feed. Figure 11.1 illustrates that a hexagon radiating elements is loaded with a slotted circle. In the place of circle slotted diamond heptagon is loaded to generate more operating frequencies. The separation between R1 and R4 is 3.7 mm and between R2 and R3 is 4.9 mm. As a way to enhance the isolation or if the mutual coupling must be decreased, the diagonal elements are placed on the same side. The conducting elements R1 and R2 are loaded the dielectric's upper surface and the remaining two radiators R3 and R4 are loaded near the base of the substrate. Loading the diagonal elements about the same side, we can get the separation between the elements is more. This causes the mutual coupling to decrease. The Hx of Hx-HpDs elements (R1, R2, R3, and R4) are used to achieve the first operational band at 22.23 GHz from (12.77–25) GHz the remaining bands because of HpDs. In the 6G spectrum, the 7 to 24 GHz and 24 to 60 GHz millimetre wave frequency bands are the initial operational frequency spectrum.

Figure 11.1 presents the designed mmWave MIMO. Dimensions of the top layer are the same as the bottom layer. The second working band at 34 GHz from (32.85–36.54) GHz is obtained by loading a heptagon with diamond slot. The second band covers mmWave frequency band in 6G spectrum. In every single hexagon, radiating elements exists a heptagon with a diamond slot. The separation among the two polygons is 0.5 mm. The elements R1, R3 and R2, R4 are spaced apart at a separation of 1.2 mm. The radiating elements R1 and R4 separated at a distance 3.7 mm and R2, R3 separated at a distance 4.9 mm. The dimensions of the microstrip line feed is 4.5 mm × 1 mm. In a separate antenna element, there are three radiating elements like hexagon, heptagon with diamond slot and rectangular feed. The dimensions of each element are adjusted such in which the antenna resonate at four different frequencies and achieve ultra wideband, which cover the 6G spectrum. It is recommended to decrease the mutual coupling among the antenna elements. Because a diagonal radiating elements (R1, R2) and (R3, R4) are loaded on the same side. This configuration leads to achieve more gain and improved performance of the antenna in relation to mutual coupling.

The essential parameter for 6G is large bandwidth, which is achievable for this type of configuration.

Figure 11.1 shows the systematic design process. The pattern is divided into four phases, the first of which is the hexagonal patch with a tiny rectangular shape patch at the intersection. The next stage is to create a circular slot with a radius of 3.5 mm in order to make a heptagonal ring. A heptagon with diamond slot (a seven-sided polygon) is loaded into the deleted circular part in the third stage. A rectangular parasite route stacked on the opposite side of the material finally excites an antenna. There is an electromagnetic connection between the heptagon and the parasitic patch. Bandwidth is increased by using the parasitic patch and the hexagonal ring at the vertex feed.

The final design of MIMO antennas at mmWave frequency is presented in Figure 11.2. According to the design, antennas one and two are loaded on top of the

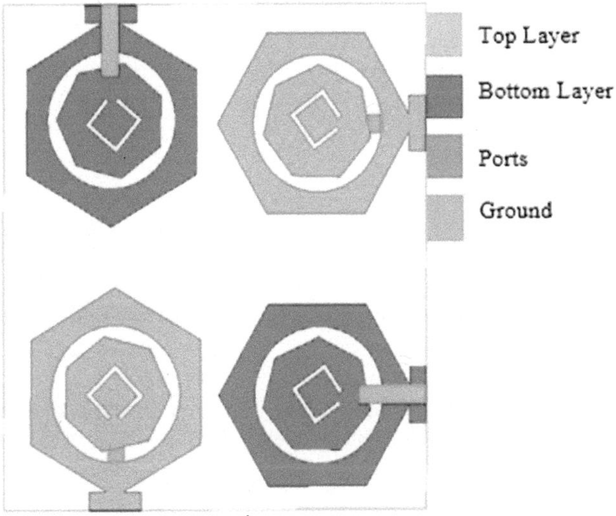

FIGURE 11.2 Four element MIMO antenna design. The single element antenna is positioned as illustrated in this figure. The four elements do not all lie on the same level. The substrate's bottom side has elements in the color red, and its top side has elements in the color brown. Separate colors are used to identify ports and the ground plane.

dielectric, and antennas three and four are filled up on the bottom. On the conducting element opposite the radiator, parasitic patches are stacked. Antennas at the upper and very lower of the patch developed a parasitic relationship with one another. Even though they are stacked on the same plane (upper part of the substrate), antenna one and antenna two are farther apart, which reduces how the mutual interaction among the elements. Antennas three and four (at the substrate's base) exhibit the same result. Now, by loading antennas one and three and placing antenna one on the upper part of the substrate and antenna three on the lower side of the substrate, the mutual coupling among them is suppressed. A similar result is visible between antennas two and four. To see the clear view of the proposed antenna the 3D version is exhibited in Figure 11.3.

Since the nearby antenna elements loaded on the substrate's upper and lower, the overall mutual coupling is minimized. The dimensions and spacing between the antenna elements are showed in the Figure 11.4. The MIMO antenna resonates at various frequencies because of the slot design. Figure 11.4 gives the proposed design's physical measurements. In the High-Frequency Structural Simulator (HFSS-19), the suggested antenna configuration, simulation, optimization, and analytical evaluations are carried out. The suggested arrangement is straightforward, simple to use, single-layer, and low profile.

11.3 RESULTS AND DISCUSSION

The impedance characteristics of a four element mmWave MIMO antenna are expressed as Sii, where i = 1, 2, 3, and 4 or S11, S22, S33, and S44 These variables

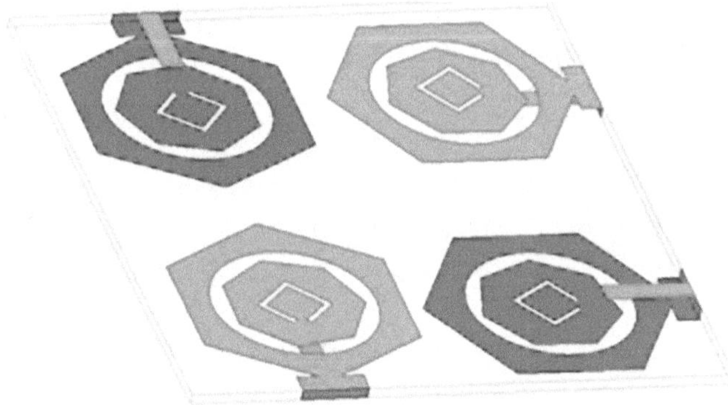

FIGURE 11.3 3D image of the proposed antenna. The three-dimensional view clearly shows which element is loaded on the substrate's top layer and bottom layer.

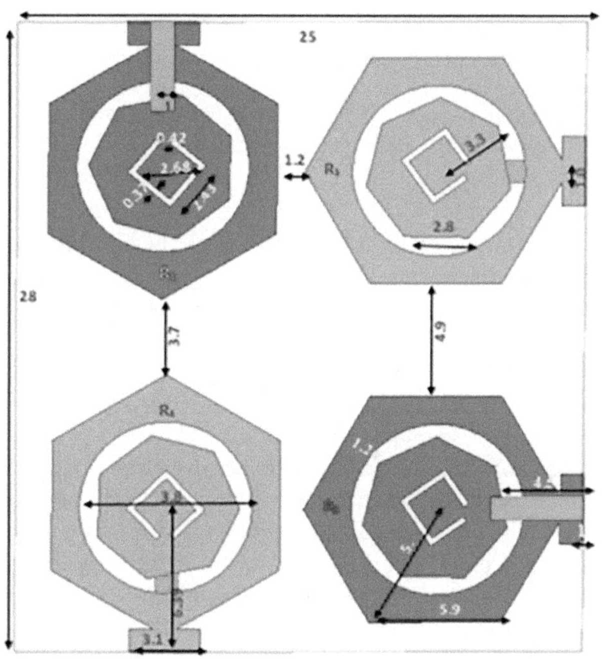

FIGURE 11.4 Dimensions of the suggested MIMO antenna. The overall dimensions and separation between the elements are indicated in the figure. All the dimensions are in mm.

are also referred to as return loss (dB) or reflection coefficient. Sij (ij), where i = 1, 2, 3, 4 and j = 1, 2, 3, 4 i.e. S12, S13, S14, S21 and so on up to S43, describes the isolation or mutual coupling properties of a mmWave antenna. The return loss (Sii) and mutual coupling (Sij) values for all applicable parameters microstrip antennas theoretical limit should be −10dB and −15dB respectively. The impedance (Sii) and isolation characteristics (Sij) of the disclosed MIMO are depicted as shown in Figure 11.5 and Figure 11.6 in relation to Sii and Sij parameters. The first operating band is from 12.77 GHz to 25 GHz with the frequency of 22.23 GHz and achieved reflection coefficient S11 = 15.43 dB. This band is due to hexagon radiating element. Using a slotted diamond-shaped element loaded on the heptagon, the second workable band is achieved at 34 GHz from (32.85–36.54) GHz with an S11 of

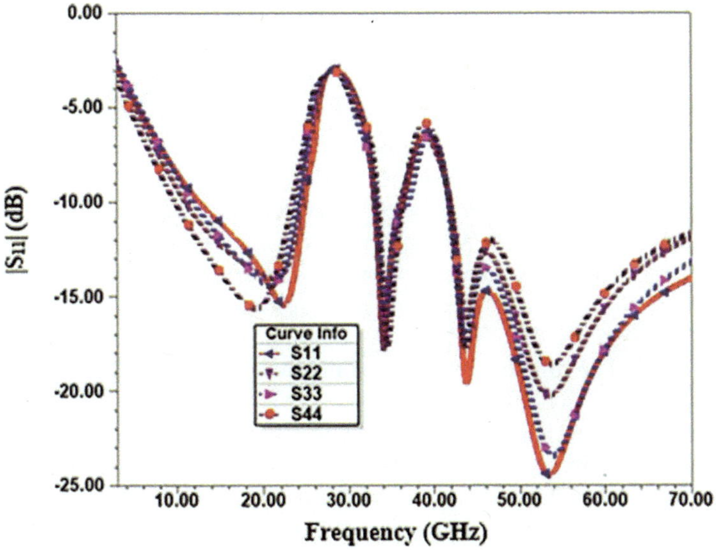

FIGURE 11.5 Return loss of the suggested mmWave MIMO antenna. From here, the first port is represented by S11, the second by S22, the third by S33, and the fourth by S44.

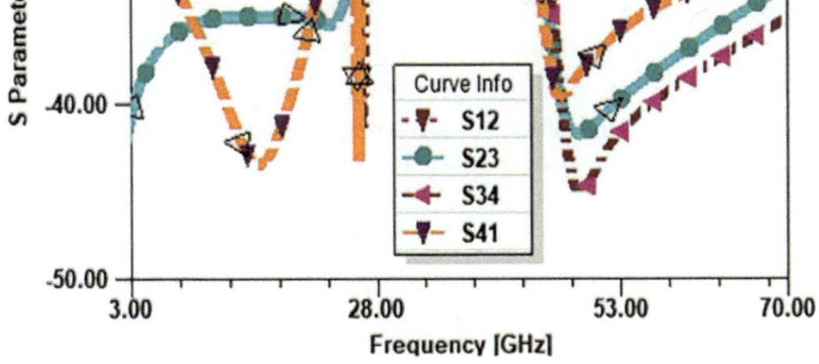

FIGURE 11.6 Transmission coefficient of mmWave MIMO. The transmission coefficient between the first port and the second port is shown by S12. S23, S34, and S41 are similar.

17.69 dB from Figure 11.5. Additionally, there are two more bands generated due to optimized heptagon loaded with slotted diamond. The third band has a resonating frequency of 43.69 GHz with S11 of 19.65 dB. The fourth operating frequency is 53.38 GHz with return loss of 24.47 dB. Out of four bands, it is achieved that three bands are ultra-wide bands. The isolation (Sij) of the mentioned mmWave antenna is achieved by diagonally positioned radiating elements loaded on the same side. It is further presented in Figure 11.6 that the mutual coupling is >25dB (or Sij –25dB). The suggested mmWave antenna has high impedance matching and improved isolation qualities, and it functions between the frequencies of 7 to 24 GHz and 30 GHz to 60 GHz. Figure 11.5 and Figure 11.6 show the impedance and isolation characteristics of the suggested MIMO according to S-parameters. Good impedance matching is included in the suggested MIMO antenna design and enhanced isolation properties.

The proposed method's effectiveness is evaluated using the envelope correlation coefficient (e) of mmWave MIMO antenna. The return loss (dB) characteristics operating in the frequency band of from 12.77 GHz to 25 GHz are depicted in Figure 11.5. From Figure 11.1 the antenna is resonate with fractional bandwidths (FBW) of 55.02% at 22.23 GHz, 10.85% at 34 GHz, and the rest of the bands are ultra wide bands. The high bandwidth at mmWave spectrum (42 GHz to more than 70 GHz) satisfies the 6G criterion.

Figure 11.5 indicates it is possible to view the simulation findings for the retrieved S_{11} and the fractional impedance bandwidth parameters for the operating frequencies. Because it negatively affects the performance of the mmWave MIMO antenna, it is important to assess the effects of reciprocal coupling between the various antenna parts while constructing a MIMO system. Figure 11.5 illustrates improved return loss values at four frequencies for a single antenna element.

Figure 11.5 shows that MIMO antenna element 1, 2, 3, and 4 functioning independently produced similar performance. The fluctuation in transmission qualities across the target frequency band is seen in Figure 11.6. Because it is less than –25 dB, the mutual coupling among two neighbouring MIMO system parts is much smaller. This demonstrates effective isolation between the MIMO components. The simulated 2D (two-dimensional) radiation pattern for the E- and H-planes at resonance frequencies is displayed in Figure 11.7 (a)–(d). Understanding an antenna's behaviour and functionality from the knowledge of its radiation patterns. These patterns are a fundamental necessity for every application involving an antenna because radiation patterns provide details about the spatial spread of electromagnetic radiation. These are used to identify the type of antenna and clarify antenna-pointed directions. With the use of these patterns, it is possible to position an antenna to any direction and observe how it responds to that particular direction. The operational frequency is determined by taking into account this pattern as the main antenna feature.

Figure 11.8 shows the mmWave MIMO 3D radiation pattern. This pattern provides all the information regarding the antenna gain. It offers the gain (dBi) in every direction that is a possibility. The suggested mmWave MIMO gain (dBi) values are mentioned in Figure 11.9. The associated mentioned MIMO antenna performance

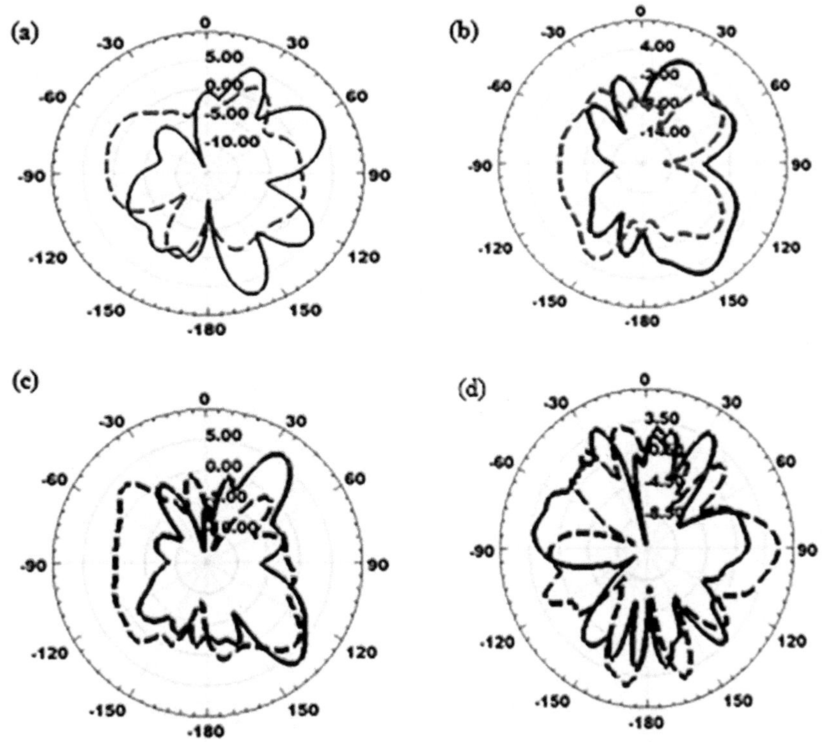

FIGURE 11.7 2D radiation pattern of the suggested MIMO antenna at (a) 22.23 GHz, (b) 34 GHz, (c) 43.69 GHz, and (d) 53.38 GHz. H-plane is indicated with a dotted line, whereas the E-plane is with a thick line.

values from Figures 11.7, 11.8 and 11.9 are included in Table 11.1. Table 11.2 lists the suggested antenna's performance statistics, including S11, fractional impedance bandwidth (FBW), antenna gain (dBi), and average gain (dBi). It is noted that the maximum gain obtained at 53.38 GHz is 9.54 dBi.

Due to surface current, the gains (dBi) for the first two operating frequencies on one side of the antenna are slightly lower than that of the third and fourth frequency. Planar shape, compactness, reasonable gain, and reasonable bandwidth are benefits of the suggested antenna, due to which it is ideal for 6G MIMO systems.

To ascertain how two antennas are independent radiating antenna elements are in their distinct performance, the correlation coefficient for envelopes (ECC) must be calculated. Equation (11.1) is used to quantitatively calculate the envelope correlation coefficient (e) throughout the suitable frequency band. S11, S12, S21, and S22 values are substituted in the equation (11.1) based on Figures 11.5 and 11.6. It should be identified that the ECC values fall below 0.003, which is within the realistically acceptable limit (0.5), for the specified frequency band.

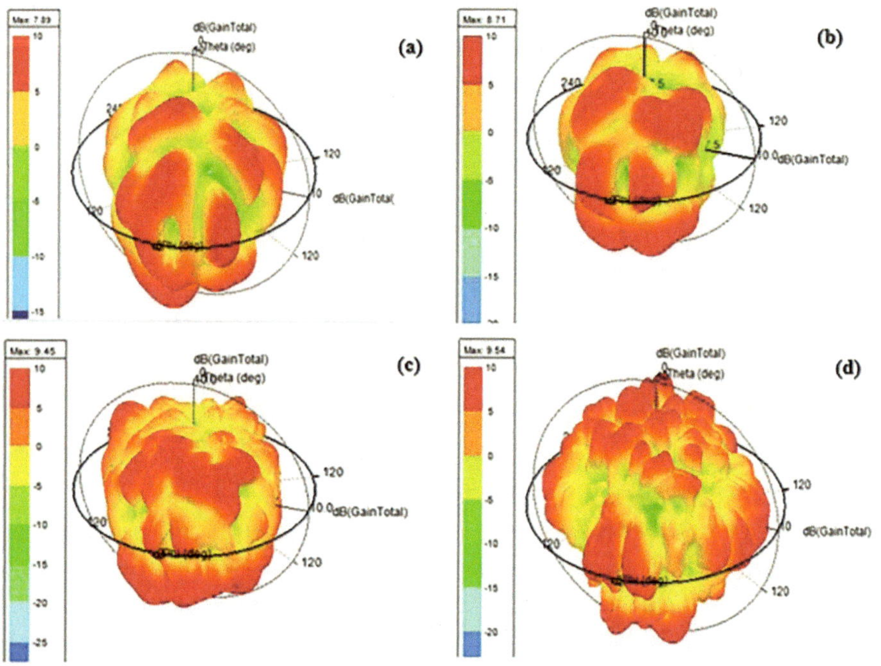

FIGURE 11.8 3D radiation of the suggested MIMO antenna at (a) 22.23 GHz, (b) 34 GHz, (c) 43.69 GHz, and (d) 53.38 GHz. E-plane is a thick line where as H-plane is a dotted line.

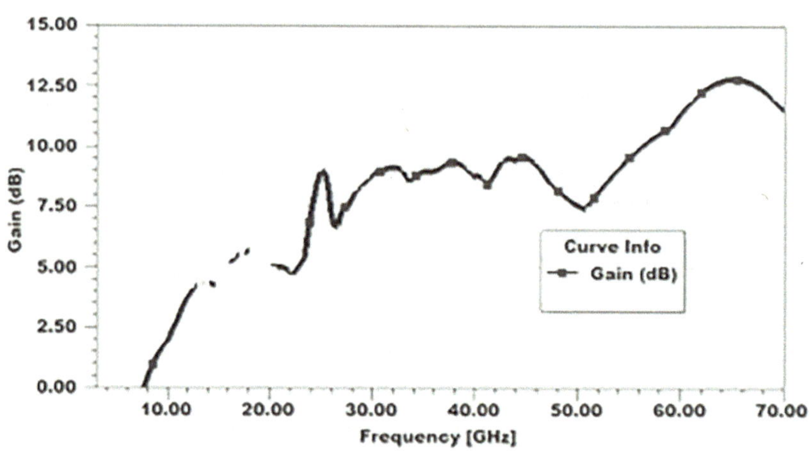

FIGURE 11.9 Gain (dBi) variation of proposed MIMO with frequency. The gain variation of the suggested MIMO antenna for the range of 3 GHz to 70 GHz.

TABLE 11.1
Proposed MIMO Antenna Performance Standards.

Parameter	Simulated			
f (GHz)	f_1 22.23	f_2 34	f_3 43.69	f_4 53.38
S_{11} (dB)	15.43	17.69	19.65	24.47
FBW	55.02%	10.85%	UWB	UWB
Gain (dBi)	7.89	8.71	9.45	9.54
Average Gain	8.898 dBi			

TABLE 11.2
Antenna Comparison between the Suggested and Existent Antennas in the Literature.

Author	Frequency (bands) GHz (*fr*)	Size (mm²)	Avg gain (dBi)	FBW	Isolation (dB)
[20]	57.2–63.4 (5)	9550	7.65	11.58%	22.31
[21]	29.75–31.50 (2)	1008	8.13	5.78%	25
[22]	24.25–27.5 (1)	915	7.03	12.5%	n.a.
[23]	24–32 (9)	750	8.9	47%	n.a.
[24]	25.7–30.5 (4)	5265	8.73	18.7%	15
[25]	24.65–28.5 (2)	1729	9.94	14.81%	37.3
[26]	27.25–28.59 (1)	400	8.72	6.54%	24
[27]	26–31 (3)	1488	9.06	18.11%	21
This work	12.77–70 (5)	700	8.89	51%	25

$$\rho_e = \frac{\left|S_{12}S^*{}_{11} + S_{22}S^*{}_{21}\right|^2}{[1-(\left|S_{11}\right|^2+\left|S_{21}\right|^2)][1-(\left|S_{22}\right|^2+\left|S_{21}\right|^2)]} \tag{11.1}$$

Table 11.2 contrasts the suggested MIMO antenna with the existing MIMO antennas reported inside the written works. This comparison table shows how the suggested MIMO antenna differs from other antennas in the literature in terms of antenna overall size, gain (dBi), fractional impedance bandwidth (%), resonant frequencies, and isolation (dB). The bulk of the supplied MIMO antennas in Table 11.2 occupy a large space when in comparison to suggested MIMO antenna. All of the reported MIMO antenna designs feature antenna in comparison to elements loaded on a single plane of the substrate, as opposed to the indicated mmWave design, which leads to low isolation values at the operating frequency. The mentioned antenna thus has many uses, including very low fabrication costs because of simple and single layer, plane geometry structure, compactness, high bandwidth (BW), high gain (dBi), and better isolation between antenna elements, all of which are required for 6G mmWave MIMO systems, particularly in the 6G mobile communication systems. The basic requirement of 6G communication is large bandwidth. The proposed MIMO meets the requirement at three bands that is at first band the achieved bandwidth is 12.23 GHz. The third and fourth band also meet the requirement of 6G bandwidth. The first band is suited for future IoT applications in addition to satellite and in-flight internet services. This MIMO can be used for smart cities, autonomous vehicles, the internet of everything, artificial intelligence, self-learning mobile networks, coverage expansion, drone operation and safe control, remote medical operations, and smart cities.

11.4 CONCLUSION

The 6G mmWave MIMO antenna that is being described is compact, easy to build, low profile, and has a consistent radiation pattern with high gain values and wide coverage. The objective is to create an antenna that will function with upcoming 6G technology. A new, simple multiband-UWB MIMO antenna has been shown to be capable of reaching a maximum fractional impedance bandwidth of 51%. The 6G mmWave implementation of MIMO antennas procedure, simulation, and performance assessment are covered in this chapter. The accomplishment using the MIMO antenna enhanced and the data rate capabilities are increased by using a four-port multiple input, multiple output (MIMO) configuration. To decline mutual interaction, the elements are placed on the substrate's upper and lower layers. At 53.38 GHz, a significant gain of 9.54 dBi is attained, which is very essential for 6G communications. The performance of mmWave MIMO is improved by using a low loss dielectric material. This demonstrates the nearby antenna elements' poor mutual coupling and better isolation characteristics. The suggested multi-band antenna can result in increased bandwidth and data throughput. Due to its reasonable performance and compactness, the recommended mmWave MIMO antenna is competitive for impending 6G mmWave MIMO applications. In order to highlight the advantages of the suggested design, a comparison is also created in between the performance of the suggested antenna and the data presented in the current literature. The results and comparison with earlier studies show that the recommended technique is very competitive and of the greatest option for 6G wireless applications.

REFERENCES

[1] Zhengquan Zhang, Yue Xiao, Zheng Ma, Ming Xiao, Zhiguo Ding, Xianfu Lei, George K. Karagiannidis, and Pingzhi Fan: 6G Wireless Networks-Vision, Requirements, Architecture, and Key Technologies, *IEEE Vehicular Technology Magazine* (2019) 2–14.

[2] Mohammed Banafaa, Ibraheem Shayea, Jafri Din, Marwan Hadri Azmi, Abdulaziz Alashbi, Yousef Ibrahim Daradkeh, and Abdulraqeb Alhammadi: 6G Mobile Communication Technology: Requirements, Targets, Applications, Challenges, Advantages, and Opportunities, *Alexandria Engineering Journal* 64 (2023) 245–274.

[3] Amjaad T. Altakhaineh, Saqer S. Alja'afreh, Aser M. Almatarneh, Eqab Almajali, Luae Al-Tarawneh, and Jawad Yousaf: A Quad-Band Shared-Aperture Antenna Based on Dual-Mode Composite Quarter-Mode SIW Cavity for 5G and 6G with MIMO Capability, *Electronics* (2023) 1–20.

[4] Shuchi Tripathi, Nithin V. Sabu, Abhishek K. Gupta, and Harpreet S. Dhillon: Millimeter-wave and Terahertz Spectrum for 6G Wireless, *arXiv, IEEE* (2021) 1–17.

[5] Kandasamy Pirapaharan, W. H. Sasinda C. Prabhashana, S. P. Pramuka Medaranga, Paul R. P. Hoole, and Xavier Fernando: A New Generation of Fast and Low-Memory Smart Digital/Geometrical Beamforming MIMO Antenna, *Electronics* (2023) 1–15.

[6] Theodore S. Rappaport, Yunchou Xing, Ojas Kanhere, Shihao Ju, Arjuna Madanayake, Soumyajit Mandal, Ahmed Alkhateeb, and Georgios C. Trichopoulos: Wireless Communications and Applications above 100 GHz: Opportunities and Challenges for 6G and beyond, *IEEE Access* 7 (2019) 78729–78757.

[7] T. S. Rappaport: 6G and beyond: Terahertz Communications and Sensing, *2019 Brooklyn 5G Summit Keynote*, Apr. 2019. [Online]. Available: https://ieeetv.ieee.org/conference-highlights/keynote-tedrappaport-terahertz-communication-b5gs-2019?

[8] K. Lee, J. Kim, E. W. Jin, and K. S. Kim: Extreme Massive MIMO for Upper-mid Band 6G Communications, in: *Proceedings of the 13th International Conference on Information and Communication Technology Convergence (ICTC)*, Jeju Island, South Korea (2022), 997–999.

[9] In-Lu Wong, Yi-Rong Chen, and Wei-Yuli: Eight-Planar-Monopole MIMO Circular Array Generating Eight Uncorrelated Waves for 6G Upper Mid-Band 8 × 8 MIMO Access Points, *IEEE Access* 11 (2023) 68018–68030.

[10] G. Zhai, Z. N. Chen, and X. Qing: Mutual Coupling Reduction of a Closely Spaced Four-element MIMO Antenna System Using Discrete Mushrooms, *IEEE Transactions on Microwave Theory and Techniques* 64 (2016) 3060–3067.

[11] B. Mohamadzade and M. Afsahi: Mutual Coupling Reduction and Gain Enhancement in Patch Array Antenna Using a Planar Compact Electromagnetic Bandgap Structure, *IET Microwaves, Antennas & Propagation* 11 (2017) 1719–1725.

[12] R. Hussain, M. Abou-Khousa, N. Iqbal, A. Algarni, S. I. Alhuwaimel, A. Zerguine, and M. S. Sharawi: A Multiband Shared Aperture MIMO Antenna for Millimeter-wave and Sub-6GHz 5G Applications, *Sensors* 22 (5) (2022) 1808.

[13] A. Altaf, A. Iqbal, A. Smida, J. Smida, A. A. Althuwayb, S. Hassan Kiani, M. Alibakhshikenari, F. Falcone, and E. Limiti: Isolation Improvement in UWB-MIMO Antenna System Using Slotted Stub, *Electronics* 9 (2020) 1582.

[14] P. Porambage, G. Gür, D. P. M. Osorio, M. Liyanage, and M. Ylianttila: 6G Security Challenges and Potential Solutions, in: *Proceedings of the Joint European Conference on Networks and Communications & 6G Summit (EuCNC/6G Summit)*, Porto, Portugal (2021) 1–6.

[15] B. Mohamadzade and M. Afsahi: Mutual Coupling Reduction and Gain Enhancement in Patch Array Antenna Using a Planar Compact Electromagnetic Bandgap Structure, *IET Microwaves, Antennas & Propagation* 11 (2017) 1719–1725.

[16] Iqbal, A., O. A. Saraereh, A. W. Ahmad, and S. Bashir: Mutual Coupling Reduction Using F-shaped Stubs in UWB-MIMO Antenna, *IEEE Access* 6 (2018) 2755–2759.

[17] W. Saad, M. Bennis, and M. Chen: A Vision of 6G Wireless Systems: Applications, Trends, Technologies, and Open Research Problems, *IEEE Network* 34 (2019) 34–142.

[18] F. Tang, Y. Kawamoto, N. Kato, and J. Liu: Future Intelligent and Secure Vehicular Network toward 6G: Machine-learning Approaches, *Proceedings of the IEEE* (2019) 292–307.

[19] G. Gui, M. Liu, F. Tang, N. Kato, and F. Adachi: 6G: Opening New Horizons for Integration of Comfort, Security, and Intelligence, *IEEE Wireless Communication* (2020) 126–130.

[20] X. Cheng, Y. Yao, T. Tomura, J. Hirokawa, T. Yu, J. Yu, and X. Chen: A Compact Multi-beam End-Fire Circularly Polarized Septum Antenna Array for Millimeter-wave Applications, *IEEE Access* 6 (2018) 62784–62792.

[21] Mohammad S. Sharawi, Symon K. Podilchak, Mohamed T. Hussain, and Yahia M. M. Antar: Dielectric Resonator Based MIMO Antenna System Enabling Millimetre-wave Mobile Devices, *IET Microwaves, Antennas & Propagation* 11 (2017).

[22] Haixin Jiang, Li-Ming Si, Weidong Hu, and Xin Lv: A Symmetrical Dual-beam Bowtie Antenna with Gain Enhancement Using Metamaterial for 5G MIMO Applications, *IEEE Photonics Journal* 11 (2019).

[23] Binqi Yang, Zhiqiang Yu, Yunyang Dong, Jianyi Zhou, and Wei Hong: Compact Tapered Slot Antenna Array for 5G Millimeter-wave Massive MIMO Systems, *IEEE Transactions on Antennas and Propagation* 65 (2017) 6721–6727.

[24] M. Ikram, M. Sharawi, A. Shamim, and A-R. Sebak: A Multiband Dual-standard MIMO Antenna System Based on Monopoles (4G) and Connected Slots (5G) for Future Smart Phones, *Microwave and Optical Technology Letters* 60 (2018) 1468–1476.

[25] Shuangshuang Zhu, Haiwen Liu, Pin Wen, and Zhijiao Chen: A Compact Gain-Enhanced Vivaldi Antenna Array with Suppressed Mutual Coupling for 5G mmWave Application, *IEEE Antennas and Wireless Propagation Letters* 17 (2018) 776–779.

[26] Yin Zhang, Jing-Ya Deng, Ming-Jie Li, Dongquan Sun, and Li-Xin Guo: A MIMO Dielectric Resonator Antenna with Improved Isolation for 5G mm-Wave Applications, *IEEE Antennas and Wireless Propagation Letters* 18 (2019) 747–751.

[27] Zamir Wani, Mahesh P. Abegaonkar, and Shiban K. Koul: A 28-GHz Antenna for 5G MIMO Applications, *Progress in Electromagnetics Research Letters* 78 (2018) 73–79.

12 Intelligent Reflecting Surface Assisted D2D Communication at mm-Wave Band

Subhra Sankha Sarma

12.1 INTRODUCTION

In recent times, the incapability of the existing wireless networks in supporting the emerging and future technologies such as communication driven by artificial intelligence (AI) [1] and the Internet of Things (IoT) [2] has driven researchers to seek out for methods and techniques which could integrate these applications to the next generation (Next-Gen) 5G/B5G wireless networks. The supposed technology can be implemented to various trending domains such as virtual reality (VR) [3], holographic telepresence [4], autonomous vehicles (AVs) [5], healthcare devices [6], disaster management [7], smart networks in urban cities [8] and indoor navigation [9]. According to a report published by Cisco, it is estimated that by 2023 5G must back up 10 percent of all users at a mean speed of 575 megabits per second (Mbps) which means that the speed for information transfer must be 13 folds quicker than the existing communication speed, and also back up around 14.7 billion machine-to-machine (M2M) communication links. This implies that the number of users doubles from 6.1 billion in 2018 to 14.7 billion by 2023 [10]. It also states that the total number of mobile users will see a surge from 8.8 billion in 2018 to 13.1 billion by 2023. Out of all of these devices, 1.4 billion of them will have 5G capability. Another estimate suggests that the count of IoT users will multiply manifold to approximately 80 billion users by 2025. This would finally increase the global revenue manifold to approximately $1.1 trillion [11]. As the 5G architecture supports massive IoT (mIoT) operations which helps in enabling IoT applications, it has a widespread usage. Thus, with the advent of the bandwidth-intensive applications, such as holographic telepresence, online gaming, etc., there is an enormous call for enhanced data rate, higher bandwidth, and marginal latency. Device-to-device (D2D) communication poses to be a frontrunner in satisfying the insatiable urge of higher data rate and larger bandwidth. It is a giant leap forward for Next-Gen networks due to the characteristics in the field of communication systems and networks. In addition to this, intelligent reflecting surface (IRS) has the ability to subdue wireless channel's substandard propagation effects and is quickly becoming a popular topic in academia as well

DOI: 10.1201/9781032656021-12

as industry due to recent growth in meta-materials technology. The combination of D2D communication with the IRS technology will usher the Next-Gen networks to new heights [12]. The next sections give an elaborate idea about the fundamentals of D2D communication and IRS technology.

12.1.1 D2D Communication

D2D communication has proven itself to be a frontrunner in countering the enormous rise of mobile users worldwide. It may be showcased as the transfer of information among two devices which are at a closer distance to each other without direct control of the BS. The representation of the D2D communication is shown in Figure 12.1. Out of all advantages of D2D communication, some of them are namely, huge bandwidth, higher throughput, very low latency, and negligible interference at the unused spectrum. Furthermore, D2D communication also exhibits higher spectral efficiency (SE) as well as energy efficiency (EE) [13]. The reason for choosing D2D connectivity is due to the level of easiness that it offers to the operators in the process of offloading traffic from the main network which thereby reduces the energy requirement. With features like directional transmission, dense deployability, etc., D2D communication can be enabled at the millimetre-Wave (mm-Wave) frequency. It has the calibre to enhance the features of the existing communication network to the Next-Gen network. Also, it enhances the quality of service (QoS) as well as the sum throughput of the gross network. As a result, D2D communication may serve to be the backbone for all bandwidth intensive applications at mm-Wave band network [14]. However, implementation of D2D communication at mm-Wave frequencies has a variety of obstructions and problems which need to be handled with immediate focus [15]. The transmitted signal from the base station (BS) in the mm-Wave frequencies suffers considerable amount of degradation from propagation loss incurred by interference. A variety of blockages contribute to the interference including atmospheric and rain attenuation, human obstruction, foliage loss, etc. [16]. This in turn subsequently lowers the QoS of the received signal of the D2D user. Release 15 (Rel-15) of the 3GPP (3rd Generation Partnership Project) focused on

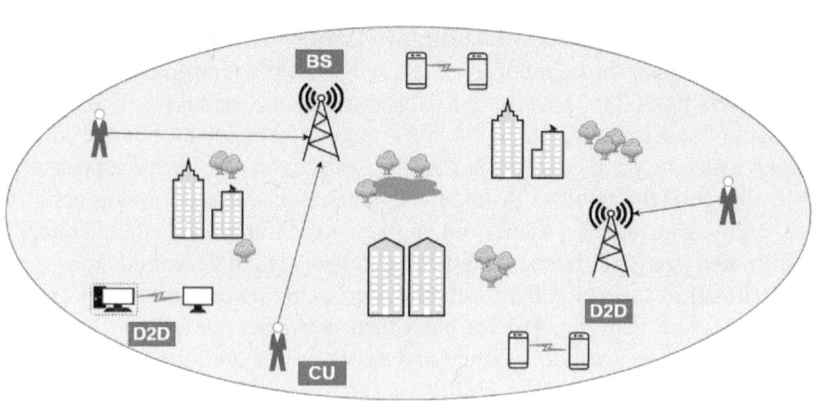

FIGURE 12.1 Representation of D2D communication.

extreme lower latency and larger throughput [17]. Release-16 (Rel-16) of the 3GPP put forward crucial characteristics for uplifting the sum rate and QoS of the gross network. Rel-16 is also known as Phase 2 of 5G.

12.1.2 Intelligent Reflecting Surface

B5G networks including the sixth generation (6G) wireless network require meeting at tougher prerequisites than 5G networks, such as increased system throughput and EE, enhanced coverage and seamless connectivity, as well as enhanced reliability in the presence of ultra-low latency. The existing technology has not fully achieved the requirements that are necessary for implementing the Next-Gen networks. The existing technology encounters some issues such as, a) shifting to higher frequencies, mm-Wave and terahertz (THz), for utilizing the higher bandwidth and data rate; and b) shortening the communication distance between the active nodes such as BS, access points (AP), relays, distributed antennas/remote radio heads (RRHs), etc., which is of paramount importance for enabling a 6G network. However, deploying active nodes also increases the energy consumption and the maintenance cost. Also, c) installing more antennas increases hardware which also increases the complexity of signal processing. Given the aforementioned challenges and constraints, it is critical to develop an innovative technology which can disrupt the existing market in order to accomplish long-term growth of the Next-Gen wireless networks to provide technology at low and reasonable cost, with lower complexity and power consumption. These issues motivated the enabling of a new paradigm, IRS to Next-Gen networks for achieving smart signal propagation channels which are also reconfigurable, for catering seamless communication between the devices [18–20]. Generally speaking, IRS is a planar surface consisting of a number of passive reflecting elements. These elements can independently produce controllable amplitude and/or phase change in the incident signal [21]. Figure 12.2 shows the representation of the IRS. Thus, by densely installing the IRSs in short distances in the wireless networks and also intelligently aligning their reflections, the propagation of signals between the transmitters and receivers can be easily reconfigured. This will help in achieving desired

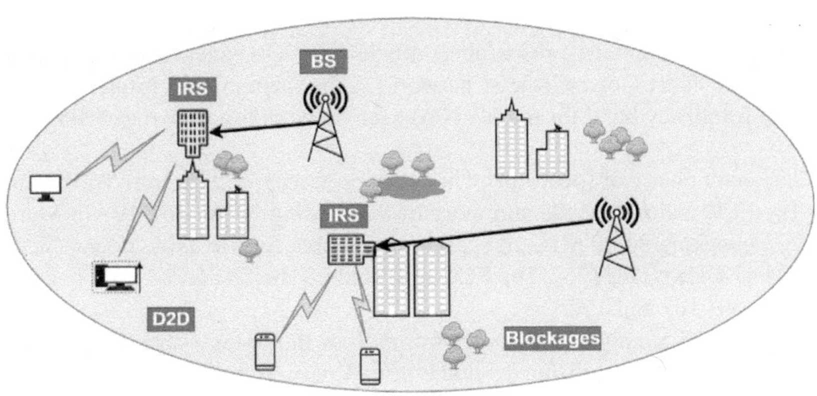

FIGURE 12.2 Representation of intelligent reflecting surface.

realizations and/or distributions, providing a new way to address the basic issues of fading impairment and interference. In addition, IRS functions using full-duplex (FD) mode with no antenna noise amplification or self-interference, giving it a competitive advantage over typical active relays. Furthermore, the IRS will achieve quantum leap improvement in providing seamless wireless communication between devices by enhancing channel capacity and reliability.

12.1.3 Mm-Wave Network

The communicating signals in a mm-Wave network operate at a frequency of 30–300 GHz with their corresponding wavelength from 10–1 mm, also known as extremely high frequency band (EHF). The operating characteristics of signal propagation in the mm-Wave network are furnished as follows.

- The mm-Wave signals exhibit higher propagation loss at 28 GHz than the microwave band. This is because the propagation loss is directly related to the square of the carrier frequency [22]. However, utilizing a highly directed antenna can significantly reduce propagation loss.
- Because of their shorter wavelength, mm-wave transmissions cannot diffract over the edges of obstacles. This has a negative impact on the signal and causes signal attenuation during line-of-sight (LoS) propagation of the broadcast signals [23]. Because mm-Wave channels use LoS transmission with Rician distribution due to a lack of multipath, outages may occur if the LoS signal is obstructed by obstructions.
- Concrete constructions and solid materials cannot be penetrated by mm-Wave signals. Nowadays, buildings in urban areas have concrete structures with highly reflecting glass, which makes it difficult for the communicating signals to penetrate through the walls undergoing signal attenuation. At 40 GHz, the signal suffers a 178 dB attenuation when passing through a concrete wall [22] severely limiting the use of communication in mm-Wave network to indoor applications. Indoor scenarios are preferred; however, outdoor implementation also finds limited application [24, 25].

Despite these disadvantages, researchers are nevertheless eager to use the mm-Wave frequency for Next-Gen wireless networks. The benefits of implementing the mm-Wave frequency band for the 5G Next-Gen network are shown as follows.

- There are plenty of spectrums which can be accessed in the mm-Wave band. The FCC announced the summary of the bidding in Auction 103 in March 2020, making 3,400 MHz of spectrum available for use in the upper 37, 39, and 47 GHz bands [26]. The FCC also sold off the 28 and 24 GHz bands in Auctions 101 and 102.
- High signal attenuation and penetration loss limits the usage of frequency reuse only for short distances in the mm-Wave communication.
- Because of the short transmission range and low beam-width of mm-Wave signals, mm-Wave communication provides enhanced security and privacy.

Radio signals travel through the medium in three ways: reflection, diffraction, and scattering. To begin, reflection takes place when a signal flowing in one media is incident into another medium with differing electromagnetic properties. The reflected signal undergoes attenuation, and some of the incident signal is absorbed [27]. Secondly, diffraction is a prominent factor in signal propagation. It is a phenomenon in which signals bend around the edges of obstacles such as large concrete buildings, walls, mountains, sky scrapers, and so on. This phenomenon is critical in modern Next-Gen wireless networks because urban areas are densely packed with buildings and monuments that hinder the path of the transmitted signal. This obstacle reduces signal strength, and the received signal at the user goes through significant attenuation. Thus, diffraction holds a crucial parameter for transmitting signals in any mobile network. Finally, scattering occurs when a signal strikes a rough surface with a size lower than the order of the signal wavelength. The incident signal on the rough surface is transmitted in all directions, thereby increasing the outage loss. All of these techniques are necessary for analysing the prevailing interferences in D2D communication at mm-Wave network. Thus, interference plays a crucial act in the propagation of signals at mm-Wave networks. The interferences that have an effect on the mm-Wave network are rain attenuation, atmospheric absorption, foliage loss, blockages, etc. It is a frequent misconception that rain and atmospheric conditions render the mm-Wave band unusable for mobile communication. This is not the case, as evidenced by the fact that an attenuation of a meagre 7 dB/km is projected for 200 m radius cell size owing to extremely severe rainfall of 25mm/hr happening at 28 GHz [16]. Thus, the rain attenuation has an insignificant toll on the mm-Wave network operating at 28 GHz band. Next, atmospheric absorption takes place only at higher frequency bands caused by water vapour and oxygen molecule [23]. The signals undergoing attenuation as a result of oxygen and water vapour absorption usually occur at frequency ranging between 57–64 GHz and 164–200 GHz bands, respectively. This infers that atmospheric attenuation has little effect on D2D communication in the mm-Wave network at 28 GHz. The channel link selected has a significant impact on signal strength at the receiver end. Communication can take place in two ways: LoS or non-line-of-sight (NLoS). The LoS is the preferred link at mm-wave frequencies. However, it suffers from its own set of issues. The signal strength attenuation varies substantially depending on whether the D2D link considered for communication is LoS or NLoS, between the transmitter and destination. Several works employed various strategies to address the problem of blockage, including the LoS Ball model [28–30], Exponential model [31], Boolean model [32, 33], Bernoulli model [34], and others.

12.2 STANDARDIZATION

D2D communication has piqued the interest of academics all around the world due to its unique features of ultra-low latency and increased SE and EE. It is necessary to have certain standardization for seamless communication between the devices to take place. Thus, this section demonstrates the standardization laid out for D2D communication operating at mm-Wave frequencies.

12.2.1 D2D Communication

According to Release-12 (Rel-12), the 3GPP has also established a basic architecture that aids in seamless integration into the Long Term Evolution-Advanced (LTE-A) ecosystem [35]. In addition, the 3GPP Rel-12 standard has been accepted by the Next-Gen public safety network. The Federal Communications Commission (FCC) has set aside 700 MHz of dedicated spectrum for the next-generation network [36]. Furthermore, LTE Proximity Services (ProSe) in the Radio Access Network (RAN) are being investigated under the 3GPP standard, which is divided into: device discovery and D2D communication [37]. Release-13 (Rel-13) added support for user equipment (UE) in network relay at layer 3, which aided in network coverage expansion. Furthermore, it introduced ProSe per packet priority (PPPP), which aided in the implementation of QoS across traffic streams. A study on D2D-based vehicle-to-vehicle (V2V) network feasibility was commenced in 3GPP Rel-13, which served as the foundation for the D2D-based vehicle-to-everything (V2X) network in 3GPP Release-14 (Rel-14) [38]. D2D synchronization, discovery, and direct communication are critical components of the 3GPP specification. As an additional component, Release-15 (Rel-15) supported vehicle platooning, driving involving automatic as well as remote feature, as well as focused on ultra-low latency and better throughput. The support for 5G radio technologies was the most important feature of Rel-15. In 2018, for the first time, 5G standardization was proposed in 3GPP Rel-15, which covered the 5G new radio (NR) technology. With the passage of time, the focus switched to Phase 2 of 5G, i.e., Release-16 (Rel-16). Rel-16 demonstrated some major distinguishing features for the 5G NR that will improve the system's data throughput and QoS. The additional components of Rel-16 included support for vertical industries such as V2X, safety of the masses, and the Industrial Internet of Things (IIoT) [39]. Some of the main areas of focus of Release-17 (Rel-17) include support for edge computing, network automation of 5G, architecture enhancement for 3GPP support of advanced V2X service, system improvement for proximity-based services, multicast-broadcast services, connectivity of unmanned aerial systems (UAS), identification of UAS, tracking including unmanned aerial vehicles (UAV) identification, authorization by UAS traffic management, authorization revocation and (re)authorization failures, UAV-UAV controller association and tracking, and User Plane connectivity for UAVs [40]. 3GPP Release 18 (Rel-18) will improve 5G performance, provide capabilities to enable more flexible and efficient spectrum use, expand device support, adapt network topology to accommodate different deployments, and deliver data-driven, intelligent network solutions [41].

12.2.2 IRS

As a network node, IRS has recently received a lot of interest among researchers. It works in a different way than the repeater. A repeater generally does not amplify the propagated signal but just relays it to the intended end user. It may be considered as a smart device that can assist to regulate the channel. It does so by reflecting the incident signal in the intended direction. It accomplishes this operation by introducing proper phase changes into the incoming signal. Researchers are focussing on IRS topologies with antenna which include both active and/or passive elements. However, from the standpoint of 3GPP standardization, the Integrated Access Backhaul (IABs) and network-controlled repeaters (NCRs) already contain active antenna components, and IRS, as a low-cost and energy-efficient device, is more likely to have just passive components. The 3GPP Rel-18 does not cover IRS as a study item, but however, more discussions are likely to happen prior to Release 19 (Rel-19) [42]. Some expected proposal discussions over IRS include,

- Deployment scenario: Examine use cases and scenarios in which the IRS will be involved. The provision of a strong NLoS path in a blockage scenario may constitute a primary use case for IRS.
- Signalling protocol: Stringent protocols for signalling between the original node and IRS controller for exchange of useful information.
- Channel modelling: IRS is considered as a cluster which is reconfigurable, within a channel model. The 38.901 channel model which is used predominantly must be investigated. This will also cover the investigation of pathloss models for signal propagation. The actual size of IRS may be considered a factor for practical deployment premise.
- Impact on existing procedures: Many techniques which are adopted nowadays include initial attach, beam alignment, beam failure retrieval, interference eradication, etc. and may be affected through the implementation of IRS. Thus, the impact of IRS on these operations also needs to be studied.

12.3 TAXONOMY OF D2D COMMUNICATION

In this sub-section we discuss the taxonomy of D2D communication as depicted in Figure 12.1. D2D communication is preferred over other technologies for operating at 28 GHz mm-Wave frequency. This is because it enhances the SE [43] and overall throughput of the network. The coverage area [44] of the network is also increased, and supports data offloading from the cellular network [45]. D2D communication may be sub-divided into two parts as depicted in Figure 12.3.

- Inband D2D: This category uses licensed spectrum for both the cellular and the D2D links. Because it is easier to manage interference in the licensed spectrum, researchers frequently employ inband D2D communication. Inband D2D is further subdivided into [46, 47],
  a) Underlay spectrum sharing: D2D users utilize the same spectrum resources as cellular users, thereby enhancing the SE and EE. But for a seamless D2D communication to take place, the interference among the users needs to be calibrated and mitigated.
  b) Overlay spectrum sharing: In this category, the spectrum is divided into two sections through the use of orthogonal channel assignment. D2D users utilize one set of channels, whereas cellular users utilize the other. This technology eliminates intra-cell interference between cellular and D2D users. But the underutilization of spectrum resources is a serious issue in the overlay D2D communication, which seriously limits its

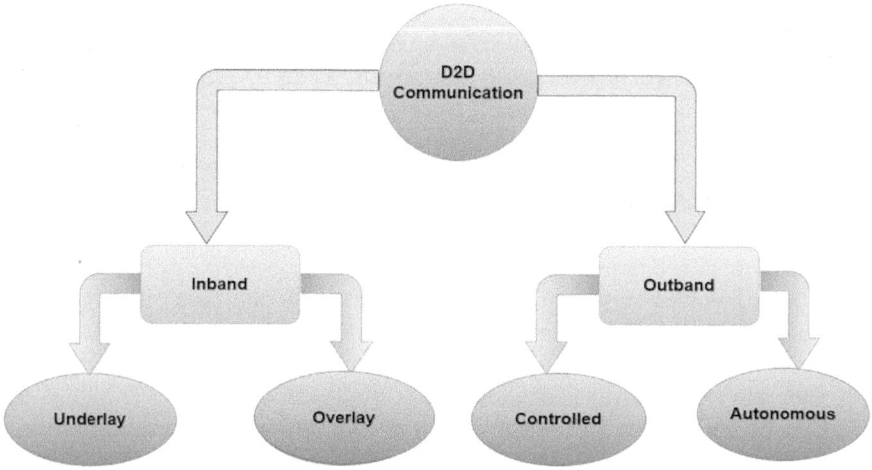

FIGURE 12.3 Classification of D2D communication.

usage. Furthermore, it has a lower SE, which reduces the data rate of D2D users. With adequate spectrum resource usage, underlay spectrum sharing has greater SE and shows better performance than overlay spectrum sharing.

- Outband D2D: This category uses unlicensed spectrum to communicate between D2D users. This technology aims to reduce or eradicate interference among cellular and D2D networks. The BS or the user controls the coordination of several radio interfaces. Outband D2D can be further subdivided as follows:
  a) Controlled: Here, BS controls the communication under the cellular network.
  b) Autonomous: D2D users can behave autonomously in the absence of cellular infrastructure. However, outband D2D communication has several key coordination management concerns for distinct operating bands.

3GPP Rel-12 [48] ideates three coverage scenarios for enabling D2D communication as depicted in the Figure 12.4.

- In-coverage: All the user equipment is within the coverage area of the evolved node B (eNB) permitting D2D communication to take place.
- Out-of-coverage: The situation in which no user equipment is within the coverage area of eNB is considered to be a case of out-of-coverage.
- Partial coverage: Again, the situation in which some of the user equipment is within the coverage area of the eNB, while some is outside is considered to be a partial coverage scenario. Hence, the user equipment in the coverage area of the eNB seeks to communicate with the user equipment which is not within the coverage area of the eNB.

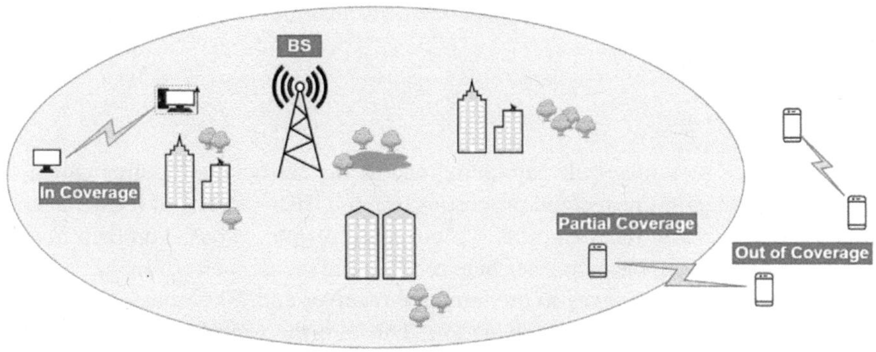

FIGURE 12.4 Coverage scenarios for D2D communication.

In general, two types of D2D communication can be enabled in a mm-Wave network [49].

- Local D2D communication: This form of communication takes place between two D2D users which comes under the same BS. In the absence of LoS path, the transferring of information can take place directly or via relays.
- Global D2D communication: Communication between two D2D users linked with distinct BSs is classified under global D2D communication through hopping via backbone networks. Also, communications including, Device-to-Base station (D2B) and Base station-to-Base station (B2B) are both part of global D2D communication.

12.4 IRS FUNDAMENTALS

An IRS has multiple unique characteristics that set it apart from other systems. They consist of passive metasurfaces and do not need any external dedicated power source to guide the incident signals to the intended end user [50]. The IRSs do not amplify the incident signals; they just relay the signals to the receiver unit. Thus, they perform without the need for any analog-to-digital converter (ADC) and digital-to-analog converter (DAC). Moreover, the IRS operates on all available frequencies due to full band response supporting full duplex transmission. Furthermore, they can be easily mounted on buildings, walls, ceilings, etc., in any environment [51]. Generally speaking, IRS is the most commonly referred term in the literature. However, the broader concept relates to any metasurfaces (passive intelligent or software-controlled) having the ability to reconfigure the incident signal, resulting in a customizable wireless environment [52]. Some of the similar terminologies [53] used for indicating IRS are Reconfigurable intelligent surface (RIS) [54], Large intelligent metasurface (LIMS) [55], Software defined metasurface (SDMS) [56], Large intelligent surface (LIS) [57, 58], Passive intelligent surface (PIS) [59], Smart reflect arrays [60], and Passive holographic MIMO (HMIMO) [61].

The next sub-section of this chapter demonstrates the structure, properties, and taxonomy of the IRS.

12.4.1 Structure

Metasurfaces or metamaterials can be defined as the artificial structures which are manufactured to gain unexpected properties [62, 63]. IRS consists of these metamaterials or metasurfaces that can modify their phase responses based on their coding. These customized tunable responses help redirect the incident electromagnetic (EM) signals from the antenna array to the intended receiver end. This solves the problem of blockage in the LoS signal propagation. Through the use of the IRS, the phase of the incident signals can be altered which further helps in interference mitigation. A response EM field [64] is generated when the EM signal is incident on the antenna surface. However, these phase responses can be altered based on perception-action system [65]. Generally, an IRS consists of three layers as depicted in Figure 12.5. The aforementioned layers can be described as shown next [50],

- Outer layer: To alter the incident signals, this layer comprises of a vast array of passive reconfigurable patches etched on a dielectric substrate.
- Intermediate layer: This layer is made up of a copper plate that lowers the signal energy leakages while reflection takes place.
- Inner layer: Finally, this layer comprises of a control circuit board capable of controlling and steering the phase and amplitude of the reflected signal instantaneously. The smart controller configures the direction of the reflected signal.

12.4.2 Salient Features

It is known that signal propagation following LoS encounters a lot of obstacles which lowers the signal strength of the propagating signal with increase in distance from

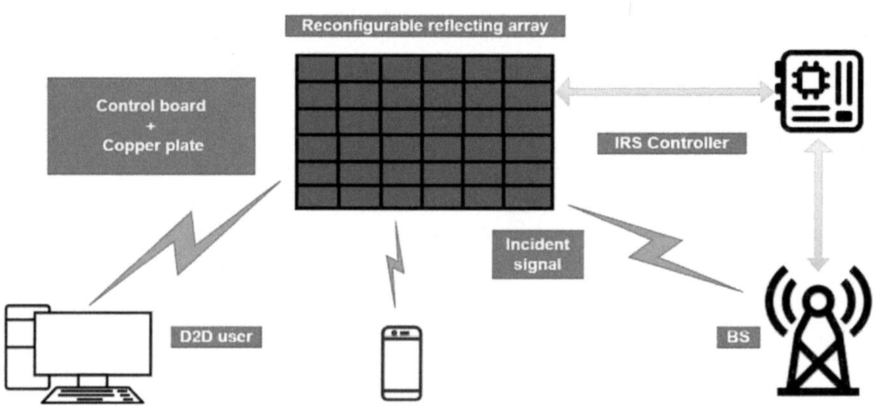

FIGURE 12.5 Layers of IRS.

the transmitter or BS [66, 67]. Thus, the development of an innovative method is necessary which helps in maintaining the signal strength of the transmitted signal and also increases the throughput of the gross network. Keeping in mind this point, the IRS can be implemented for relaying the signals from the transmitter (BS) to the receiving end users. Its main job is to append a number of programmable paths to an uncontrollable transmission channel, allowing the user to cater advantageous channel parameters to mitigate pathloss issues. However, the number of programmable paths is equal to the count of reflecting elements. It is adequate to provide passive beamforming, in which an array gain is received due to the temporal delay applied to distinct reflecting elements, resulting in constructive interference [68]. Furthermore, the array gain is set so that the received signal's signal-to-noise ratio (SNR) is directly related to the square of the total count of reflecting elements in the IRS [69]. The incident signal from the BS on the surface of the IRS follows the principle of Snell's law which suggests that the incident ray, refracted ray, and the normal lie on the same plane and the ratio of the sine of the angle of incidence and angle of refraction remains constant for a given medium.

The IRS may be implemented in providing seamless communication between two devices in mm-Wave 5G/B5G network in two ways. The IRS may make use of active or passive reflecting elements [70]. Generally, passive reflecting surfaces are preferred over active counterparts because the implementation is simple and moreover, the energy efficiency is higher when compared to active reflecting elements, albeit the higher pathloss they incur makes them unsuitable for many systems. Furthermore, active reflecting elements eliminate the consequences of multipath fading which is experienced by the passive IRS systems by frequently incorporating amplifying features such as negative resistance components [71] which permits the amplification factor to be greater than one. However, passive IRS systems obtain better weighted sum rates than active systems depending on the configurations. The cases when the amplification factor is relatively low or an extensively greater number of reflecting components are used in the IRS [72], yield better performance. In larger use situations, active IRS-aided systems outperform passive IRS systems with the same power consumption [71].

Again, a different kind of IRS can be employed for communication known as partially active IRS. They employ a single amplifying component for numerous IRS parts while still allowing for independent phase shifts. As a result of which, the system experiences a minor loss in performance in a trade-off for better power gains [73]. It is also found that for calculating channel estimation, some passive IRS systems also utilize active RF elements [74].

While comparing IRS with the other existing technologies, IRS surely poses an advantage in providing seamless communication between two devices in a mm-Wave 5G/B5G network. The IRS is compared to existing technologies such as Amplify-and-forward (AF), Decode-and-forward (DF), Massive multiple-input, multiple-output (mMIMO), and backscatter communication.

- mMIMO: The IRS possesses passive or active array configuration operating through reflect or transmit method. This makes the IRS unique in comparison to the active intelligent surface-based massive MIMO [75].

- AF relay: It is known that an AF relay amplifies and regenerates the signals during transmission, but in an IRS, the incident signals are reflected as a passive array in the absence of a transmitter. Thus, IRS gets rid of the requirement for transmit power usage. Supposedly, an IRS operates in full duplex mode which infers that it is spectrum efficient, whereas AF operates in half duplex mode. This results from the factor that AF encounters strong interference in full duplex mode, necessitating optimal interference mitigation techniques. This makes IRS a better candidate than AF relay.
- DF relay: DF relaying, like AF relaying, follows the same procedure which involves decoding and its subsequent regeneration of the transmitted signal from the source before transmitting it to the receiver end. It has a much larger complexity and uses a lot of signal processing resources because of the decoding procedure [76]. The IRS, on the other hand, undertakes no decoding. It simply performs passive reflection. As a result, it is less expensive and uses very little power.
- Backscatter communication: In the IRS system, it increases the performance of the existing communication link by simply reflecting or relaying the signal (without any decoding). However, in backscatter transmission, the path from transmitter to user end gets affected by unwanted interference and must be mitigated or eliminated at the end user. In IRS assisted communication, however, the same important information is relayed through both the direct and reflect-path signals and then they are constructively combined at the end user to enhance the total received power [77, 78].

12.5 APPLICATION OF IRS TO D2D COMMUNICATION: CHALLENGES AND SOLUTIONS

As mentioned in the preceding sections, D2D communication in the mm-Wave frequency faces a number of problems. Three specific realistic scenarios suppress the benefits of D2D communication namely, a) the space between the transmitter and the receiver end is large, as a result of which the signal strength lowers; b) weak D2D link between communicating end users; and c) the communication between the devices in the underlaying channel gets affected by severe interference from the adjacent users in the cell. Also, the communication between the D2D users takes place in the mm-Wave frequencies which cater to more obstacles from the surroundings. However, IRS poses to be a viable candidate in mitigating the interference by allowing the communication to take place LoS transmission of signals. Although, there are certain factors which tend to lower the performance of the D2D communication in the mm-Wave band. Thus, this section showcases a review of the problems faced in implementing D2D communication in the mm-Wave network through the application of IRS technology such as, resource allocation, interference management, and power management for seamless D2D communication and also their subsequent solutions thereby satisfying the required QoS of the network.

12.5.1 Resource Allocation

To counter the various issues of interference in enabling D2D communication in the mm-Wave frequency by the use of IRS technology, optimal resource management is required. Researchers throughout the world have employed various techniques for optimal allocation of resources to mitigate interference. Here, we investigate some of the techniques for effective resource management.

In [79], the authors employed joint processing coordinated multipoint cellular networks assisted by the IRS for an underlaying D2D communication using a downlink channel. The authors tend to enhance the overall throughput by collective designing of the cellular user association, the active and passive beamforming at the BS and IRS, respectively; and the transmit power of every single D2D transmitter. The scheme also employs big-M framework, penalty scheme, and successive convex approximation, guaranteeing to converge to a Karush-Kuhn-Tucker (KKT) solution. The authors in [80] allocate the resources through employing block coordinate descent algorithm in addition with semi-definite relaxation technique. It helps in managing the beamforming vector, power optimization, and phase shift matrix. As a result of the optimization technique used, the overall throughput of the network gets enhanced. Later in [81], the authors studied the framework for beamforming and resource management on mixed timescales. Here, an algorithm known as mixed-integer stochastic successive convex approximation (MISSCA) is employed which helps in minimizing the system latency, in addition to the heavy overhead in trade-off with CSI.

Next in [82], the authors implemented a multivariable resource optimization function for mitigating the consequences of interference on the IRS-aided D2D communication network. The article proposed game theory based hybrid sub-optimal algorithms for solving the issues of resource allocation, power optimization, and enhancing throughput of the network. M. Forouzanmehr et al. in [83] designed a resource allocation framework which helps in alleviating interference between the cellular along with the D2D users and also enhances the system throughput. The article also employed Fenchel's duality approach for designing the passive beamforming matrices for enhancing the signal strength at IRS. Again, it is known that blockages tend to lower the signal strength of the transmitting signal from the BS or access points resulting in losing the LoS path of transmission. Thus in [84], the authors implemented IRS to boost off the signal strength for a seamless communication. Rotational mirrors are also used to increase the signal strength along with the IRS in an indoor network. The application of the IRS in uplifting the signal strength and providing a better way of communication among the devices is not only limited to D2D communication. In [85], the authors implemented the IRS technology in drone-terrestrial networks for providing LoS links between the drones and the devices at the surface. To this end, the authors allocated the resources for the IRS assisted drone-terrestrial networks by implementing the blockchain and AI enabled configurable reflection mechanism. Furthermore, IRS can effectively provide better network coverage to the users present in blockages and at the cell edges for exchange of information between the devices. Thus in [86], L. Ferdouse et al. implemented a reinforcement learning-based policy optimization scheme for optimal resource allocation in the downlink channel. This scheme boosts off enhancing the energy efficiency and the overall data rate of the network. Next in [87], the authors employed both IRS passive reflection and active relaying for allocation of resources used for exchange of valuable information between the devices. For active relaying, they have implemented DF relays. The timings of these relays are controlled by the IRS controller. Through this scheme, the authors tried to maximize the network coverage and thereby enhance the data rate. The article in [88] modelled a multi-agent Markov decision problem for resource management to solve the issues of implementing IRS in three different use cases namely, ideal IRS, non-ideal IRS with continuous, as well as discreet phase shifts. The article also solved the problem of power allocation dynamically.

Later in [89], the authors demonstrated a scheme wherein a BS exchanges information with two users through the help of energy harvesting relay network in the presence of IRS employing a power-splitting architecture. Particle swarm optimization (PSO) algorithm is utilized in order to enhance the EE and the overall throughput of the network abiding by individual power constraints for each of the users, thereby enhancing the performance of the network. M. Hua et al. [90] proposed a dynamic IRS framework wherein the IRS phase-shift vectors change instantly for both wireless energy as well as information transmission. A penalty-based algorithm is employed for allocating the resources and in the process enhancing the data rate of the network. In [91], the authors utilized a hybrid deep reinforcement learning technique. Here, the overall data rate is improved by employing intelligent resource allocation of subcarrier, beamforming, and phase shift of the IRS. Interference is mitigated by employing dueling double deep Q networks (D3QN) along with twin delayed deep deterministic policy gradient (TD3). A joint optimization framework [92] is formulated for optimal resource allocation of the IRS assisted non-orthogonal multiple access (NOMA) system. Many-to-one matching algorithm is employed for channel assignment. The IRS aided system outperforms the conventional system in terms of system throughput and performance gains as depicted by their simulation findings. Now in [93], the authors investigated a multi-user system based on IRS for finding the fundamental capacity limits. It deals in allocation of resources employing Lagrange duality method. The simulation results demonstrated the enhancement of the rate region and capacity gain by dynamically adjusting the IRS reflection matrix. Next in [94], the authors designed a framework for IRS assisted coordinated multi-point (CoMP) system employing genetic algorithm (GA). The work tried to enhance the EE by jointly calibrating BS clustering, user association, channel assignment, power management, and optimal configuration of the IRS, while abiding by the user QoS. Finally in [95], a minorization-maximization (MM)-based algorithm is implemented for optimal resource allocation in an IRS based system thereby increasing the achievable sum finite block-length rate. This scheme showcased lower complexity in comparison to existing schemes. Thus, in this section detailed challenges along with their respective solutions for implementing IRS in the existing wireless communication system are demonstrated as shown in table 12.1. However, designing and controlling of the IRS needs immediate attention for enabling it into the Next-Gen wireless networks thereby providing seamless communication between devices.

TABLE 12.1
Summary of Related Works for Resource Management

Year	Objective	Technique	Performance
2022 [79]	Transmit power optimization, active and passive beamforming	Big-M formulation, penalty method, and successive convex approximation	Enhanced overall data rate
2021 [81]	Beamforming and resource management on mixed timescales	Mixed-integer stochastic successive convex approximation	System latency is reduced
2022 [82]	Multivariable resource optimization for interference mitigation	Game theory based hybrid sub-optimal algorithms	Enhanced throughput of the network
2022 [85]	Facilitating LoS links between the drones and the devices at the surface	Blockchain technology	Better network coverage to the users present in blockages and at the cell edges
2023 [86]	Optimal resource allocation in downlink channel	RL-based policy optimization scheme	Enhanced the EE and the overall capacity of the network
2021 [87]	Joint optimization of time allocations and passive beamforming to maximize the user's achievable rate	DF relaying	Maximized user achievable rate and IRS network coverage
2022 [88]	Dynamic power allocation	Multi-agent deep reinforcement learning	Enhanced network sum rates
2023 [91]	Intelligent resource allocation of subcarrier, beamforming and phase shift of the IRS	Hybrid deep reinforcement learning technique	Improved overall data rate
2020 [92]	Optimal resource allocation	Many-to-one matching algorithm	Enhanced system throughput and performance gains
2021 [93]	Finding the fundamental capacity limits	Lagrange duality method	Enhancement of the rate region and capacity gain
2023 [95]	Optimal resource allocation	MM-based algorithm	Increased achievable sum finite block-length rate

12.5.2 Interference Management

Next, interference management plays a pivotal role in enabling seamless communication between the D2D devices in mm-Wave network aided by IRS technology. Several studies have been conducted to improve the D2D network's performance metrics through interference management. Thus, this section investigates the various strategies used to mitigate or eradicate the interference affecting the users, thereby enhancing the performance of the network.

In [96], the authors employed rate-splitting multiple access (RSMA) approach for dead-zone users in the presence of IRS. The problem was divided into multiple sub-problems which were then unified by implementing block-coordinate descent (BCD) algorithm. The simulation results depicted the superiority over other existing schemes in terms of achievable data rate. Here in the RSMA technique [66], the message which is to be transmitted is sub-divided into multiple messages (common and private). Later, the messages are encoded and then transmitted to the receiver end users. This technique helps in enhancing the capacity of the gross communication system. Next in [97], the authors derived the closed form expressions of outage probability, cumulative distribution function (CDF), and probability density function (PDF) of the network wherein both NOMA and IRS assisted communications exist in a B5G wireless network. Again, human blockage also poses to be a leading factor contributing to interference which affects the D2D communication and subsequently degrades the signal strength. This issue has been discussed in [98]. Here, the authors designed an UAV network with IRS placed on top of it to provide LoS links for D2D transmission. This work explored on the effect of height of the UAV on the performance of the signal transmission. B. Han et al. [99] employed genetic algorithm for the minimization of interference in an indoor scenario for mm-Wave D2D network. The co-channel interference was reduced through implementation of IRS aided by anti-phase signal reflection. Again in [100], the authors analysed the performance of a comprehensive framework in a cellular network aided by IRS and subsequently derived the closed form expressions of the coverage probability, ergodic capacity, and EE by employing stochastic geometry approach. Also, the aggregate interference of the IRS and BS is analysed.

The authors in [101] implemented deep reinforcement learning (DRL) approach in a D2D communication assisted by IRS for eradication of interference. The transmit power and phase shift matrix at the IRS were optimized which helps in maximizing the gross data rate of the network. The work also employed Markov decision process to formulate the proximal policy optimization in order to work out the maximization game. Later in [102], the authors designed a mm-Wave D2D network aided by IRS. An IRS activation and selection problem was also formulated which was used to maximize the data rate of the network. This work mainly dealt in minimization of interference due to blockage. In [103], the authors employed precoder design based on block diagonalization technique for reduction of interference in a MIMO network aided by IRS in the presence of multiple users. The interference rendered by the IRS was mitigated in this framework. Next in [104], the authors designed a framework for D2D communication in B5G network assisted by IRS employing reinforcement learning (RL) approach for maximizing the SE of the network.

The IRS reflection coefficients are also optimized for enhancing the performance of the network structured by Markov Decision Process (MDP). Furthermore, in [105], the authors developed a framework where multiple IRSs are deployed in an underlaying D2D network. Block coordinate descent (BCD) framework was implemented for enhancing the sum-rate of the D2D users. Finally in [106], the authors modelled a scheme where the D2D network is facilitated by IRS aided UAV systems. The implementation of reinforcement algorithm solved the passive beamforming optimization problem. A detailed summary of all the interference management techniques is shown in table 12.2.

12.5.3 Power Management

This sub-section describes the impact of power management on the performance of the D2D communication aided by IRS technology in 5G/B5G networks. Power

TABLE 12.2
Summary of Related Works for Interference Management

Year	Objective	Technique	Performance
2023 [96]	Enhance performance for dead-zone users	RSMA approach	Enhanced achievable data rate
2023 [97]	Integration of NOMA and IRS assisted communication	Stochastic geometry	Increased outage probability, CDF, and PDF
2023 [98]	Provide LoS link and analyse the effect of human blockages	Geometrical-Empirical human bodies blockage model	Probability of blockage and system SE
2022 [99]	Minimization of interference for indoor scenario	Genetic algorithm	Data rate and blocking probability
2022 [100]	Performance of a comprehensive framework in a cellular network aided by IRS	Stochastic geometry	Coverage probability, ergodic capacity, and EE
2021 [101]	Optimize transmit power and phase shift matrix	DRL approach	Maximized overall data rate
2022 [102]	Minimization of interference due to blockage	IRS activation and selection problem	Maximized the data rate of the network
2023 [103]	Reduction of interference in a MIMO network	Block diagonalization technique	Improved system capacity and higher SE
2022 [104]	Optimization of IRS reflection coefficients	RL and MDP approach	Maximized the SE of the network
2021 [105]	Power allocation and receive beamforming	BCD technique	Sum-rate enhancement of the D2D users
2023 [106]	Passive beamforming optimization problem	Reinforcement algorithm	Lower latency and greater robustness

optimization is necessary to control the transmit power from the BS as well as the D2D users. Thus in this sub-section, the issues and their corresponding solutions of the power management are presented in a detailed manner.

In [109], the authors modelled an IRS aided D2D communication system in the presence of imperfect CSI. The work optimized the transmit beamforming as well as phase shifts of the IRS by employing S-Procedure to enhance the SE and EE of the overall network. The closed-form expressions for the optimized number of reflecting elements for bound analysis are derived in [107]. Here, more focus is given to the configuration of the number of reflecting elements in the IRS considering power consumption constraints. In [108], the authors proposed a framework where the IRS is implemented to extend network coverage to the blind spots. To enhance the overall throughput of the network, power is optimized. The sequential fractional programming (SFP) method is implemented for optimizing the passive beamforming at IRS, and power optimization is carried on by the use of a standard convex optimization method. This scheme showcases better data rate in comparison to the other existing AF relay methods. The introduction of IRS to the D2D communication at a 5G mm-Wave network increases the interference due to the addition of new links. Thus, a particle swarm optimization (PSO) algorithm is employed in [109] to optimize the transmit power and the phase shift of the IRS. The simulation results suggest that the data rate gets enhanced, whereas the transmit power is reduced considerably. Next in [110], the authors, similar to the previous work, also jointly optimized the transmit power and the phase shift of the IRS. The authors employed alternating optimization and successive convex approximation techniques for enhancing the throughput of the users. It is seen that co-channel interference has large scale effect on the D2D users. Thus, the authors in [111] proposed an RIS assisted D2D communication employing two-timescale optimization scheme. This scheme adapts power optimization thereby increasing the D2D ergodic weighted sum-rate for a certain predefined outage probability. Again, the authors in [112] designed an IRS based framework for meeting the requirements of vehicular communication. The scheme optimized the IRS reflection coefficients, power, and spectrum allocation increasing the vehicle-to-infrastructure (V2I) link capacity. Again in [113], the authors adopted a RIS based D2D communication scenario for maximizing the sum rate of the network and minimizing the mutual interference. This scheme performs joint optimization of transmit power and phase shifts of all RIS links. The problem of power optimization is solved by the application of the gradient descent method. Z. Yang et al. [114] combined the IRS, backscatter, and wireless powered technologies for the advancement of ultra-low-power applications. The work proposed a self-sufficient scenario for the D2D communication to take place. Then, the closed form expressions of coverage probability and average data rate are obtained. Finally, the work in [115] demonstrated a communication system for IoT and IRS assisted UAV. Power optimization and passive beamforming are done for enhancing the EE of the network. Dinkelbach algorithm, SCA, and BCD methods are implemented alternately for optimizing the parameters. This work significantly reduces the energy consumption of the network. A detailed summary of all the power management techniques is shown in table 12.3.

TABLE 12.3
Summary of Related Works for Power Management

Year	Objective	Technique	Performance
2022	Optimization of transmit beam-forming and phase shifts of the IRS	S-Procedure	Enhanced SE and EE
2022 [107]	Minimize number of reflecting elements in the IRS	Bound analysis	Optimal number configuration
2022 [108]	Extend network coverage to the blind spots	SFP method	Enhanced overall throughput
2023 [109]	Optimize transmit power and phase shift of the IRS	Particle swarm optimization	Enhanced data rate and reduced transmit power
2020 [110]	Jointly optimize transmit power and phase shift of the IRS	Alternating optimization and successive convex approximation	Enhanced the throughput of the users
2020 [111]	Power optimization	Two-timescale optimization scheme	D2D ergodic weighted sum-rate
2020 [112]	Optimize the IRS reflection coefficients, power and spectrum allocation	Alternating optimization algorithm	Sum V2I link capacity
2021 [113]	Joint optimization of transmit power and phase shifts of all RIS links	Gradient descent method	Increased sum rate and lower mutual interference
2021 [114]	Self-sufficient scenario for the D2D communication	Stochastic geometry	Enhanced coverage probability and average data rate
2022 [115]	Power optimization and passive beamforming	Dinkelbach algorithm, SCA, and BCD methods	Enhanced the EE of the network

12.6 IMPLEMENTATION OF IRS

The IRS poses to be a viable candidate for the enabling of D2D communication in the 5G/B5G network. Theoretically, D2D communication can harness the advantages of the IRS technology for uplifting the data rate and also increasing the network coverage. But in the practical scenario, there are some issues which hinder the successful deployment of the IRS technology in the 5G network. These issues are furnished as follows,

- Large IRS: Since the transmitted signals from the BS are incident on the IRS surface consisting of an array of antenna, a large surface area of the IRS is required to cater to a significant large area. As most of the existing works deals in higher frequency of the transmitting signal, with an increase in distance of the D2D user from the BS, the signal strength decreases significantly. Thus, this issue needs to be worked upon.

- Channel state information: In a realistic scenario, the CSI of the network is unknown for various reasons. The configuration of the IRS phase-shift matrix is dependent on the CSI of the IRS based network. Also, multiple interference affects the reflected signals from the IRS antenna array.
- Higher complexity: Also, the signals incident on the IRS surface from the BS get reflected by an array of antennas which tend to interfere among themselves. This will increase the complexity of the network. This also proves to be a hindrance in the successful implementation of the IRS technology at a D2D communication in the 5G mm-Wave network.

12.7 POTENTIAL RESEARCH DIRECTIONS

Although this chapter demonstrated the implementation of IRS for D2D communication in a 5G mm-Wave network, it is not restricted to a mm-Wave network only. There are other verticals where the IRS technology can be used for enhancing the performance of the network such as visible light communication (VLC), terahertz communication (THz), UAV communication, M2M communication, etc. Thus, the IRS technology has a lot of potential for implementation at Next-Gen networks thereby enhancing the performance of the respective networks. Some of the potential research directions of IRS technology are as follows,

- Phase shift controller: The IRS mounted on the walls of a building or over a drone may experience time varying channel conditions which will surely hamper the transmission of data between the devices. It will also get affected by the fading and shadowing. Thus, a phase shift controller is needed to mitigate the effects of fading, thereby improving the channel conditions of the network.
- Interference dynamics: The IRS consists of an array of antennas which reflect the impinging signal to the desired device in the mm-Wave network. Thus, interference due to various factors as described in the earlier sections may prove to be a hindrance to the seamless communication to take place. Thus, optimization techniques are required to minimize the effects of interference. Also, the dynamics of the interference need to be considered for a practical network scenario.
- Channel modelling: As the transmitted signal undergoes severe signal degradation due to the effect of the interference of the mm-Wave network, a competent channel model is required for the propagation of the transmitted signals which will strengthen the required SNR of the received signal at the end user.

12.8 CONCLUSION

The IRS technology provides a lot of advantages which makes it a front runner in the implementation of D2D communication at the mm-Wave 5G network. It helps in improving the coverage capacity and data rate, which would be essential for providing seamless communication among devices in the 5G network. Bandwidth intensive applications such as online gaming, holographic telepresence, and so on can make use of the huge data rate and network coverage provided by the IRS technology. However, there are various issues which hinder the incorporation of IRS technology into a 5G network, including co-channel interference, blockages, severe pathloss attenuation, channel impairment, etc. These issues can be eradicated or minimized through various methods including optimal resource allocation, power optimization, mode selection, and interference mitigation. Thus, this chapter broadly demonstrated and critically evaluated the various techniques used to enable IRS technology to D2D communication in 5G mm-Wave network. This chapter also showcased the basics of the D2D communication, mm-Wave network, and the IRS technology. The salient features of IRS technology are also discussed. Furthermore, this chapter also explored the realistic issues of deploying IRS in a 5G network.

REFERENCES

[1] Tang, F., Kawamoto, Y., Kato, N., and Liu, J. Future intelligent and secure vehicular network toward 6G: Machine-learning approaches. *Proceedings of the IEEE,* 2020, 108, 292–307.

[2] Al-Fuqaha, A., Guizani, M., Mohammadi, M., Aledhari, M., and Ayyash, M. Internet of things: A survey on enabling technologies, protocols, and applications. *IEEE Communications Surveys & Tutorials,* 2015, 17, 2347–2376.

[3] Chen, M., Challita, U., Saad, W., Yin, C., and Debbah, M. Artificial neural networks-based machine learning for wireless networks: A tutorial. *IEEE Communications Surveys & Tutorials,* 2019, 21, 3039–3071.

[4] He, L., Liu, K., He, Z., and Cao, L. Three-dimensional holographic communication system for the metaverse. *Optics Communications*, 2023, 526, 128894.

[5] Ahangar, M.N., Ahmed, Q.Z., Khan, F.A., and Hafeez, M. A survey of autonomous vehicles: Enabling communication technologies and challenges. *Sensors,* 2021, 21, 706.

[6] Islam, S.M.R., Kwak, D., Kabir, M.H., Hossain, M., and Kwak, K.S. The internet of things for health care: A comprehensive survey. *IEEE Access,* 2015, 3, 678–708.

[7] Jeyaseelan, J., Sriram Kumar, D., and Caroline, B.E. Disaster management using free space optical communication system. *Photonic Network Communications*, 2020, 39, 1–14.

[8] Zanella, A., Bui, N., Castellani, A., Vangelista, L., and Zorzi, M. Internet of things for smart cities. *IEEE Internet of Things Journal,* 2014, 1, 22–32.

[9] Ahmed, Q.Z., Hafeez, M., Khan, F.A., and Lazaridis, P. Toward beyond 5G future wireless networks with focus toward indoor localization. In *Proceedings of the 2020 IEEE Eighth International Conference on Communications and Networking (ComNet)*, IEEE, Hammamet, Tunisia, 27–30 October 2020, pp. 1–5.

[10] Cisco, U. *Cisco Annual Internet Report (2018–2023) White Paper*. 2020. Available online: www.cisco.com/c/en/us/solutions/collateral/executive-perspectives/annual-internet-report/white-paper-c11-741490.html (accessed on 18 October 2023).

[11] Minoli, D., and Occhiogrosso, B. Practical aspects for the integration of 5G networks and IoT applications in smart cities environments. *Wireless Communications and Mobile Computing*, 2019, Article ID 5710834, 30 pages.

[12] Huang, C., Mo, R., and Yuen, C. Reconfigurable intelligent surface assisted multiuser MISO systems exploiting deep reinforcement learning. *IEEE Journal on Selected Areas in Communications,* 2020, 38, 1839–1850.

[13] Qamar, F., Hindia, M.H.D., Dimyati, K., Noordin, K.A., Majed, M.B., Abd Rahman, T., and Amiri, I.S. Investigation of future 5G-IoT millimeter-wave network performance at 38 GHz for urban microcell outdoor environment. *Electronics*, 2019, 8(5), 495.

[14] Venugopal, K., Valenti, M.C., and Heath, R.W. Device-to-device millimeter wave communications: Interference, coverage, rate, and finite topologies. *IEEE Transactions on Wireless Communications*, 2016, 15(9), 6175–6188.

[15] Mitra, R.N., and Agrawal DP. 5G mobile technology: A survey. *ICT Express*, 2015, 1(3), 132–137.

[16] Rappaport, T.S., Sun, S., Mayzus, R., Zhao, H., Azar, Y., Wang, K., Wong, G.N., Schulz, J.K., Samimi, M., and Gutierrez, F. Millimeter wave mobile communications for 5G cellular: It will work! *IEEE Access*, 2013, 1, 335–349.

[17] Ghosh, A., Maeder, A., Baker, M., and Chandramouli, D. 5G evolution: A view on 5G cellular technology beyond 3GPP release 15. *IEEE Access*, 2019, 7, 127639–127651.

[18] Wu, Q., and Zhang, R. Intelligent reflecting surface enhanced wireless network: Joint active and passive beamforming design. In *Proceedings of the IEEE GLOBECOM*, IEEE, December 2018.

[19] Wu, Q., and Zhang, R. Intelligent reflecting surface enhanced wireless network via joint active and passive beamforming. *IEEE Transactions on Wireless Communications*, 2019, 18(11), 5394–5409.

[20] Wu, Q., and Zhang, R. Towards smart and reconfigurable environment: Intelligent reflecting surface aided wireless network. *IEEE Communications Magazine*, 2020, 58(1), 106–112.

[21] Wu, Q., Zhang, S., Zheng, B., You, C., and Zhang, R. Intelligent reflecting surface-aided wireless communications: A tutorial. *IEEE Transactions on Communications*, 2021, 69(5), 3313–3351.

[22] Qiao, J., Shen, X.S., Mark, J.W., Shen, Q., He, Y., and Lei, L. Enabling device-to-device communications in millimeter-wave 5G cellular networks. *IEEE Communications Magazine*, 2015, 53(1), 209–215.

[23] Wei, L., Hu, R.Q., Qian, Y., and Wu, G. Key elements to enable millimeter wave communications for 5G wireless systems. *IEEE Wireless Communications*, December 2014, 21(6), 136–143.

[24] Ju, S., Kanhere, O., Xing, Y., and Rappaport, T.S. A millimeter-wave channel simulator NYUSIM with spatial consistency and human blockage. In *2019 IEEE Global Communications Conference (GLOBECOM)*, IEEE, Hawaii, USA 2019, pp. 1–6.

[25] Rangan, A., Rappaport, T.S., and Erkip, E. Millimeter-wave cellular wire-less networks: Potentials and challenges. *Proceedings of the IEEE*, March 2014, 102(3), 366–385.

[26] FCC Auction 103. Available online: www.fcc.gov/millimeter-wave-708090-ghz-service (accessed on 20 December 2021).

[27] Garg, V.K. Chapter 3—Radio propagation and propagation path-loss models. In *The Morgan Kaufmann Series in Networking*, Wireless Communications Networking. Morgan Kaufmann, Amsterdam: Netherlands 2007, pp. 47–84.

[28] Singh, S., Kulkarni, M.N., Ghosh, A., and Andrews, J.G. Tractable model for rate in self-backhauled millimeter wave cellular networks. *IEEE Journal on Selected Areas in Communications*, 2015, 33(10), 2196–2211.

[29] Turgut, E., and Gursoy, M.C. Uplink performance analysis in D2D-enabled Millimeter Wave cellular networks with clustered users. *IEEE Transactions on Wireless Communications*, 2019, 18(2), 1085–1100.

[30] Deng, N., Haenggi, M., and Sun, Y. (2018b). Millimeter-wave device-to-device networks with heterogeneous antenna arrays. *IEEE Transactions on Communications*, 66(9), 4271–4285.

[31] Giatsoglou, N. et al. D2D-aware device caching in mmWave-cellular networks. *IEEE Journal on Selected Areas in Communications*, 2017, 35(9), 2025–2037.

[32] Baccelli, F. et al. *Stochastic Geometry and Wireless Networks–Part I: Theory*. Delft: Now Publishers, Inc., 2009.

[33] Wu, S., Atat, R., Mastronarde, N., and Liu, L. Improving the coverage and spectral efficiency of millimeter-wave cellular networks using device-to-device relays. *IEEE Transactions on Communications*, 2018, 66(5), 2251–2265.

[34] Wei, N., Lin, X., and Zhang, Z. Optimal relay probing in millimeter-wave cellular systems with device-to-device relaying. *IEEE Transactions on Vehicular Technology*, 2016, 65(12), 10218–10222.

[35] Kar, U.N., and Sanyal, D.K. A critical review of 3GPP standardization of device-to-device communication in cellular networks. *SN Computer Science,* 2020, 1, 37.

[36] Lien, S.Y., Chien, C.C., Liu, G.S.T., Tsai, H.L., Li, R., and Wang, Y.J. Enhanced LTE device-to-device proximity services. *IEEE Communications Magazine,* 2016, 54(12), 174–182.

[37] Mumtaz, S., and Rodriguez, J. Introduction to D2D communication. In Mumtaz, S., and Rodriguez, J. (eds) *Smart Device to Smart Device Communication*. Cham: Springer, 2014.

[38] ETSI 3GPP: Evolved Universal Terrestrial Radio Access (E-UTRA); User Equipment (UE) radio transmission and reception; (Release 14), TS 36.102 V14.5.0.2017.

[39] National Instruments. *3GPP Release 15 Overview.* Available online: https://spectrum.ieee.org/telecom/wireless/3gpp—release – 15 – overview (accessed on 21 December 2021).

[40] *Release 17 Update from SA2*. Available online: https://www.3gpp.org/news –events/2200 – sa2artcle (accessed on 21 December 2021).

[41] *Release 18.* Available online: www.3gpp.org/specifications-technologies/releases/release-18. (accessed on 20 October 2023).

[42] *Evolution of Network Nodes: From IAB to IRS.* Available online: www.3gpp.org/news-events/partner-news/iab-irs (accessed on 21 October 2023).

[43] Phunchongharn, P., Hossain, E., and Kim, D.I. Resource allocation for device-to-device communications under-laying LTE-advanced net-works. *IEEE Wireless Communications*, August 2013, 20(4).

[44] Pei, Y., and Liang, Y. Resource allocation for device-to-device communications overlaying two-way cellular networks. *IEEE Transactions on Wireless Communications,* July 2013, 12(7), 3611–3621.

[45] Yang, M.J. et al. Solving the data overload: Device-to-device bearer control architecture for cellular data offloading. *IEEE Transactions on Vehicular Technology*, March 2013, 8(1), 31–39.

[46] Sarma, S.S., Hazra, R., and Mukherjee, A. Symbiosis between D2D communication and Industrial IoT for Industry 5.0 in 5G mm-Wave cellular network: An interference management approach. *IEEE Transactions on Industrial Informatics*, 2021, 18(8), 5527–5536.

[47] Sarma, S.S., Khuntia, P., and Hazra, R. Power control scheme for device-to-device communication using uplink channel in 5G mm-Wave network. *Transactions on Emerging Telecommunications Technologies*, 2022, 33(6), e4267.

[48] Kar, U.N., and Sanyal, D.K. A critical review of 3GPP standardization of device-to-device communication in cellular networks. *SN Computer Science,* 2020, 1, 37.

[49] Sarma, S.S., and Hazra, R. Interference mitigation methods for D2D communication in 5G network. In *Cognitive Informatics and Soft Computing*. Springer, Berlin: Germany 2020, pp. 521–530.

[50] Okogbaa, F.C., Ahmed, Q.Z., Khan, F.A., Abbas, W.B., Che, F., Zaidi, S.A.R., and Alade, T. Design and application of intelligent reflecting surface (IRS) for beyond 5g wireless networks: A review. *Sensors,* 2022, 22(7), 2436.

[51] Sarma, S.S., and Hazra, R. Intelligent reflecting surface at mm-wave band for D2D communication: An insight. In *Emerging Electronics and Automation: Select Proceedings of E2A 2021*. Singapore: Springer Nature Singapore, 2022, pp. 307–3141.

[52] Basar, E., Di Renzo, M., De Rosny, J., Debbah, M., Alouini, M.S., and Zhang, R. Wireless communications through reconfigurable intelligent surfaces. *IEEE Access*, 2019, 7, 116753–116773.

[53] Zhao, J. A survey of intelligent reflecting surfaces (IRSs): Toward 6G wireless communication networks. *arXiv,* arXiv:1907.04789, 2019.

[54] Di Renzo, M. et al. Reconfigurable intelligent surfaces vs. relaying: Differences, similarities, and performance comparison. *IEEE Open Journal of the Communications Society,* 2020, 1, 798–807.

[55] He, Z.Q., and Yuan, X. Cascaded channel estimation for large intelligent metasurface assisted massive MIMO. *IEEE Wireless Communications Letters,* 2019, 9, 210–214.

[56] Liaskos, C., Tsioliaridou, A., Nie, S., Pitsillides, A., Ioannidis, S., and Akyildiz, I. An interpretable neural network for configuring programmable wireless environments. In *Proceedings of the 2019 IEEE 20th International Workshop on Signal Processing Advances in Wireless Communications (SPAWC)*, Cannes, France, IEEE, 2–5 July 2019, pp. 1–5.

[57] Hu, S., Rusek, F., and Edfors, O. Beyond massive MIMO: The potential of data transmission with large intelligent surfaces. *IEEE Transactions on Signal Processing*, 2018, 66, 2746–2758.

[58] Jung, M., Saad, W., Jang, Y., Kong, G., and Choi, S. Performance analysis of large intelligent surfaces (LISs): Asymptotic data rate and channel hardening effects. *IEEE Transactions on Wireless Communications*, 2020, 19, 2052–2065.

[59] Mishra, D., and Johansson, H. Channel estimation and low-complexity beamforming design for passive intelligent surface assisted MISO wireless energy transfer. In *Proceedings of the ICASSP 2019–2019 IEEE International Conference on Acoustics, Speech and Signal Processing (ICASSP)*, Brighton, UK, IEEE, 12–17 May 2019, pp. 4659–4663.

[60] Tan, X., Sun, Z., Koutsonikolas, D., and Jornet, J.M. Enabling indoor mobile millimeter-wave networks based on smart reflect-arrays. In *Proceedings of the IEEE INFOCOM 2018-IEEE Conference on Computer Communications*, Honolulu, HI, USA, IEEE, 16–19 April 2018, pp. 270–278.

[61] Huang, C., Hu, S., Alexandropoulos, G.C., Zappone, A., Yuen, C., Zhang, R., Di Renzo, M., and Debbah, M. Holographic MIMO surfaces for 6G wireless networks: Opportunities, challenges, and trends. *IEEE Wireless Communications,* 2020, 27, 118–125.

[62] Cui, T.J., Qi, M.Q., Wan, X., Zhao, J., and Cheng, Q. Coding metamaterials, digital metamaterials and programmable metamaterials. *Light: Science & Applications*, 2014, 3(10), e218–e218.

[63] Chen, H.-T., Taylor, A.J., and Yu, N. A review of metasurfaces: Physics and applications. *Reports on Progress in Physics*, 2016, 79(7), 076401.

[64] Liaskos, C., Nie, S., Tsioliaridou, A., Pitsillides, A., Ioannidis, S., and Akyildiz, I. A new wireless communication paradigm through software-controlled metasurfaces. *IEEE Communications Magazine*, 2018, 56(9), 162–169.

[65] Di Renzo, M. et al. Smart radio environments empowered by reconfigurable intelligent surfaces: How it works, state of research, and the road ahead. *IEEE Journal on Selected Areas in Communications*, 2020, 38(11), 2450–2525.

[66] Sarma, S.S., Hazra, R., and Goswami, P. Power optimization in a multi-cell D2D communication for smart city in a mm-Wave cellular network: An mIoT perspective. *IEEE Internet of Things Journal*, 1 November 2023, 10(21), 18686–18694.

[67] Sarma, S.S., Sachan, A., Hazra, R., Talukdar, F.A., Mukherjee, A., Chatterjee, P., and Al-Numay, W. D2D communication in a 5G mm-Wave cellular network for wireless sensor networks. *IEEE Sensors Journal*, 1, 1–19. doi: 10.1109/JSEN.2023.3300586.

[68] Bjornson, E., Ozdogan, O., and Larsson, E.G. Reconfigurable intelligent surfaces: Three myths and two critical questions. *IEEE Communications Magazine,* 2020, 58(12), 90–96.

[69] Wu, Q., and Zhang, R. Towards smart and reconfigurable environment: Intelligent reflecting surface aided wireless network. *IEEE Communications Magazine*, 2019, 58(1), 106–112.

[70] Sharma, A., Bhambri, A., and Huilgol, P.S. Intelligent reflective surfaces and its application across various communication techniques: A review. *Physical Communuication*, 1, 2023 pp. 1–28. doi: 10.36227/techrxiv.23677509.v1.

[71] Long, R., Liang, Y.C., Pei, Y., and Larsson, E G. Active reconfigurable intelligent surface-aided wireless communications. *IEEE Transactions on Wireless Communications*, 2021, 20(8), 4962–4975.

[72] You, C., and Zhang, R. Wireless communication aided by intelligent reflecting surface: Active or passive? *IEEE Wireless Communications Letters*, 10(12), 2659–2663, 2021.

[73] Liu, K., Zhang, Z., Dai, L., Xu, S., and Yang, F. Active reconfigurable intelligent surface: Fully-connected or sub-connected? *IEEE Communications Letters*, 2021, 26(1), 167–171.

[74] Alexandropoulos, G.C., and Vlachos, E. A hardware architecture for reconfigurable intelligent surfaces with minimal active elements for explicit channel estimation. In *ICASSP 2020–2020 IEEE International Conference on Acoustics, Speech and Signal Processing (ICASSP)*, IEEE, Barcelona, Spain, 2020, pp. 9175–9179.

[75] Long, W., Chen, R., Moretti, M., Zhang, W., and Li, J. A promising technology for 6G wireless networks: Intelligent reflecting surface. *Journal of Communications and Information Networks*, 2021, 6, 1–16.

[76] Sarma, S.S., Hazra, R., and Chong, P.H.J. Performance analysis of DF relay-assisted D2D communication in a 5G mmWave network. *Future Internet*, 2022, 14(4), 101.

[77] Chen, Z., Ma, X., Han, C., and Wen, Q. Towards intelligent reflecting surface empowered 6G terahertz communications: A survey. *China Communications,* 2021, 18, 93–119.

[78] Di Renzo, M., Zappone, A., Debbah, M., Alouini, M.S., Yuen, C., De Rosny, J., and Tretyakov, S. Smart radio environments empowered by reconfigurable intelligent surfaces: How it works, state of research, and the road ahead. *IEEE Journal on Selected Areas in Communications*, 2020, 38, 2450–2525.

[79] Wang, W., Yang, L., Meng, A., Zhan, Y., and Ng, D.W.K. Resource allocation for IRS-aided JP-CoMP downlink cellular networks with underlaying D2D communications. *IEEE Transactions on Wireless Communications*, June 2022, 21(6), 4295–4309.

[80] Zhang, C., Chen, W., He, C., and Li, X. Throughput maximization for intelligent reflecting surface-aided device-to-device communications system. *Journal of Communications and Information Networks*, December 2020, 5(4), 403–410

[81] Liu, Y., Hu, Q., Cai, Y., and Juntti, M. Latency minimization in intelligent reflecting surface assisted D2D offloading systems. *IEEE Communications Letters*, September 2021, 25(9), 3046–3050.

[82] Xu, F., Ye, Z., Cao, H., and Hu, Z. Sum-rate optimization for IRS-aided D2D communication underlaying cellular networks. *IEEE Access*, 2022, 10, 48499–48509.

[83] Forouzanmehr, M., Akhlaghi, S., and Khalili, A. Resource allocation for multi-IRS-aided D2D communication underlying cellular networks. *IET Communications*, 2022, 16(13), 1628–1641.

[84] Hamad, A.N., Adnan Qidan, A., El-Gorashi, T E.H., and Elmirghani, J.M.H. Resource allocation in IRS-aided optical wireless communication systems. In *2023 23rd International Conference on Transparent Optical Networks (ICTON),* IEEE, Bucharest, Romania, pp. 1–5, 2023.

[85] Pan, Q., Wu, J., Bashir, A.K., Li, J., Vashisht, S., and Nawaz, R. Blockchain and AI enabled configurable reflection resource allocation for IRS-aided coexisting drone-terrestrial networks. *IEEE Wireless Communications*, December 2022, 29(6), 46–54.

[86] Ferdouse, L., Woungang, I., Anpalagan, A., and Yamamoto, K. A resource allocation policy for downlink communication in distributed IRS aided multiple-input single-output systems. *IEEE Transactions on Communications*, April 2023, 71(4), 2410–2424.

[87] Zheng, B., and Zhang, R. IRS Meets relaying: Joint resource allocation and passive beamforming optimization. *IEEE Wireless Communications Letters*, September 2021, 10(9), 2080–2084.

[88] Chen, J., Guo, L., Jia, J., Shang, J., and Wang, X. Resource allocation for IRS assisted SGF NOMA transmission: A MADRL approach. *IEEE Journal on Selected Areas in Communications*, April 2022, 40(4), 1302–1316.

[89] Thi, H.N., Kieu, T.X., Truong, L.H., and Le Thi, A. Resource allocation for NOMA/IRS network with energy harvesting in presence of Hardware Impairment. In *2023 IEEE 3rd International Conference in Power Engineering Applications (ICPEA)*, IEEE, Putrajaya, Malaysia, 2023, pp. 169–174.

[90] Hua, M., and Wu, Q. Joint dynamic beamforming design and resource allocation for IRS-aided FD-WPCN. In *2021 IEEE Global Communications Conference (GLOBECOM)*, IEEE, Madrid, Spain, 2021, pp. 01–06.

[91] Wu, W., Yang, F., Zhou, F., Wu, Q., and Hu, R.Q. Intelligent resource allocation for IRS-enhanced OFDM communication systems: A hybrid deep reinforcement learning approach. *IEEE Transactions on Wireless Communications*, June 2023, 22(6), 4028–4042.

[92] Zuo, J., Liu, Y., Qin, Z., and Al-Dhahir, N. Resource allocation in intelligent reflecting surface assisted NOMA systems. *IEEE Transactions on Communications*, November 2020, 68(11), 7170–7183.

[93] Mu, X., Liu, Y., Guo, L., Lin, J., and Al-Dhahir, N. Capacity and optimal resource allocation for IRS-assisted multi-user communication systems. *IEEE Transactions on Communications*, June 2021, 69(6), 3771–3786.

[94] Chen, J., Xie, Y., Mu, X., Jia, J., Liu, Y., and Wang, X. Energy efficient resource allocation for IRS assisted CoMP systems. *IEEE Transactions on Wireless Communications*, July 2022, 21(7), 5688–5702.

[95] Li, Z., Shen, H., Xu, W., Zhu, P., and Zhao, C. Resource allocation for IRS-assisted uplink URLLC systems. *IEEE Communications Letters*, June 2023, 27(6), 1540–1544.

[96] Katwe, M., Singh, K., Clerckx, B., and Li, C.-P. Rate splitting multiple access for sum-rate maximization in IRS aided uplink communications. *IEEE Transactions on Wireless Communications*, April 2023, 22(4), 2246–2261.

[97] Selim, M.M., and Tomasin, S. Performance analysis of RIS-assisted downlink NOMA wireless systems under D2D interference. *Digital Signal Processing*, 2023, 104269.

[98] Nor, A.M., Fratu, O., and Halunga, S. The human blockage impact on ARIS assisted D2D communication systems. In *12th International Conference on Modern Circuits and Systems Technologies (MOCAST)*, IEEE, Athens, Greece, 2023, pp. 1–4.

[99] Han, B., Chen, X., Jiao, L., and Guo, D. Interference elimination in IRS-enabled indoor mmw-D2D communication. In *2022 IEEE 24th Int Conf on High Performance Computing & Communications; 8th Int Conf on Data Science & Systems; 20th Int Conf on Smart City; 8th Int Conf on Dependability in Sensor, Cloud & Big Data Systems & Application (HPCC/DSS/SmartCity/DependSys)*, IEEE, Hainan, China, 2022, pp. 1374–1381.

[100] Shafique, T., Tabassum, H., and Hossain, E. Stochastic geometry analysis of IRS-assisted downlink cellular networks. *IEEE Transactions on Communications*, February 2022, 70(2), 1442–1456.

[101] Nguyen, K.K., Masaracchia, A., Yin, C., Nguyen, L.D., Dobre, O.A., and Duong, T.Q. Deep reinforcement learning for intelligent reflecting surface-assisted D2D communications. *arXiv preprint*, arXiv:2108.02892, 2021.

[102] Han, B., Chen, X., Guo, D., and Jiao, L. Intelligent reflecting surface enabled in D2D millimeter wave communication. In *2022 IEEE International Conference on Smart Internet of Things (SmartIoT)*, IEEE, Suzhou, China, 2022, pp. 86–93.

[103] Feitosa, W.D.O., Guerreiro, I.M., Cavalcanti, F., Rodrigo, P., Maciel, T.F., Lobão, M.C.R., Makki, B., and Fodor, G. Interference mitigation with block diagonalization for IRS-aided MU-MIMO communications. *arXiv preprint*, arXiv:2306.17307, 2023.

[104] Sultana, A., and Fernando, X. Intelligent reflecting surface-aided device-to-device communication: A deep reinforcement learning approach. *Future Internet*, 2022, 14(9), 256.

[105] Cao, Y., Lv, T., Ni, W., and Lin, Z. Sum-rate maximization for multi-reconfigurable intelligent surface-assisted device-to-device communications. *IEEE Transactions on Communications*, 2021, 69(11), 7283–7296.

[106] Hou, Z. et al. Joint IRS selection and passive beamforming in multiple IRS-UAV-enhanced anti-jamming D2D communication networks. *IEEE Internet of Things Journal*, 2023, 10(22), 19558–19569.

[107] Li, D. Bound analysis of number configuration for reflecting elements in IRS-assisted D2D communications. *IEEE Wireless Communications Letters*, October 2022, 11(10), 2220–2224.

[108] Zhao, D., Lu, H., Wang, Y., Sun, H., and Gui, Y. Joint power allocation and user association optimization for IRS-assisted mmWave systems. *IEEE Transactions on Wireless Communications*, January 2022, 21(1), 577–590.

[109] Wang, R., Wen, X., Xu, F., Ye, Z., Cao, H., Hu, Z., and Yuan, X. Joint particle swarm optimization of power and phase shift for IRS-aided D2D underlaying cellular systems. *Sensors,* 2023, 23(11), 5266.

[110] Guan, X., Wu, Q., and Zhang, R. Joint power control and passive beamforming in IRS-assisted spectrum sharing. *IEEE Communications Letters*, July 2020, 24(7), 1553–1557.

[111] Cai, C., Yang, H., Yuan, X., Zhang, Y.-J.A., and Liu, Y. Reconfigurable intelligent surface assisted D2D underlay communications: A two-timescale optimization design. *Journal of Communications and Information Networks*, December 2020, 5(4), 369–380.

[112] Chen, Y., Wang, Y., Zhang, J., and Li, Z. Resource allocation for intelligent reflecting surface aided vehicular communications. *IEEE Transactions on Vehicular Technology*, 2020, 69(10), 12321–12326.

[113] Chen, Y. et al. Reconfigurable intelligent surface assisted device-to-device communications. *IEEE Transactions on Wireless Communications*, 2021, 20(5), 2792–2804.

[114] Yang, Z., Feng, L., Zhou, F., Qiu, X., and Li, W. Analytical performance analysis of intelligent reflecting surface aided ambient backscatter communication network. *IEEE Wireless Communications Letters*, December 2021, 10(12), 2732–2736.

[115] Ji, Z., Tu, J., Guan, X., Yang, W., Yang, W., and Wu, Q. Energy efficient design in IRS-assisted UAV data collection system under malicious jamming. *arXiv preprint*, arXiv:2208.14751, 2022.

13 Machine Learning for Advanced Wireless Communication

Applications, Challenges, Problems, and Open Research Questions

Wasswa Shafik

ABBREVIATIONS

5G Fifth generation
AGC Automatic gain control
AI Artificial intelligence
ANNs Artificial neural networks
AR Augmented reality
CNNs Convolutional neural networks
CSNNs Compressed sensing neural networks
DL Deep Learning
IoT Internet of Things
LSTM Long short-term memory
MIMO Multiple Input Multiple Output
ML Machine Learning
PCA Principal component analysis
RNNs Recurrent neural networks
RSS Received signal strength
SDN Software-defined networking
SNR Signal-to-noise ratio
SVMs Support vector machines
TOA Time-of-arrival
VR Virtual reality

DOI: 10.1201/9781032656021-13

13.1 INTRODUCTION

The rapid adoption of wireless communication technologies has completely changed how we communicate, share information, and conduct business. The demand for higher data rates, ultra-reliable connections, and low-latency communications has increased to previously unheard-of levels with the arrival of 5G and beyond (Sonia & Singh, 2022). In addition, as the number of connected devices, Internet of Things (IoT) applications, and wireless service demand grow, current wireless networks are under much stress. Researchers and industry professionals have considered ML a promising solution to address these issues and realize the full potential of wireless communication systems (Nwachukwu et al., 2022; Shafik, 2023).

A subset of artificial intelligence (AI) called ML allows computers to infer patterns from data and make deft decisions without explicit programming. It provides a new paradigm for wireless communication that enables communication systems to adapt dynamically and improve their performance in response to user demands and changing environmental conditions (Bindle et al., 2022). ML's transformative potential has sparked much interest in the various facets of advanced wireless communication: Investigate cross-domain approaches integrating ML in wireless communication and networking realms, offering insights into achieving more efficient and intelligent wireless communication systems; and identify existing challenges and open problems throughout the literature review, calling for further research to address these limitations. The ability to manage communication requirements that are both diverse and increasingly complex defines advanced wireless communication systems (Anil Kumar et al., 2022). These systems must support a wide range of applications, including, among others, autonomous vehicles, smart cities, augmented reality, virtual reality, industrial automation, and streaming ultra-high-definition video (Deepika et al., 2022). Wireless networks must function effectively and quickly adapt to changing circumstances, such as channel conditions, traffic loads, and diverse user behaviors.

Modern communication demands are dynamic and unpredictable, and traditional wireless communication methods that rely on predetermined algorithms and fixed parameters find it challenging to meet these demands. Conventional methods frequently call for manual adjustments and recurrent updates, which result in inefficiencies and subpar performance (Zhang et al., 2022). This limitation is especially noticeable when wireless environments are changing quickly, and traditional approaches might not be able to provide the desired level of service. Advanced wireless communication systems present several challenges, and ML emerges as a game-changing solution (Shafik, Matinkhah, Afolabi, et al., 2021). Wireless networks can adapt, optimize, and self-configure with the help of ML, which gives them the ability to do so and thus realize their full potential. ML algorithms can continuously learn from past and in-the-moment observations, improving system performance by making wise decisions and predictions in contrast to predefined rules (Shafik, Matinkhah, Shokoor, et al., 2021).

The ability of ML to handle the enormous amounts of data produced by connected devices and users is one of the technology's key advantages in wireless communication. Wireless networks are being deluged with massive data streams due to the expansion of IoT devices and the spread of data-centric applications (Jun et al., 2021). ML algorithms are superior at sifting through and making sense of this flood of data, while traditional methods struggle. Moreover, ML models can adjust and advance as they come across fresh data and scenarios (Scherer et al., 2022). This adaptability is invaluable in wireless communication, where network conditions can quickly change due to user mobility, interference, and environmental variations. Adaptive algorithms based on ML can continuously learn and change, resulting in improved and optimized network operations (Zenasni et al., 2022).

A further promising development is incorporating DL methods into wireless communication systems. Deep neural networks can take on complex tasks that were previously difficult for conventional methods because they can automatically learn hierarchical representations from raw data (Sivasundari & Ananthajothi, 2022). DL has already shown that it can completely change the way wireless communication works by being used to do things like wireless channel prediction, link adaptation, and massive MIMO beamforming. Despite the significant advancements in applying ML to cutting-edge wireless communication, some problems and research directions still need to be addressed (Kannadhasan et al., 2022). The urgent issues are ensuring ML models' robustness and dependability in real-world situations, addressing mobile devices' constrained power and computational resources, and ensuring privacy and security in ML-driven communication systems (Wu et al., 2022). Figure 13.1 demonstrates the application of ML at different computational layers, including perception, communication, blockchain composite, and at industrial layer.

Interdisciplinary collaborations between experts in ML, wireless communication, signal processing, and networking are becoming increasingly important as the field of ML for advanced wireless communication develops (Tashan et al., 2022). By bridging the gap between these domains, creative solutions that use ML's advantages to get around the limitations of conventional wireless communication techniques may result (Nwachukwu et al., 2022). The combination of ML and advanced wireless communication can ultimately change how we interact with connectivity and data transfer. ML enables the development of truly intelligent and dynamic communication ecosystems by providing wireless networks with intelligence and adaptability (Nwachukwu et al., 2022).

13.1.1 The Contribution of the Chapter

The following are the chapter contributions:

- Overview of the role of Machine Learning in wireless communication, emphasizing its potential benefits and challenges.
- Exploration of Machine Learning techniques for signal processing, including denoising, equalization, and modulation recognition.
- Review of research on channel estimation using ML models to predict channel characteristics.
- Analysis of Machine Learning applications in interference management and spectrum sensing for optimizing spectrum utilization.
- Investigation of Machine Learning techniques for resource allocation and beamforming to enhance network performance.

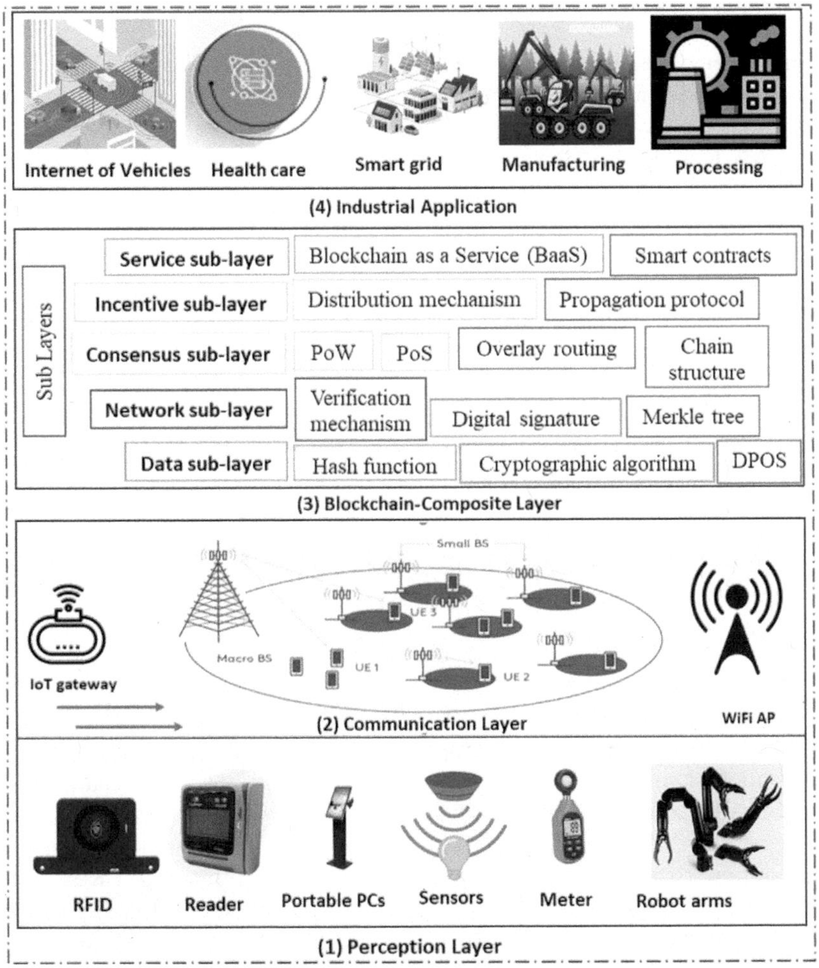

FIGURE 13.1 Application of Machine Learning in Different Computational Layers.

- Examination of Deep Learning and neural network architectures in wireless communication.
- Exploration of cross-domain approaches integrating Machine Learning in wireless communication and networking.
- Identification of challenges and open problems in applying Machine Learning to wireless communication.
- Comparative analysis of different Machine Learning techniques and their suitability for wireless communication tasks.
- Highlighting the significance of Machine Learning in advancing wireless communication technologies.
- Finally, it provides valuable insights for future research in the field by demonstrating open research questions and lessons learned.

13.1.2 The Organization of the Chapter

The remainder of this chapter is structured as follows: Section 13.2. Overview of ML in Wireless Communication provides a brief overview of the role of ML in the field of wireless communication. It highlights the potential benefits and challenges of applying ML techniques in this domain. Section 13.3 presents the ML Techniques for Signal Processing, discussing the different ML algorithms employed and their applications in signal denoising, equalization, and modulation recognition. Section 13.4 demonstrates Applications of ML in Channel Estimation, entailing interference management and spectrum sensing, resource allocation, and beamforming, among others. Section 13.5 illustrates the advantages and challenges of using DL models for complex wireless communication tasks. Section 13.6 presents Cross-Domain Approaches focusing on wireless networking, addressing how cross-domain approaches can lead to more efficient and intelligent wireless communication systems. Section 13.7 illustrates the challenges, problems, and open research questions. Finally, Section 13.8 presents the lessons learned and the conclusion.

13.2 OVERVIEW OF MACHINE LEARNING IN WIRELESS COMMUNICATION

We imagine a world in which ML and cutting-edge wireless communication are combined to create self-organizing, self-optimizing, and self-healing wireless networks. These intelligent networks will continuously learn from user behaviors, network conditions, and environmental factors to dynamically adapt their configurations and resource allocations (Bindle et al., 2022). This flexibility will be essential in high-user mobility situations, congested urban areas, and various device types.

Implementing ML algorithms as edge intelligence at the network edge will lower latency and improve the ability to make decisions in real-time (Anil Kumar et al., 2022). Edge intelligence will enable devices and base stations to process data locally to reduce the load on the core network and enable quick response times for time-sensitive applications like autonomous vehicles and industrial automation (Tashan et al., 2022). ML holds great promise for supporting emerging communication paradigms like the massive IoT, mission-critical communications, and traditional wireless communication applications. ML's scalability and data-driven nature can effectively manage the enormous IoT device population, ensuring dependable connectivity and using available network resources best (Henrique & Prasad, 2022a). Furthermore, ML can help with anomaly detection, network failure prediction, and ensuring uninterrupted service delivery in mission-critical applications like public safety and healthcare (Christalin Nelson et al., 2022; Krishnamoorthy, 2022).

There are difficulties in implementing ML in advanced wireless communication. ML models' interpretability and transparency remain significant concerns, particularly in safety-critical applications where human accountability is crucial (Li et al., 2022). To prevent discriminatory decision-making, paying close attention to any potential bias and fairness issues in ML algorithms is essential. Furthermore, power efficiency becomes a significant concern for battery-operated devices, such as smartphones and IoT sensors, as ML algorithms become more complex and resource-intensive (Elsayed & Zincir-Heywood, 2022). To enable ML models to be widely used in wireless communication, it will be essential to develop them in a lightweight and energy-efficient way for devices with limited resources (Bertalanic et al., 2022).

Partnerships between academia, industry, and regulatory bodies will significantly impact the future of ML for advanced wireless communication. Researchers must collaborate closely with wireless communication engineers, data scientists, and domain experts to identify practical use cases, validate algorithms, and ensure the seamless integration of ML into wireless communication standards (Chattopadhyay, 2022). In this future landscape of advanced wireless communication empowered by ML, we envision a plethora of practical applications that will revolutionize various industries and enhance the quality of life for individuals worldwide (Boas et al., 2022). In healthcare, ML-driven wireless communication can support telemedicine and remote patient monitoring, enabling healthcare professionals to monitor patients' vital signs and health conditions in real-time (Wang et al., 2022). Wearable devices with ML capabilities can detect anomalies and predict potential health issues, leading to early intervention and improved patient outcomes.

ML can improve traffic management in smart cities, resulting in less congestion and greater transportation effectiveness. Intelligent traffic lights can dynamically change signal timing based on traffic flow, minimizing delays and enhancing overall urban mobility. This is made possible by ML algorithms (Venkatesh et al., 2022; Shafik, 2024). In addition, ML can be used in smart energy management, where wireless communication networks support the monitoring and optimizing energy usage in buildings, resulting in lower energy consumption and a greater emphasis on sustainability. ML-powered wireless communication can potentially revolutionize industries like manufacturing and agriculture (Lopez-Perez et al., 2022). Wireless sensor networks powered by ML can improve irrigation and boost crop yield by monitoring soil moisture, temperature, and crop health in precision agriculture. In manufacturing, ML and wireless communication can enable predictive maintenance, enabling early equipment fault detection and averting expensive unplanned downtime (Arun Kumar, 2022).

The adoption of ML in 5G and subsequent networks will facilitate the development of cutting-edge services like augmented reality (AR) and virtual reality (VR) encounters. For seamless and immersive user experiences, wireless networks with intelligent traffic steering can optimize data delivery for AR and VR applications (Bahache et al., 2022). The amalgamation of ML and wireless communication will also foster groundbreaking advancements in autonomous vehicles. ML algorithms will play a pivotal role in enhancing the decision-making capabilities of self-driving cars, making them safer, more efficient, and capable of navigating complex urban environments (Chander, 2022). Despite the remarkable potential of ML for advanced wireless communication, researchers and practitioners must remain attentive to the ethical implications of these technologies (Rohini et al., 2022). Issues surrounding data privacy, security, and algorithmic bias demand rigorous attention and proactive measures to ensure that ML-driven wireless communication benefits society without compromising fundamental rights and values (Waqas et al., 2022).

The fusion of ML and advanced wireless communication heralds an era of transformative technological progress. This synergy promises to deliver unprecedented levels of connectivity, intelligence, and efficiency in wireless networks, empowering various industries and enabling innovative applications that will positively impact billions of lives (Alzaidi et al., 2022; Kamruzzaman & Alruwaili, 2022). As researchers and innovators embark on this transformative journey, they must prioritize ethical considerations and seek ways to maximize the societal benefits of this disruptive technology (Lamar et al., 2022). By fostering an ecosystem of collaboration, innovation, and responsibility, we can unlock the full potential of ML for advanced wireless communication, ushering in a future where seamless connectivity and intelligent networks are the backbone of an increasingly interconnected world.

13.3 MACHINE LEARNING TECHNIQUES FOR SIGNAL PROCESSING

ML signal processing techniques have revolutionized the extraction of valuable information from noisy and complex signals. Traditionally, signal-processing tasks were approached with rule-based algorithms and mathematical transformations. However, ML methods offer a data-driven approach, enabling systems to learn from examples and adapt to various signal characteristics.

13.3.1 Denoising

Denoising is a crucial signal processing operation that enhances the quality of signals and the accuracy of subsequent analysis and processing. Traditional denoising methods frequently rely on statistical and custom filtering methods, which may not capture the complex patterns present in real-world signals (X. Liu et al., 2023). ML techniques, which directly denoise signals from data, offer an alternative approach. Denoising tasks have been successfully carried out using an autoencoder, an unsupervised DL model. An autoencoder, which consists of an encoder that converts the input signal to a lower-dimensional representation, reconstructs the clean signal from the reduced representation (Kumar et al., 2022). While being trained, autoencoders develop the capacity to map noisy signals to their clean counterparts. In light of this, autoencoders perform better than traditional denoising methods and effectively denoise signals with varying noise levels, mainly when the noise characteristics are complex and non-linear (Okafor & Longe, 2022).

13.3.2 Equalization

In wireless communication systems, where signals travel through various channel conditions, equalization is a crucial signal-processing task. Intersymbol interference results from the channel's introduction of distortion and attenuation of various frequency components of the transmitted signal (Varatharaj & Ramalingam, 2022). The purpose of equalizers is to accurately recover the original transmitted symbols and make up for this distortion. Adaptive equalizers that adjust to changing channel conditions have been created using ML algorithms (Haji Mirzaee et al., 2022). Decision trees and support vector machines (SVMs) have categorized the channel state based on features extracted from received signals. The equalization coefficients are then adjusted per this information (Masur et al., 2022). Additionally, it has been shown that DL techniques, such as deep feedforward neural networks, can directly learn complex equalization functions from data. By training on a wide range of channel conditions, DL-based equalizers can adapt to different channel problems and do a better job of equalizing than traditional methods (Y. Liu et al., 2022).

13.3.3 Modulation Recognition

In wireless communication systems, where various modulation schemes are used to encode information into the transmitted signals, modulation recognition is a crucial task. The design of the receiver and the control of interference depend on accurate modulation recognition (Xie et al., 2022). A data-driven method for automatically categorizing modulation types based on samples of the received signal is provided by ML techniques. Artificial neural networks (ANNs), random forests, and k-nearest neighbor algorithms have all been used for this (Hassan et al., 2022). Deep neural networks, in particular, have demonstrated exceptional performance in modulation recognition tasks. Deep neural networks can discover even the most negligible variations between modulation schemes because they can automatically learn intricate patterns and features from the raw signal samples (Alanya-Beltran et al., 2022). Deep neural networks can also outperform conventional hand-crafted feature-based classifiers and generalize well to various modulation types with enough training data.

13.3.4 Feature Extraction

The fundamental signal processing step of feature extraction entails choosing pertinent and instructive features from unprocessed signal data for further analysis and classification. Traditional feature extraction techniques rely on predefined transformations, such as wavelet and Fourier transforms, which might not capture the most discriminative features for particular tasks (Zeb et al., 2022). A popular ML method for dimensionality reduction and feature extraction is principal component analysis (PCA) (Gupta et al., 2023). PCA, which identifies the principal components that account for the majority of the variance in the data and projects the raw signal onto these components, produces a more succinct and valuable representation (Salahdine et al., 2023). DL architectures like convolutional neural networks (CNNs) and recurrent neural networks (RNNs) have also been used to extract features directly from raw data to make end-to-end signal processing pipelines in which the model learns feature extraction and subsequent analysis tasks (Ometov et al., 2023).

13.3.5 Time Series Prediction

Time series prediction is a crucial signal-processing task in various fields, including finance, weather forecasting, and stock market prediction, as it entails projecting future values based on historical observations (Shafi et al., 2023). Time series data may contain complex temporal dependencies and non-linear patterns that conventional techniques like autoregressive models and moving average models may find difficult to detect (Isabona et al., 2023). In particular, RNNs and long short-term memory (LSTM) networks have become effective time series prediction tools. RNNs and LSTMs can effectively capture long-term dependencies in time series because they are built to handle sequential data (K. Wang et al., 2023). These architectures' recurrent connections enable them to keep track of previous inputs and make educated predictions based on the past. As a result, RNNs and LSTMs are well suited for tasks like forecasting the weather, the stock market, and demand, where past trends significantly impact present results (Sonia & Singh, 2022).

13.3.6 Sparse Signal Recovery

Reconstructing sparse signals from a small number of noisy measurements is the goal of the signal processing task known as sparse signal recovery. This issue is common in applications like compressive sensing, where it is impractical or expensive to acquire complete measurements of signals (Nwachukwu et al., 2022). Traditional signal recovery techniques, such as compressed sensing with L1-norm minimization, might need many measurements to recover sparse signals accurately. DL is used by ML techniques like compressed sensing neural networks (CSNNs) to recover sparse signals from under-sampled measurements effectively (Bindle et al., 2022). CSNNs are made up of many layers of neurons, each of which represents a signal measurement. The network uses backpropagation to learn how to extract the original sparse signal from the undersampled measurements during training (Anil Kumar et al., 2022). Compared to conventional methods, this approach significantly lowers the number of measurements needed while achieving comparable or even better performance.

13.3.7 Real-Time Signal Processing and Edge Computing

Many applications, including real-time communication systems, autonomous vehicles, and medical equipment, require real-time signal processing. Real-time applications may struggle to meet the low-latency requirements of conventional signal processing techniques (Scherer et al., 2022). ML algorithms have efficiently completed real-time signal processing tasks, especially lightweight models like MobileNets and EfficientNets. These compact models are made to be deployed on devices with limited resources, which makes them perfect for use in edge computing scenarios (Zenasni et al., 2022). ML enables low-latency decision-making and lessens the workload on centralized servers by carrying out signal-processing tasks at the network edge.

13.3.8 Cross-Modal Signal Processing

Integrating data from various modalities, such as audio, image, and text, is called cross-modal signal processing. Cross-modal signal processing tasks like audio-visual speech recognition and audio-text sentiment analysis have been carried out using ML techniques (Kannadhasan et al., 2022). For instance, to increase the precision of speech recognition in noisy environments, multimodal DL models, which combine CNNs and RNNs, can successfully learn representations from audio and visual data (Tashan et al., 2022). Cross-modal processing has much potential for more immersive and context-aware applications, such as human-computer interaction, virtual reality, and smart home systems.

13.4 APPLICATIONS OF MACHINE LEARNING IN CHANNEL ESTIMATION

Channel estimation is a crucial task that involves predicting the characteristics of the wireless channel to enable efficient and reliable data transmission. ML techniques have shown great promise in improving channel estimation in wireless communication systems.

13.4.1 Adaptive Channel Estimation

Adaptive channel estimation can be done using ML algorithms such as neural networks and SVMs (Henrique & Prasad, 2022b). These algorithms can accurately predict channel characteristics in real-time by learning from previous channel responses and observed data. Adaptive channel estimation techniques can increase the accuracy of data transmission and boost system performance by continuously adjusting to changing channel conditions.

13.4.2 Non-Linear Channel Estimation

Traditional channel estimation techniques might use linear channel models, which might not accurately represent the intricate non-linearities found in real-world wireless channels. ML techniques, particularly DL models like neural networks, are well suited for non-linear channel estimation tasks (Krishnamoorthy, 2022). In order to estimate non-linear channels more precisely and robustly, DL models can learn complex mappings between input data (such as transmitted signals) and channel characteristics.

13.4.3 Time-Variant Channel Estimation

User mobility, multipath propagation, and environmental changes are a few examples of factors that cause wireless channels to frequently display time-variant behavior. Traditional estimation techniques that assume static channels face difficulties when dealing with time-variant channels. Kalman filters and RNNs are two examples of ML algorithms capable of handling time-variant channel estimation (Li et al., 2022). Kalman filters can recursively update channel estimates based on time-varying observations, whereas RNNs can capture temporal dependencies and patterns in channel variations.

13.4.4 Multi-User MIMO Channel Estimation

Inter-user interference makes it difficult to accurately estimate the channels for multiple users in multi-user Multiple-Input Multiple-Output (MIMO) systems. ML algorithms can perform joint channel estimation by considering the correlations and interactions between the channels of multiple users (Boas et al., 2022; Lopez-Perez et al., 2022). Multi-user MIMO channel estimation can achieve improved spatial multiplexing and better interference mitigation by utilizing the power of ML.

13.4.5 Low-SNR Channel Estimation

Accurate channel estimation becomes more complex in low signal-to-noise ratio (SNR) scenarios because the received signal is masked by noise. DL models with denoising capabilities are one example of a ML technique that can improve channel estimation performance in low-SNR situations (Bahache et al., 2022). These models can improve the accuracy of channel estimates and learn to denoise received signals, making communication under challenging environments more dependable.

13.4.6 Channel Estimation with Limited Feedback

Due to channel dimensionality and bandwidth restrictions in massive MIMO and millimeter-wave communication systems, providing explicit feedback for channel estimation may not be feasible. The amount of feedback needed can be decreased while maintaining the channel estimation accuracy using ML-based feedback compression techniques (Lopez-Perez et al., 2022). These methods use the ML models' capacity for learning to compress feedback data and effectively use the feedback resources available.

13.4.7 Hybrid Beamforming Channel Estimation

In mmWave MIMO systems, hybrid beamforming is a technique used to keep beamforming capabilities while reducing hardware complexity. Channel estimation for the analog and digital beamforming stages can be done using ML, and the hybrid beamforming matrix can be optimized (Bahache et al., 2022). The system can adaptively adjust the beamforming parameters based on the channel conditions and improve beam alignment and communication performance by using ML for hybrid beamforming channel estimation.

13.4.8 Interference Management

In wireless communication, interference management deals with the problems brought on by overlapping signals from various sources in the same frequency band. In dynamic environments, traditional methods rely on fixed power control and frequency planning, which may not be the best approach (Ding et al., 2022). ML techniques make a data-driven approach to interference management possible, in which models learn from historical data and in-the-moment measurements to make wise choices about power allocation, frequency reuse, and scheduling. By adapting to changing interference patterns, ML models can improve network performance, reduce interference, and increase overall throughput (Alzaidi et al., 2022).

13.4.9 Spectrum Sensing

Spectrum sensing is essential in cognitive radio systems, where secondary users must find and access underutilized spectrum bands without interfering with primary users. Traditional sensing techniques could be inaccurate on fading and noisy channels (Kamruzzaman & Alruwaili, 2022). A better method is provided by ML, which trains models like support vector machines and DL architectures on labeled data to identify primary user signals among noise and interference precisely. Even in complex propagation environments, ML models achieve high detection accuracy because of their capacity to learn pertinent features from received signals (Okafor & Longe, 2022).

13.4.10 Resource Allocation and Beamforming

Beamforming and resource allocation are crucial methods for improving wireless communication systems. In order to maximize system throughput and user satisfaction, resource allocation involves managing constrained network resources like bandwidth and transmit power (Shafik et al., 2016). For receiver signal strength to be increased and interference to be reduced, beamforming aims to concentrate transmitted energy in particular directions. Based on observed network states and performance feedback, ML techniques like reinforcement learning enable dynamic and adaptive resource allocation strategies (Shafik et al., n.d.). Additionally, beamforming vectors can be learned using ML models while considering the channel and user locations, resulting in flexible and effective solutions.

Wireless networks can achieve higher spectral efficiency, less interference, and better user experiences by optimizing resource allocation and beamforming with ML. These applications demonstrate the adaptability and efficiency of ML techniques in channel estimation for various wireless communication scenarios. By fusing the strengths of data-driven learning with conventional signal processing techniques, ML increases channel estimation's accuracy, adaptability, and efficiency. This enhances the effectiveness and dependability of wireless communication systems. The following section explains which current technologies ML supports.

13.5 DEEP LEARNING AND NEURAL NETWORK APPLICATIONS IN WIRELESS COMMUNICATION

DL (currently known as transfer learning) and neural networks have revolutionized various aspects of wireless communication systems, enabling significant advancements in performance, efficiency, and adaptability. The following are some of the key applications of DL and neural networks in wireless communication are explained below.

13.5.1 Modulation Classification

CNNs, in particular, have proven exceptional at classifying the modulation schemes used in wireless communication signals. By learning from unprocessed signal samples, CNNs can automatically extract hierarchical features and patterns that distinguish different modulation types (K. Wang et al., 2023). DL for modulation classification allows for more precise and reliable modulation scheme identification, even in noise and interference.

13.5.2 Channel Estimation

DL methods like LSTM and RNNs networks have tackled channel estimation tasks. These models are well suited for time-variant and multi-user MIMO channel estimation because they can capture temporal dependencies and variations in channel conditions (Isabona et al., 2023). DL-based channel estimation has higher accuracy and adaptability than conventional methods because it learns from previous channel responses.

13.5.3 Signal Denoising and Enhancement

Autoencoders with DL and denoising capabilities are used to reduce noise and improve the quality of received signals. These models can reconstruct clean versions and increase the dependability of wireless communication in difficult environments by learning the underlying structure of noisy signals (Shafi et al., 2023). DL-based signal denoising and augmentation improve system performance and lower error rates.

13.5.4 Beamforming and Antenna Array Optimization

The configurations of antenna arrays and beamforming have been optimized using DL. Deep neural networks can learn intricate relationships between channel parameters and ideal beamforming weights, producing adaptive and effective beam forming solutions (Lakshmikantha Reddy & Meena, 2023; Ometov et al., 2023). This application is especially beneficial in millimeter-wave communication systems, where beamforming is essential for achieving high data rates and dependable connections.

13.5.5 Network Optimization and Resource Allocation

Dynamic resource allocation and network optimization have both been accomplished using reinforcement learning and other DL techniques. Based on observed network states and feedback on system performance, deep reinforcement learning models learn to allocate bandwidth, transmitting power, and other resources in the most efficient way possible in real-time (Salahdine et al., 2023). With this strategy, network resources are used more adaptably and effectively, increasing system capacity and the user experience.

13.5.6 Spectrum Sensing and Dynamic Spectrum Access

Available spectrum bands for dynamic spectrum access can be found and identified in cognitive radio networks using DL models. DL models can effectively sense and use underutilized frequency bands without damaging primary users by learning from diverse spectrum data (Gupta et al., 2023). This makes spectrum sharing more flexible and effective, which improves spectrum utilization and boosts network capacity.

13.5.7 Wireless Channel Prediction

Based on prior measurements, DL models, such as LSTM networks, have been used to forecast wireless channel conditions. Channel prediction makes it possible to allocate resources more wisely and adapt communication parameters in advance, improving the performance of time-sensitive applications and lowering latency in wireless communication systems (Zeb et al., 2022).

13.5.8 MIMO Detection and Decoding

Multiple-Input Multiple-Output (MIMO) systems increase data rates and link reliability by transmitting and receiving signals using multiple antennas. However, computational difficulties arise with MIMO detection and decoding, particularly when many antennas exist (Sarkar, 2022). DL models have been used to effectively perform MIMO detection and decoding, including deep neural networks and graph-based neural networks. In massive MIMO scenarios, these models can learn intricate mappings between received signals and transmitted symbols, enabling faster and more precise decoding (Lopez-Perez et al., 2022).

13.5.9 Interference Mitigation and Self-Interference Cancellation

Wireless communication frequently encounters interference issues, particularly in crowded urban areas with overlapping networks. DL techniques have been used to reduce interference and cancel self-interference in full-duplex communication systems (Nwachukwu et al., 2022). DL models enable more efficient interference cancellation by learning to recognize interference patterns and gauge self-interference levels, which improves spectral efficiency and increases data rates.

13.5.10 Radio Environment Mapping

Radio environment maps have been made using DL, which offer useful details about signal strengths, channel conditions, and interference patterns in various locations. DL models can create precise and dynamic radio environment maps by learning from crowdsourced data or in-the-moment measurements (Anil Kumar et al., 2022). These maps can be used for resource allocation, network planning, and optimization, improving the user experience and overall network performance.

13.5.11 Localization and Positioning

DL models like convolutional neural networks and RNNs have been used for wireless localization and positioning. With fewer localization errors, these models can more accurately estimate mobile device positions by learning from received signal strength (RSS) measurements or time-of-arrival (TOA) data (Sonia & Singh, 2022). Location-based services, indoor navigation, and asset tracking are all possible uses for wireless localization with DL.

13.5.12 Dynamic Channel Selection

DL can be used to carry out dynamic channel selection in situations involving dynamic spectrum access where several wireless technologies coexist. Based on real-time observations of channel conditions and interference levels, deep reinforcement learning models learn to choose the best communication channel among those that are available (Scherer et al., 2022). This dynamic channel selection makes seamless handover between various wireless technologies possible.

13.5.13 Automatic Gain Control and Power Control

In appreciation of DL models, wireless communication systems can now automatically adjust gain and power levels. These models can improve automatic gain control (AGC) and power control algorithms, resulting in improved link quality, less interference, and more stable communication performance by learning from feedback on signal quality and channel conditions (Scherer et al., 2022).

13.5.14 Energy-Efficient Communication

Energy-efficient communication systems can be designed using DL techniques like model compression and neural architecture search. DL models can be deployed on devices with limited resources, such as edge devices and IoT sensors, thanks to model compression, which reduces the size of the models (Meng et al., 2020). Energy-efficient wireless communication solutions are made possible by neural architecture search, which optimizes the architecture of DL models to reduce computational complexity and energy consumption (Shafik et al., n.d.).

13.6 CROSS-DOMAIN APPROACHES: WIRELESS AND NETWORKING

Cross-domain approaches refer to integrating and collaborating concepts, techniques, and methodologies from different domains to address challenges and improve performance in wireless communication and networking systems. These approaches leverage the strengths and knowledge of each domain to create innovative solutions that transcend traditional boundaries. The following are some vital cross-domain approaches in wireless communication and networking:

13.6.1 Machine Learning and Wireless

A potent cross-domain strategy is the fusion of wireless communication and ML. Wireless systems can use ML algorithms for various purposes, including channel estimation, interference management, resource allocation, and beamforming (Shafik et al., n.d.). ML models improve wireless communication systems' effectiveness, adaptability, and intelligence by learning from data and adapting to shifting network conditions. An advantageous relationship between the two fields can be established by using wireless communication data to train ML models and enhance the efficiency of ML tasks.

13.6.2 Networking and Software-Defined Networking (SDN)

A networking paradigm known as SDN separates the control plane from the data plane to allow centralized network management and programmability. SDN integration with conventional networking enables resource optimization, traffic engineering, and dynamic and flexible network configuration (Zenasni et al., 2022). SDN can be extended to wireless access networks in the context of wireless communication, enabling effective resource management and orchestration in environments with multiple technologies and operators. Wireless networks can be made more scalable, manageable, and responsive to changing traffic patterns and user demands by combining networking and SDN principles (Kannadhasan et al., 2022).

13.6.3 Cloud Computing and Edge Computing

The challenges of latency, scalability, and data processing in wireless networks are addressed by the convergence of cloud computing and edge computing, which is a cross-domain strategy. Cloud computing offers enormous computational resources and storage capabilities in centralized data centers (Tashan et al., 2022). However, the latency introduced by sending data to far-off data centers can be troublesome in time-critical applications like the Internet of Things, autonomous vehicles, and augmented reality. On the other hand, edge computing reduces latency and improves real-time processing capabilities by bringing computation and data storage closer to the end users at the network edge. Wireless networks can balance centralized processing and local data handling by combining cloud and edge computing advantages, ensuring efficient resource utilization and responsiveness to user needs (Christalin Nelson et al., 2022).

13.6.4 Security and Privacy in Wireless Networks

Wireless networks must incorporate security and privacy principles from the cybersecurity domain to protect data, user identities, and communication integrity (Jun et al., 2021). Wireless communication systems can benefit from adapting and applying cybersecurity methods like intrusion detection, authentication, and encryption (Matinkhah & Shafik, 2019). Cross-domain strategies protect wireless networks from threats like data tampering, denial-of-service attacks, and eavesdropping while preserving the confidentiality and integrity of sensitive data (Shafik et al., 2022).

13.6.5 Wireless Sensing and the Internet of Things

Wireless sensing technologies and the Internet of Things can be combined to create intelligent, interconnected systems that collect and analyze data in real-time (Tashan et al., 2022). IoT platforms can process and analyze this data to make informed decisions and launch automated actions. Wireless sensors can gather environmental data, health parameters, and other information (Shafik et al., 2022). Applications like environmental monitoring, precision agriculture, healthcare, and smart cities are made possible by cross-domain approaches between wireless sensing and IoT, where wireless networks play a crucial role in facilitating seamless data exchange and analysis.

13.6.6 Green Networking and Sustainable Wireless Communication

Green networking aims to lower wireless communication systems' energy usage and carbon footprint. By incorporating sustainability and environmental science concepts, cross-domain strategies can be developed to optimize power consumption, network coverage, and energy-efficient communication protocols (Christalin Nelson et al., 2022; Li et al., 2022). Sustainable wireless networks can be created using methods, for example, dynamic power allocation, energy harvesting, and sleep mode operation.

13.6.7 Wireless and Internet of Things Analytics

Wireless communication and IoT analytics make real-time data analysis and decision-making possible. IoT devices produce massive amounts of data; wireless communication is the foundation for sending this data to cloud platforms or edge servers. Wireless communication and IoT analytics work together to help businesses and industries make the most of the data that IoT devices collect (Chattopadhyay, 2022). This integration makes IoT deployments more efficient and cost-effective by facilitating predictive maintenance, anomaly detection, and optimization.

13.6.8 Wireless and Blockchain Technology

Wireless communication can be combined with blockchain technology, known for its decentralized and tamper-resistant nature, to improve wireless networks' security, privacy, and trust. Blockchain-based identity management and authentication systems can guarantee secure and dependable access to wireless networks by preventing man-in-the-middle attacks and unauthorized access (W. Wang et al., 2022). Blockchain can also be utilized to monitor and confirm data transmissions, ensuring data immutability and integrity in wireless communication.

13.6.9 Wireless and Augmented/Virtual Reality

User experiences that are immersive and interactive are made possible by the combination of wireless communication and augmented and virtual reality technologies. Wireless networks that transmit data at high speeds and with low latency are necessary for AR and VR applications (Venkatesh et al., 2022). Users can interact in real-time with AR and VR applications like remote collaboration, virtual tours, and immersive gaming by combining wireless communication with AR and VR.

13.6.10 Wireless and Renewable Energy Integration

Integrating wireless communication with renewable energy sources can offer sustainable and dependable connectivity in off-grid and energy-restricted environments. Reducing reliance on conventional power grids, wireless networks can be powered by solar, wind, or other renewable energy systems (Arun Kumar, 2022). By supporting applications like disaster response, environmental monitoring, and rural connectivity, this integration makes it easier to deploy wireless communication in remote and rural areas.

13.6.11 Wireless and Artificial Intelligence for Network Optimization

Advanced network optimization and autonomous management are made possible by fusing wireless communication and AI techniques. AI algorithms can optimize wireless network configuration and parameters in real-time based on network conditions, traffic patterns, and user demand (Sarkar, 2022). This multi-domain strategy guarantees effective spectrum use, reduced interference, and enhanced overall network performance.

13.6.12 Wireless and Big Data Analytics

The processing and analysis of massive amounts of data produced by wireless networks and connected devices is made possible by the integration of big data analytics with wireless communication. By utilizing big data analytics, wireless networks can learn more about user behavior, network performance, and new trends (Rohini et al., 2022). This data-driven strategy makes it easier to make wise decisions and create individualized wireless communication services and applications.

13.6.13 Wireless and Edge Intelligence

Edge computing and ML are combined to create edge intelligence, which enables real-time data analysis and decision-making at the network edge. Edge intelligence and wireless communication can be combined to process data closer to the source, requiring less latency and bandwidth (Kamruzzaman & Alruwaili, 2022). Real-time insights are essential in time-sensitive applications like industrial automation, autonomous vehicles, and healthcare, where this method is beneficial.

13.7 CHALLENGES, PROBLEMS, AND OPEN RESEARCH QUESTIONS

While ML has shown great promise in advancing wireless communication, it also faces several challenges and problems that must be addressed to fully realize its potential.

13.7.1 Challenges and Problems

Some of the key challenges are as follows:

13.7.1.1 Data Collection and Labeling

For efficient training, ML algorithms heavily rely on broad and varied datasets. Getting representative and high-quality datasets can be difficult in wireless communication. Gathering real-world wireless data in various settings and situations takes a lot of time and money. Labeling the data for supervised learning tasks like modulation classification and channel estimation can also be time-consuming and expensive (Sonia & Singh, 2022). Unbalanced datasets and a lack of labeled data can produce biased models with poor generalization capabilities.

13.7.1.2 Limited Training Data for Dynamic Environments

User mobility, fading, and environmental changes are just a few examples of the factors that cause wireless communication channels to behave in a time-variant and dynamic way. ML models may become overfitted due to training on small data samples, which prevents the models from generalizing to brand-new, unforeseen channel conditions (Nwachukwu et al., 2022). It is difficult to continuously gather and update training data to match the dynamic nature of wireless channels.

13.7.1.3 Energy Efficiency and Resource Constraints

Energy efficiency is essential in wireless communication systems, especially for battery-powered devices and IoTs like sensors. However, DL models can be computationally demanding and require a lot of power and resources for training and inference, of incredibly complex ones (Bindle et al., 2022). Due to their constrained processing power and memory, these models are difficult to deploy on devices with limited resources or edge computing environments.

13.7.1.4 Latency and Real-Time Processing

Many wireless communication applications, including augmented reality and self-driving cars, demand real-time processing and quick responses. Real-time applications may be hampered by the significant inference time introduced by DL models with many layers and parameters (Anil Kumar et al., 2022). It is important to carefully weigh the trade-offs involved in reducing model complexity and optimizing for low-latency operations while maintaining accuracy.

13.7.1.5 Explainability and Trustworthiness

It can be challenging to comprehend how ML models, particularly deep neural networks, make particular decisions because they are frequently called "black boxes." Transparency and explainability are essential in wireless communication systems for fostering reliability (Zhang et al., 2022). Developing interpretability techniques to offer insights into model choices is necessary, particularly in safety-critical applications.

13.7.1.6 Overfitting and Generalization

There is a risk of overfitting in wireless communication tasks, especially when working with limited data, where models memorize noise or peculiarities in the training data instead of learning useful patterns (Scherer et al., 2022; Zenasni et al., 2022). Reliable performance in real-world scenarios depends on models that generalize well to unknown data and various channel conditions.

13.7.1.7 Security and Robustness

Wireless communication ML models are susceptible to data tampering and adversarial attacks. Attackers can fool models by faking wireless signals or input data, compromising security and reducing performance (Sivasundari & Ananthajothi, 2022). Defense mechanisms and adversarial training methods must be incorporated when creating robust and secure ML models for wireless communication.

13.7.1.8 Spectrum Efficiency and Interference

In order to optimize spectrum efficiency and prevent harmful interference to primary users or nearby networks, ML applications in spectrum sensing, interference management, and resource allocation must strike a balance (Kannadhasan et al., 2022). In dynamic and crowded wireless environments, ensuring fair and effective spectrum sharing continues to be challenging.

13.7.1.9 Regulatory and Standardization Issues

To ensure interoperability and adherence to industry standards, integrating ML into wireless communication systems may raise regulatory concerns and necessitate standardization efforts. It can be challenging to uphold current regulations while utilizing ML's full potential in wireless communication (Krishnamoorthy, 2022).

13.7.1.10 Scalability and Complexity

Deploying ML models on a large scale is challenging as wireless communication systems become more complex and larger. Training and maintaining models across numerous devices and network nodes may call for distributed and federated learning approaches to handle the computational load. These issues and challenges must be resolved for advanced wireless communication to fully benefit from ML (Krishnamoorthy, 2022). ML can significantly enhance wireless communication systems' effectiveness, adaptability, and reliability by creating reliable algorithms, effective training strategies, and domain-specific optimization techniques.

13.7.2 Open Research Questions

The following are some of the technical open research questions that need to be addressed: Continued research and innovation in this field will pave the way for future wireless technologies and applications. Several open research questions remain to be addressed in the rapidly evolving field of ML for advanced wireless communication. These questions drive the exploration of new frontiers and pave the way for innovative solutions and advancements.

13.7.2.1 Generalization in Dynamic Environments

How can ML models be trained to generalize effectively in dynamic and time-variant wireless communication environments? Addressing this question involves developing novel approaches for continual learning and transfer learning to enable models to adapt and maintain performance in changing channel conditions and network settings.

13.7.2.2 Federated Learning for Edge Devices

How can federated learning be effectively applied to wireless edge devices with limited resources and privacy constraints? Exploring new techniques for federated learning and model aggregation in resource-constrained environments is crucial to enabling collaborative learning without compromising user privacy or energy efficiency.

13.7.2.3 Explainable Artificial Intelligence for Wireless Communication

How can ML models used in wireless communication tasks be made more explainable and interpretable? Developing methods to provide meaningful insights into model decisions will enhance trust, facilitate debugging, and ensure compliance with regulatory requirements in safety-critical applications.

13.7.2.4 Robustness and Security

How can ML models in wireless communication systems be made more robust and resilient against adversarial attacks and data poisoning? Research on developing robust training techniques and adversarial defense mechanisms will be vital to safeguarding wireless networks from security breaches.

13.7.2.5 Efficient Model Design

How can model compression and quantization techniques be optimized to reduce the computational and memory requirements of DL models while maintaining performance in wireless communication applications? Developing efficient model architectures tailored to wireless communication scenarios is essential for enabling deployment on resource-constrained devices.

13.7.2.6 Joint Optimization of Communication and Learning

How can communication and ML tasks be optimized jointly to maximize the system's performance? Investigating novel methods for integrating communication objectives into the model training process and vice versa will enable more efficient and adaptive wireless communication systems.

13.7.2.7 Resource Allocation in Heterogeneous Networks

How can ML be leveraged to optimize resource allocation in dense and heterogeneous wireless networks, considering various technologies like 5G, Wi-Fi, and IoT devices? Research on intelligent resource management strategies that adapt to network conditions and user demands will be crucial for efficient spectrum utilization and improved user experiences.

13.7.2.8 Cross-Modal Signal Processing

How can ML techniques be extended to handle cross-modal signal processing tasks in wireless communication, such as audio-visual signal analysis and multi-modal data fusion? Exploring new architectures and learning techniques to integrate information from diverse modalities will enable more advanced and context-aware applications.

13.7.2.9 Energy-Aware Learning

How can ML models be designed and optimized for energy efficiency in wireless communication systems, particularly battery-powered devices and IoT sensors? Research on developing energy-aware learning algorithms and model architectures will contribute to sustainable and eco-friendly wireless networks.

13.7.2.10 Standardization and Deployment

How can the adoption and integration of ML in wireless communication systems be standardized to ensure interoperability and scalability? Collaborative efforts between academia, industry, and standardization bodies will be essential to defining best practices and guidelines for incorporating ML into wireless technologies. By addressing these open research questions, the field of ML for advanced wireless communication will continue to advance, unlocking new possibilities for more intelligent, adaptive, and efficient wireless networks. These research directions will shape the future of wireless communication technologies and pave the way for transformative applications and services.

13.8 LESSONS LEARNED AND CONCLUSION

The application of ML to advanced wireless communication has led to valuable lessons learned, driving the evolution of both fields.

13.8.1 Lessons Learned from the Chapter

Within this section, some significant vital lessons that have emerged are discussed:

13.8.1.1 Data Is the Backbone of Technology Success

Time-variant and dynamic communication channels are inherent to wireless technology. For machine learning (ML) applications in wireless communication to be successful, models must be able to adapt to changing channel conditions, user mobility, and network circumstances. Strategies like transfer learning and continuous learning are needed to maintain model performance in dynamic situations.

13.8.1.2 Adaptability to Dynamic Environments

By their very nature, wireless communication networks are time-varying and dynamic. ML applications in wireless communication need models that can adapt to shifting channel circumstances, user mobility, and network scenarios. Transfer learning and continuous learning are two methodologies that are necessary to maintain model performance in changing contexts.

13.8.1.3 Interdisciplinary Collaboration Is Key

Safety-critical applications have experienced issues as a result of DL models' lack of interpretability. Transparency regarding how models generate judgments and a dedication to both are necessary to establish confidence in ML-based wireless communication systems. Explainable AI strategies have been created in order to get over this issue and give a richer knowledge of model behavior.

13.8.1.4 Trade-Offs in Model Complexity

In safety-critical applications, DL models' interpretability issues have been a difficulty. It takes dedication to both openness and commitment to both to establish confidence in ML-based wireless communication networks. Explainable AI strategies have been created as a solution to this issue and to give users a clearer knowledge of model behavior.

13.8.1.5 Explainability and Trustworthiness

Safety-critical applications have experienced issues as a result of DL models' interpretability limitations. Transparency regarding how models decide what to do and a dedication to both are necessary for establishing confidence in ML-based wireless communication systems. Explainable AI strategies have been created in order to get over this issue and give a deeper knowledge of model behavior.

13.8.1.6 Data Privacy and Security

The application of ML may help to increase the energy efficiency of wireless communication networks. Research on energy-conscious learning algorithms and the use of renewable energy sources has been stimulated by the growing importance of sustainability and green networking.

13.8.1.7 Real-World Deployment Challenges

It may be advantageous to utilize ML to increase the energy efficiency of wireless communication networks. Research on energy-conscious learning algorithms and incorporating renewable energy sources has been stimulated by the growing importance of sustainability and green networking.

13.8.1.8 The Power of Edge Computing

The application of ML may benefit the enhancement of the energy effectiveness of wireless communication systems. Research on energy-aware learning algorithms and the use of renewable energy sources has been sparked by the growing relevance of sustainability and green networking.

13.8.1.9 Sustainable and Green Wireless Technologies

The application of ML may help to increase the energy efficiency of wireless communication networks. Research on energy-conscious learning algorithms and the incorporation of renewable energy sources has been stimulated by the growing importance of sustainability and green networking.

13.8.1.10 ML as a Catalyst for Innovation

In wireless communication, ML has sparked innovation by opening up new applications, services, and use cases that were difficult or impractical in the past. Wireless networks have evolved into intelligent and adaptive systems thanks to their capacity to draw insightful conclusions from massive datasets and make data-driven decisions. The application of ML in advanced wireless communication has taught us several important lessons, all of which highlight the necessity of ongoing research, collaboration, and innovation to utilize these technologies fully. Exciting possibilities are on the horizon as ML and wireless communication continue to be integrated.

13.8.2 Conclusion

High-tech wireless communication using ML has resulted in spectacular successes and crucial insights. The use of ML has fundamentally altered a number of elements of wireless communication, from signal processing and channel prediction to resource allocation and network optimization, through interdisciplinary collaboration and a deeper understanding of data-driven methodologies. Ongoing research concentrates on model complexity, understandability, and security as well as data accessibility. The key to furthering wireless communication technology development is to embrace the lessons learned, such as the value of high-quality data, the necessity of flexibility in dynamic contexts, and the role of edge computing. As the development of ML and wireless communication progresses, it is anticipated that these two fields will combine seamlessly to produce a world of wireless networks and services that are more intelligent, efficient, and linked.

REFERENCES

Alanya-Beltran, J., Raut, R., Patil, S., Christobel, Y. A., Vats, P., Nagaprasad, S., & Kumar, Y. P. (2022). Machine Learning-Based Intelligent Wireless Communication System for Solving Real-World Security Issues. *Security and Communication Networks, 2022.* https://doi.org/10.1155/2022/7978822

Alzaidi, M. S., Subbalakshmi, C., Roshini, T. V., Shukla, P. K., Shukla, S. K., Dutta, P., & Alhassan, M. (2022). 5G-Telecommunication Allocation Network Using IoT Enabled Improved Machine Learning Technique. *Wireless Communications and Mobile Computing, 2022.* https://doi.org/10.1155/2022/6229356

Anil Kumar, B., Goyal, S. B., & Islam, S. M. N. (2022). Cognitive Computing Models in Communication Systems. In *Cognitive Computing Models in Communication Systems.* https://doi.org/10.1002/9781119865605

Arun Kumar, K. A. (2022). RF Fingerprinting of Software Defined Radios Using Ensemble Learning Models. *Journal of Communications, 17*(4). https://doi.org/10.12720/jcm.17.4.287-293

Bahache, M., Tahari, A. E. K., Herrera-Tapia, J., Lagraa, N., Calafate, C. T., & Kerrache, C. A. (2022). Towards an Accurate Faults Detection Approach in Internet of Medical Things Using Advanced Machine Learning Techniques. *Sensors, 22*(15). https://doi.org/10.3390/s22155893

Bindle, A., Gulati, T., & Kumar, N. (2022). Exploring the Alternatives to the Conventional Interference Mitigation Schemes for 5G Wireless Cellular Communication Network. *International Journal of Communication Systems, 35*(4). https://doi.org/10.1002/dac.5059

Chander, B. (2022). Artificial Neural Networks and Support Vector Machine for IoT. In *Internet of Things.* https://doi.org/10.1007/978-3-030-87059-1_3

Christalin Nelson, S., Singh, R. K., & Prakash, G. L. (2022). Hybrid Deep Learning Model Based on Intelligent Microbat Routing (IMR) and Popularity Content Caching (PCC) for an Effective Caching and Routing in Vehicular Edge Networks. *Computers and Electrical Engineering, 103.* https://doi.org/10.1016/j.compeleceng.2022.108353

Ding, D., Zou, M., Wang, X., Qin, G., Zhang, S., Chan, S. Y., Meng, Q., Liu, Z., Zhang, Q., & Chen, Y. (2022). Thermal Conductivity of Polydisperse Hexagonal BN/Polyimide Composites: Iterative EMT Model and Machine Learning Based on First Principles Investigation. *Chemical Engineering Journal, 437.* https://doi.org/10.1016/j.cej.2022.135438

Gupta, D., Ragab, M., Mansour, R. F., Khamparia, A., & Khanna, A. (2023). AI-Enabled 6G Networks and Applications. In *AI-Enabled 6G Networks and Applications.* https://doi.org/10.1002/9781119812722

Haji Mirzaee, P., Shojafar, M., Cruickshank, H., & Tafazolli, R. (2022). Smart Grid Security and Privacy: From Conventional to Machine Learning Issues (Threats and Countermeasures). *IEEE Access, 10.* https://doi.org/10.1109/ACCESS.2022.3174259

Hassan, S., Tariq, N., Naqvi, R. A., Rehman, A. U., & Kaabar, M. K. A. (2022). Performance Evaluation of Machine Learning-Based Channel Equalization Techniques: New Trends and Challenges. *Journal of Sensors, 2022.* https://doi.org/10.1155/2022/2053086

Henrique, P. S. R., & Prasad, R. (2022a). A Glimpse of the Future beyond 2020—B5G. In *6G The Road to the Future Wireless Technologies 2030.* https://doi.org/10.1201/9781003336884-3

Henrique, P. S. R., & Prasad, R. (2022b). The Road for the Future Wireless Network. In *6G The Road to the Future Wireless Technologies 2030.* https://doi.org/10.1201/9781003336884-4

Isabona, J., Imoize, A. L., Ojo, S., Do, D.-T., & Lee, C.-C. (2023). Machine Learning-Based GPR with LBFGS Kernel Parameters Selection for Optimal Throughput Mining in 5G Wireless Networks. *Sustainability, 15*(2). https://doi.org/10.3390/su15021678

Jun, Y., Craig, A., Shafik, W., & Sharif, L. (2021). Artificial Intelligence Application in Cybersecurity and Cyberdefense. In *Wireless Communications and Mobile Computing* (Vol. 2021). https://doi.org/10.1155/2021/3329581

Kamruzzaman, M. M., & Alruwaili, O. (2022). AI-based Computer Vision Using Deep Learning in 6G Wireless Networks. *Computers and Electrical Engineering, 102.* https://doi.org/10.1016/j.compeleceng.2022.108233

Kannadhasan, S., Nagarajan, R., Karthick, A., & Ghosh, A. (2022). Smart Antennas, Electromagnetic Interference and Microwave Antennas for Wireless Communications. In *Smart Antennas, Electromagnetic Interference and Microwave Antennas for Wireless Communications.* https://doi.org/10.1201/9781003373230

Kumar, L. J. S., Krishnan, P., Shreya, B., & M. S., S. (2022). Performance Enhancement of FSO Communication System Using Machine Learning for 5G/6G and IoT Applications. *Optik, 252.* https://doi.org/10.1016/j.ijleo.2021.168430

Lakshmikantha Reddy, S., & Meena, M. (2023). Performance of Cooperative Spectrum Sensing Techniques in Cognitive Radio Based on Machine Learning. In *Cognitive Science and Technology.* https://doi.org/10.1007/978-981-19-8086-2_25

Li, Q., Cheng, H., Yang, Y., Tang, H., Wang, J., Luo, G., & Sun, W. (2022). Deep Reinforcement Learning-based Resource Allocation for 5G Machine-type Communication in Active Distribution Networks with Time-varying Interference. *Mobile Networks and Applications, 27*(6). https://doi.org/10.1007/s11036-022-02006-5

Liu, X., Deng, Y., & Mahmoodi, T. (2023). Wireless Distributed Learning: A New Hybrid Split and Federated Learning Approach. *IEEE Transactions on Wireless Communications, 22*(4). https://doi.org/10.1109/TWC.2022.3213411

Liu, Y., Zhang, S., Ding, Z., Schober, R., Al-Dhahir, N., Hossain, E., & Shen, X. (2022). Special Issue on Next Generation Multiple Access—Part I. In *IEEE Journal on Selected Areas in Communications* (Vol. 40, Issue 4). https://doi.org/10.1109/JSAC.2021.3139485

Lopez-Perez, D., De Domenico, A., Piovesan, N., Xinli, G., Bao, H., Qitao, S., & Debbah, M. (2022). A Survey on 5G Radio Access Network Energy Efficiency: Massive MIMO, Lean Carrier Design, Sleep Modes, and Machine Learning. *IEEE Communications Surveys and Tutorials, 24*(1). https://doi.org/10.1109/COMST.2022.3142532

Meng, H., Shafik, W., Matinkhah, S. M., & Ahmad, Z. (2020). A 5G Beam Selection Machine Learning Algorithm for Unmanned Aerial Vehicle Applications. *Wireless Communications and Mobile Computing, 2020.* https://doi.org/10.1155/2020/1428968

Okafor, K. C., & Longe, O. M. (2022). ASM-ROBOT: A Cyber-Physical Home Automation Controller with Memristive Reconfigurable State Machine. *International Journal of Advanced Computer Science and Applications, 13*(1). https://doi.org/10.14569/IJACSA.2022.01301103

Ometov, A., Mezina, A., & Nurmi, J. (2023). On Applicability of Imagery-Based CNN to Computational Offloading Location Selection. *IEEE Access*, *11*. https://doi.org/10.1109/ACCESS.2022.3232469

Salahdine, F., Han, T., & Zhang, N. (2023). 5G, 6G, and Beyond: Recent Advances and Future Challenges. In *Annales des Telecommunications/Annals of Telecommunications*. https://doi.org/10.1007/s12243-022-00938-3

Shafik, W. (2023). Cyber Security Perspectives in Public Spaces: Drone Case Study. In *Handbook of Research on Cybersecurity Risk in Contemporary Business Systems* (pp. 79–97). IGI Global.

Shafik, W. (2024). Wearable Medical Electronics in Artificial Intelligence of Medical Things. In *Handbook of Security and Privacy of AI-Enabled Healthcare Systems and Internet of Medical Things* (pp. 21–40). https://doi.org/10.1201/9781003370321-2

Shafik, W., Matinkhah, S. M., Afolabi, S. S., & Sanda, M. N. (2021). A 3-dimensional Fast Machine Learning Algorithm for Mobile Unmanned Aerial Vehicle Base Stations. *International Journal of Advances in Applied Sciences*, *10*(1). https://doi.org/10.11591/ijaas.v10.i1.pp28-38

Shafik, W., Matinkhah, S. M., Shokoor, F., & Sharif, L. (2022). A Reawakening of Machine Learning Application in Unmanned Aerial Vehicle: Future Research Motivation. *EAI Endorsed Transactions on Internet of Things*, *8*(29). https://doi.org/10.4108/eetiot.v8i29.987

Shafik, W., Matinkhah, S. M., Shokoor, F., & Nur Sanda, M. (2021). Internet of Things-based Energy Efficiency Optimization Model in Fog Smart Cities. *International Journal on Informatics Visualization*, *5*(2). https://doi.org/10.30630/joiv.5.2.373

Shafik, W., Mobin, M., & Matinkhah, S. M. (n.d.). *A Networked 5G Unmanned Aerial Vehicle Communication Pricing and Challenges*.

Sivasundari, A., & Ananthajothi, K. (2022). Impact of AI in 5G Wireless Technologies and Communication Systems. In *Artificial Intelligent Techniques for Wireless Communication and Networking*. https://doi.org/10.1002/9781119821809.ch2

Tashan, W., Shayea, I., Aldirmaz-Colak, S., Aziz, O. A., Alhammadi, A., & Daradkeh, Y. I. (2022). Advanced Mobility Robustness Optimization Models in Future Mobile Networks Based on Machine Learning Solutions. *IEEE Access*, *10*. https://doi.org/10.1109/ACCESS.2022.3215684

Varatharaj, N., & Ramalingam, S. T. (2022). Enabling Continuous Connectivity Services for Ambrosus Blockchain Application by Incorporating 5G-multilevel Machine Learning Orchestrations. *Journal of Intelligent and Fuzzy Systems*, *42*(4). https://doi.org/10.3233/JIFS-211745

Venkatesh, V., Raj, P., & Anushia Devi, R. (2022). Smart Supportable Framework by Internet of Things (IoT) and Machine Learning for Future Agriculture. In *Advances in Computers* (Vol. 127).

Wang, K., Song, T., Wang, Y., Fang, C., He, J., Nirmalathas, A., Lim, C., Wong, E., & Kandeepan, S. (2023). Evolution of Short-Range Optical Wireless Communications. *Journal of Lightwave Technology*, *41*(4). https://doi.org/10.1109/JLT.2022.3215590

Wang, W., Zhu, L., Yao, C., & Yu, L. (2022). Machine Learning Facilitating TDOA Passive Positioning Based on Optimization Theory. In *Information and Control* (Vol. 51, Issue 4). https://doi.org/10.13976/j.cnki.xk.2022.1416

Waqas, M., Tu, S., Halim, Z., Rehman, S. U., Abbas, G., & Abbas, Z. H. (2022). The Role of Artificial Intelligence and Machine Learning in Wireless Networks Security: Principle, Practice and Challenges. *Artificial Intelligence Review*, *55*(7). https://doi.org/10.1007/s10462-022-10143-2

Xie, Y., Wang, Y., Kandeepan, S., & Wang, K. (2022). Machine Learning Applications for Short Reach Optical Communication. In *Photonics* (Vol. 9, Issue 1). https://doi.org/10.3390/photonics9010030

Zeb, S., Mahmood, A., Hassan, S. A., Piran, M. J., Gidlund, M., & Guizani, M. (2022). Industrial Digital Twins at the Nexus of NextG Wireless Networks and Computational Intelligence: A Survey. In *Journal of Network and Computer Applications* (Vol. 200). https://doi.org/10.1016/j.jnca.2021.103309

Zhang, S., Gu, H., Chi, K., Huang, L., Yu, K., & Mumtaz, S. (2022). DRL-Based Partial Offloading for Maximizing Sum Computation Rate of Wireless Powered Mobile Edge Computing Network. *IEEE Transactions on Wireless Communications*, *21*(12). https://doi.org/10.1109/TWC.2022.3188302

14 Estimation of Channel for Multipath Transmission in Wireless Communication Using OFDM

Mahapriya C, Shweta B Thomas, and S. Sangeetha

14.1 INTRODUCTION

In 2019, 5G was developed with wireless technology to become the next generation. It's providing 10 gigabits per second of speed. 5G uses higher frequency bands than previous versions, operating at the range of low band, midband or millimetre wave frequencies from 24 to 66 GHz. The latest generation of cellular technology has had a significant impact on the advancement of 6G communication for several years. The 6G technology operates in a more extensive range of 30 GHz to 3000 GHz frequencies which are capable of providing significantly superior bandwidths, compared with 5G. The 6G will be able to deliver 206.25 gigabits per second after 2030. In order to automatically manage, monitor and optimise networks, machine learning technology will be a critical factor for the deployment of 6G networks. In order to meet the very challenging and strict requirements for future applications, current methods of signal processing and communications design are not sufficient [1]. This will make everything, like self-driven cars, automated production. The aim of the new mobile communication system is to maximise data speeds with more mobility, requiring a wide range of frequencies and higher levels of integration. OFDM plays a major role in wireless communication for transmitting high speed data over indoor and outdoor applications [2]. Considering the factors such as energy efficiency and privacy and security, future wireless communication systems should serve many applications that can be effectively supported by an intelligent cyber physical infrastructure based on novel network structures, spectrum access schemes or resource allocation solutions. Highly innovative technical solutions are required to address a wide range of wireless communications challenges, so as to be able to achieve an effective delivery of ultra-high data rates, dense connectivity and robust coverage while respecting significantly divergent requirements on the quality of services [3].

DOI: 10.1201/9781032656021-14

14.2 ORTHOGONAL FREQUENCY DIVISION MULTIPLEXING (OFDM)

From the last decade, the application of OFDM in wireless communication systems (especially for mobile communication) has attracted interests due to its various advantages and reduction of effects of frequency selective fading [4]. Nevertheless, estimating the fading channel parameters and monitoring the same are required to perform a high rate of demodulation coherently and spectrum efficient OFDM systems that utilise M-array modulation structures with no constant amplitudes. For OFDM systems, it is possible to use a correlation between the response of channel frequency at various time and frequency for estimating the channels. Based on the filtering in both time and frequency domain, the estimation of channel for the OFDM systems has been implemented in [5–7]. A singular-value decomposition-based estimation of channel was also presented in [8]. No assumptions can be made about the channel model of these methods, which means there may be a large degree of estimation difficulty. Multipath propagation is, however, often used to describe a radio channel in wireless communication system. The multipath propagation has been modelled by a series of dominant specular lines, usually 2 to 6 in large cell with platforms having high base station antenna [9]. In addition, the possibility of a low multipath fade channel is present due to fast data transmission in wireless communications. Further, the ratio of length of OFDM sample spanned by the multipaths to the number of multipaths can be termed as the sparsity of the multipath channel [10].

14.2.1 Estimation of OFDM Channel Using Singular Value Decomposition

The OFDM baseband model is depicted in Figure 14.1 wherein modulation of data is performed by IDFT and demodulation is performed by DFT with low computational complexity [11]. The assumption of the preservation of elimination of inter-symbol interference, i.e., ISI and the tone orthogonality are preserved by initiating the cyclic prefix (CP). Also, slow fading is assumed and hence its value is one OFDM symbol. Consider L is the length of the CP and N is the number of tones in the OFDM system respectively and thus the system is defined as a parallel set of Gaussian Channels with the correlation attenuation, *Sb* and mathematically, the OFDM system with sampling period (*Ts*) is given by

$$g(t) = \sum_m \beta_m \delta\left(t - T_m T_s\right) \tag{14.1}$$

where *g*(*t*) is the impulse response of the channel and *n*(*t*) is white gaussian noise of the estimated channel [13].

14.3 ESTIMATION OF CHANNEL IN OFDM

As the OFDM has the advantages of better transmission speed and bandwidth efficiency and good robustness in relation to the multiple path fade and delay, OFDMM has become a widespread application of telecom systems due to their wide

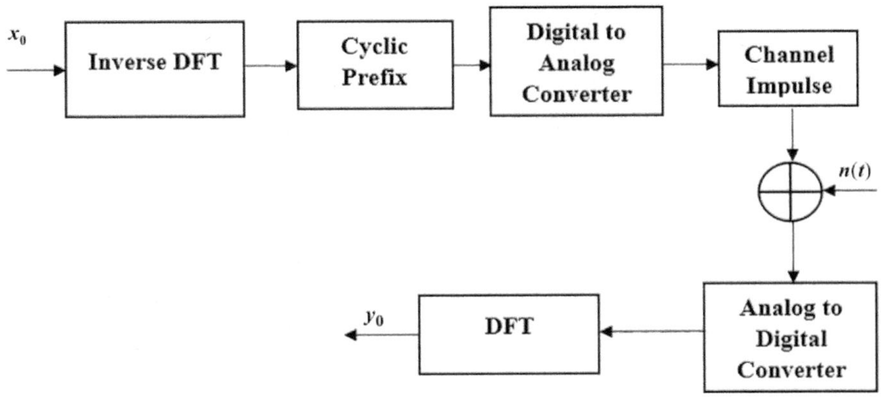

FIGURE 14.1 Mentioning OFDM channel response using DFT and IDFT.

applicability. For example, using it in a Digital Audio Broadcasting (DAB) system, digital subscriber line (DSL) standard, ISDN Line or Wireless LAN standard such as 802.11 Wi-Fi of IEEE (US), and the HIPRLAN/2 (European). For the broadband access, the standards such as 802.1WiMAX (IEEE), and for 4th Generation mobile communications, it is also proposed. The advantage of differential phase-shift keying (DPSK) is also utilised in OFDM systems to avoid the requirement of tracking a time varying channel, but it confines the number of bits per symbol and the loss of 3 dB signal-to-noise ratio (SNR) is occurred. Coherent modulation allows arbitrary constellation of signals, but it requires the efficient channel estimation strategies for detection and decoding. The development of channel estimation networks for OFDM systems is characterised by two major problems. The arrangement of pilot information is the major concern wherein a reference signal is used by both transmitters and receivers. The development of an estimator with reduced complexity and excellent channel tracking ability is the second concern. Since the fading channel of OFDM systems may be considered as a 2D signal (time and frequency), the 2Dsquare Wiener interpolation filter is used to determine the best channel for estimation of mean square exponential error. It is unfortunate that, for practical purposes, this 2D estimator structure is too difficult to implement. In OFDM systems, where the two constraints work in concert and a good trade-off between them is required, users of estimators which have low complexity and higher accuracy need to take advantage of these data rates mixed with low bit error rate. In OFDM systems, one dimensional channel projections are normally taken for the purposes of achieving a trade-off between complexity and accuracy [7, 12–17]. Block-type and comb-type pilot channel estimates, where the pilots are inserted in the frequency direction with respect to time, shall be two of the basic 1D channel estimations. The estimate of block pilot arrangement depends on the errors such as LSE, MMSE and modified MMSE. The estimation for the comb pilot arrangement includes maximum likelihood (ML) estimation, LS with 1D interpolation and the parametric channel modelling-based (PCMB) estimation. Other methods for estimation of channel, such as the method of linear 2D interpolations, an estimator approach that uses inverse filtering and decoding, which applies to OFDM systems with multiple transmitters and receiver antennas, etc. have also been investigated [18–22].

14.3.1 A Simple Method for Multi Antenna OFDM System

Even though the temporal channel is estimated simply using optimum training sequences to achieve the best performance during training, the relative phases of the OFDM systems could not be determined when the systems are in transmit mode. This is because, at different antennas, the transmitted signal, t_i [q, p] are predicted using transmitted random data. Therefore, a simple but effective technique is still needed for OFDM systems to lower the complexity in the parameter of channel's variation. An algorithm to estimate the channel for OFDM system with multiple transmit antenna is discussed in this section.

$$\begin{aligned} S_{11}[q]\,\widetilde{h1}[q] &= R_1[q] - S_{21}[q]\,\widetilde{h2}[q] \\ S_{22}[q]\,\widetilde{h2}[q] &= R_2[q] - S_{12}[q]\,\widetilde{h1}[q] \end{aligned} \tag{14.2}$$

For systems with constant modulation, the discussion in the previous part was as follows:

$S_{11}[q] = S_{22}[q] = KT$ and, therefore

$$\begin{aligned} \widetilde{h1}[q] &= \frac{1}{K}\left(R_1[q] - S_{21}[q]\widetilde{h}_2[q]\right) \\ \widetilde{h2}[q] &= \frac{1}{K}\left(R_2[q] - S_{12}[q]\widetilde{h}_1[q]\right) \end{aligned} \tag{14.3}$$

where, if $\widetilde{h2}$ $[q]$ evaluates to be known, then $\widetilde{h1}$ $[q]$ is calculated without the need of matrix inversion. However, anyone among $\widetilde{h1}$ $[q]$ or $\widetilde{h2}$ $[q]$ evaluates to be known. $\widehat{hi}$ $[q]$'s for $i = 1,2$ denotes the parameter of channel variation at time t, i.e.,

$$\widehat{hl}[q] = \sum\nolimits_{b\geq 0} f_p\,\widetilde{hl}[q-b] \tag{14.4}$$

where f_b (b ≥ 0) is the coefficient of the channel estimator [23, 24] and the corresponding Fourier transform is Ø_w. At time, $q - 1$, the estimation of channel parameters robustly, $\widetilde{h1}$ $[q]$ and $\widetilde{h2}$ $[q]$ are replaced respectively with $\widehat{h1}$ $[q-1]$ and $\widehat{h2}$ $[q-1]$ in (14.5),

$$\begin{aligned} \widetilde{h1}[q] &= \frac{1}{K}\left(R_1[q] - S_{21}[q]\widehat{h}_2[q-1]\right) \\ \widetilde{h2}[q] &= \frac{1}{K}\left(R_2[q] - S_{12}[q]\widehat{h}_1[q-1]\right) \end{aligned} \tag{14.5}$$

Hence, the complexity in computation of the system is reduced, but it may cause degradation of the system's performance. However, it does not affect the performance of the OFDM system significantly.

14.3.2 Data Transmission over Fading Channel by Multiple Antenna System

A transmitter antenna having M-element array transmitter and S-element array for receiver is considered for narrowband transmission system in this study. Each element of S-array can be considered as one distinct receiver. A broadcast scene is assumed wherein a common message is attempted to transmit, irrespective of the number of receivers with different symbols for M-array antenna elements. The lth receiver received the complex baseband signal. This received signal is the superposition of noisy M-transmitted symbols $X_1, X_2, \ldots X_M$. Each signal is attenuated and phase-shifted by a complex coefficient $\propto_i$. *li* indicates the fading occurred between the two antennas involved in the transmission, i.e.,

$$Y_l = \sum_{i=1}^{M} \propto_{i,l}^{*} X_i + V_l \qquad 1 = 1,2,\ldots.,S \tag{14.6}$$

With the reflections off by the objects present in the transmission path, this fading channel is derived using the effect of multiple signal copies arrived at a receiver in different times [**25**].

14.3.3 Channel Model and Optimal Training Sequences

If the transmission system consists of multiple antennas for transmitting and one antenna for receiving, then the analysis for multiple antennas can be generalised. Consider the signals s_1 and s_2 are propagated at a time over the channels $a_1 = \left[a_1(0)\ldots a_1(S-1)\right]^T$ and $a_2 = \left[a_2(0)\ldots a_2(S-1)\right]^T$, which are frequency-selective and wherein *T* denotes the transpose of the vector. The finite-impulse response (FIR) filter is used to model each channel with S taps. Therefore, at time 'b' the expression of the received signal is

$$y(b) = \sum_{i=0}^{S-1} a_1(i) s_1(b-i) + \sum_{i=0}^{S-1} a_2(i) s_2(b-i) + z(b) \tag{14.7}$$

where $z(b)$ represents the white Gaussian noise (AWGN). The inputs s_1 and s_2 are finite-signal constellation and propagated at the training data and data symbols that maximise the channel capacity respectively as *Ni* and *Nt*. The receiver has the known 2*Nt* symbols in training for estimating the unknown 2*S* channel coefficients. For this, the channels a_1 and a_2 are assumed to be unchanged for the data block transmission [26].

14.3.4 Pilot arrangement Based Estimation of Channel in OFDM Systems

The N_p carrier signals are inserted uniformly with the $x_q(m)$ number of pilots, then the binary data is represented by:

$X(b) = X(mS + 1)$

$$\begin{cases} x_{q(m)}, & l = 0 \\ inf.data & l = 1,\ldots S-1 \end{cases} \tag{14.8}$$

Also, at pilot sub-carriers, the channel frequency response is the $H_p(b)$, $b=0,1,\ldots..N_p$. The LS based channel estimation in pilot sub-carriers with input data $X_p(b)$ and output data $Y_p(b)$ at b^{th} pilot sub-carrier is given by:

$$H_e = \frac{Y_p}{X_p} \quad b=0,1,N_p-1 \tag{14.9}$$

To remove the ICI, guard time provides OFDM symbol. Since LS estimate is prone to ICI and noise, MMSE is used for complexity compromising. In view of the matrix inversion being included in every iteration, a simplified MMSE estimator was proposed by [27]. The advantage of this simplified system is that performance of the inverse calculations occurs once. In [28] reduce the complexity, a low approximation of rank is used to decompose single values.

14.3.5 Interpolation Techniques for Comb-Type Pilot Arrangement

At pilot sub-carrier, the channel is estimated at data sub-carriers using channel information in comb-type pilot estimation of channels. The linear interpolation has been shown for better performance of the piecewise-constant interpolation in [29]. At b^{th} data-carrier, $mS < b < S(m+1)$, the channel is estimated using linear interpolation by:

$$\begin{aligned} H_e(b) &= H_e(mS+l) \qquad 0 \le 1 < S \\ &= \left(H_p(m+1) - H_p(m)\right)\frac{1}{S} + H_p(m) \end{aligned} \tag{14.10}$$

where $H_e(b)$ is the estimated channel and $H_p(b)$ is the frequency response of the channel pilot sub-carriers. However, in [27], it is proved that the II order interpolation has better performance than the linear one and is given by:

$$H_e(b) = H_e(mS+l)$$

$$c_1 H_p(m-1) + c_0 H_p(m) + c_{-1} H_p(\mathrm{m}+1)\beta$$

$$\text{where } \begin{cases} c_1 = \dfrac{\beta(\beta-1)}{2} \\ c_0 = -(\beta-1)(\beta+1), \quad \beta = \dfrac{1}{N} \\ c_{-1} = \dfrac{\beta(\beta+1)}{2} \end{cases} \tag{14.11}$$

The original data sequence is converted to low pass interpolated data by inserting zeros followed by low pass FIR filtering. It passes the original data without changing and it interpolates to minimise the MSE between the interpolated points. Further, it is known that spline cubic interpolation results a continuous and smooth polynomial. The zero padded spatial interpolation and DFT/IDFT is a high-resolution interpolation [30]. After channel estimation $\{H_p(b), b = 0, 1, N_p - 1\}$, first using IDFT time domain has been converted:

$$G(q) = \sum_{b=0}^{N_p-1} H_p e^{(2\pi bq/N_p)}, q = 0, 1, \ldots\ldots N_p - 1. \tag{14.12}$$

The channel estimation of all frequencies is calculated by using the following equation (14.13)

$$H(b) = \sum_{n=0}^{N-1} G_N(q) e^{-j(2\pi/N)nb}, \quad 0 \leq \mathrm{b} \leq \mathrm{N} - 1. \tag{14.13}$$

14.4 GRAND CHALLENGES AND OPPORTUNITIES FOR WIRELESS COMMUNICATION

14.4.1 Security, Secrecy, and Privacy

Wireless technology provides tremendous information. The information is being transmitted via wireless communication system. Transmission of the information via radio, broadcasting antenna and radar is lost due to external noise in free space. In order to transmit the information efficiently without interference, new technology should be implemented. For instance, examining the propagation characteristics of networks might improve their safety performance as compared to physically layer security technologies [31].

14.4.2 Communication Infrastructure

Improving the utilisation of resources by new system architecture is another way to cope with an increasing demand for wireless services. The drawbacks of non-dynamic resource allocation of wireless cellular system are eliminated by large-scale cell free network structure [32]. By using array of antenna, multiple base station serves a greater number of users in order to get the better efficiency. In order to get the better trade-off performance, the system has to be designed with low complexity, low cost. To avoid this issue the connectivity of network is important in communication, and research efforts are being carried out. In addition, the integration of satellite, air and sea network is a new trend in channel communication which will bias major attention and study as part of an overall aim to provide heterogenous broadband connectivity with next generation wireless systems.

14.4.3 Coding and Modulation

In future wireless networks, it is crucial to introduce new coding and modulation schemes so that the possible gains predicted by theories can be achieved in practice. Modulation scheme, channel coding, integrated terrestrial and submarine network can be useful for connecting the network [33]. Furthermore, a good strategy is the adoption of more efficient modulation and channel programming schemes or new modulation sources in order to enable much higher frequency efficiency for future mobile networks. Each generation has been dominated by a new pattern of coding along the technological evolution. In the 2G, in 3G and 4G, as well as in 5G, convolution coding has been taken into account, turbo codes have been replaced with more efficient ones, LDPC was used along with Polar Codes. The performance of the codes has been optimised for certain code rates, as a majority of these considered codes have been designed with low levels of capabilities in this regard. In such a case, it is expected that performance improvements and robustness to various environments would be achieved through the introduction of enhanced codes in current systems. In addition, to prevent the information loss that would result if modulation and channel schemes were created separately, e.g., when detection and decryption are performed in isolation, it may be possible to combine their design. In this case, an exhaustive examination of how to design the channels coding, modulation, detection and decoding, low latency, high reliability/higher bandwidths in order to ensure its successful implementation on a new generation of wireless networks needs to be carried out [34].

14.5 CONCLUSION

In order to meet the tough and diverging requirements for next wireless systems, there are many technological problems that must be addressed such as excellent spectral efficiency, very slow lag times, an enormous amount of device connectivity, fast maximum possible data rate, exceptionally reliable user experience, a diverse range of various levels of quality of service, energy efficiency and sharply reduced costs. In the next few years, a lot of technical breakthroughs will take place in wireless communications. In order to achieve these breakthroughs, a key factor will be cross disciplinary dialogue and collaboration between academics and industry. To promote these multidisciplinary activities as well as contribute to the development of a Globally Widespread Cyber Physical Infrastructure that can be fully interrogated around the world, the "Wireless Communication" section in the Frontier plans to provide an area where ideas are generated, polished and perfected.

REFERENCES

[1] www.frontiersin.org/articles/10.3389/frsip.2021.664331/full

[2] www.techrepublic.com/article/5g-vs-6g/

[3] S. Dang, O. Amin, B. Shihada, and M.-S. Alouini, "What should 6G be?" *Nat. Electron.*, vol. 3, pp. 20–29, 2020.

[4] H. Sari, G. Karam, and I. Jeanclaude, "Transmission techniques for digital terrestrial TV broadcasting," *IEEE Commun. Mag.*, vol. 33, pp. 100–109, Feb. 1995.

[5] P. Hoeher, S. Kaiser, and I. Robertson, "Two-dimensional pilot-symbol-aided channel estimation by wiener filtering," in *Proc. ICASSP'97*. IEEE, Munich, Germany, 1997, pp. 1845–1848.

[6] M. Speth, S. Fechtel, G. Fock, and H. Meyr, "Broadband transmission using OFDM: System performance and receiver complexity," in *Proc. Int. Zurich Seminar on Broadband Communications*. IEEE, Munich, Germany, 1998, pp. 99–104.

[7] Y. Li, L. J. Cimini, and N. R. Sollenberger, "Robust channel estimation for OFDM systems with rapid dispersive fading channels," *IEEE Trans. Commun.*, vol. 46, pp. 902–915, July 1998.

[8] O. Edfors, M. Sandell, J. J. van de Beek, S. K. Wilson, and P. O. Börjesson, "OFDM channel estimation by singular value decomposition," *IEEE Trans. Commun.*, vol. 46, pp. 931–939, July 1998.

[9] J. T. Chen, A. Paulraj, and U. Reddy, "Multichannel maximum-likelihood sequence estimation (MLSE) equalizer for GSM using a parametric channel model," *IEEE Trans. Commun.*, vol. 47, pp. 53–63, Jan. 1999.

[10] I. Kang, M. P. Fitz, and S. B. Gelfand, "Blind estimation of multipath channel parameters: A modal analysis approach," *IEEE Trans. Commun.*, vol. 47, pp. 1140–1150, Aug. 1999.

[11] A. Peled and A. Ruiz, "Frequency domain data transmission using reduced computational complexity algorithms," in *Proc. IEEE Int. Conf. Acoustics, Speech, and Signal Processing*. Denver, CO, IEEE, 1980, pp. 964–967.

[12] O. Edfors, M. Sandell, J.-J. Van de Beek, D. Landstrom, and F. Sjöberg, *An Introduction to Orthogonal Frequency Division Multiplexing*. Lulea, Sweden: Lulea Tekniska Universitet, 1996, pp. 1–58.

[13] J.-J. Van de Beek, O. S. Edfors, M. Sandell, S. K. Wilson, and O. P. Börjesson, "On channel estimation in OFDM systems," *45th IEEE Vehicular Technology Conference*. Chicago, IL, IEEE, July 1995, vol. 2, pp. 815–819.

[14] P. Strobach, "Low-rank adaptive filters," *IEEE Trans. Signal Process.*, vol. 44, pp. 2932–2947, Dec. 1996.

[15] S. Coleri, M. Ergen, A. Puri, and A. Bahai, "channel estimation techniques based on pilot arrangement in OFDM systems," *IEEE Trans. Broadcast.*, vol. 48, pp. 223–229, Sept. 2002.

[16] J. Wu and W. Wu "A comparative study of robust channel estimators for OFDM systems," *Proceedings of ICCT*, pp. 1932–1935, 2003.

[17] B. Yang, K. B. Letaief, R. S. Cheng, and Z. Cao, "Channel estimation for OFDM transmission in multipath fading channels based on parametric channel modelling," *IEEE Trans. Commun.*, vol. 49, pp. 467–479, Mar. 2001.

[18] B. Ottersten, M. Viberg, and T. Kailath, "Performance analysis of the total least squares ESPRIT algorithm," *IEEE Trans. Signal Process.*, vol. 39, pp. 1122–1135, May 1991.

[19] X. Hou, S. Li, D. Liu., C. Yin, and G. Yue, "On two-dimensional adaptive channel estimation in OFDM systems," in *60th IEEE Vehicular Technology Conference*. Los Angeles, CA, IEEE, Sept. 2004, vol. 1, pp. 498–502.

[20] F. Sanzi, J. Sven, and J. Speidel, "A comparative study of iterative channel estimators for mobile OFDM systems," *IEEE Trans. Wireless Commun.*, vol. 2, pp. 849–859, Sept. 2003.

[21] Y. Li, "Simplified channel estimation for OFDM systems with multiple transmit antennas," *IEEE Trans. Commun.*, vol. 1, pp. 67–75, Jan. 2002.

[22] G. Auer, "Channel estimation in two dimensions for OFDM Systems with multiple transmit antennas," *GLOBECOM*, pp. 322–326, 2003.

[23] Y. (G.) Li, N. Seshadri, and S. Ariyavisitakul, "Channel estimation for OFDM systems with transmitter diversity in mobile wireless channels," *IEEE J. Select. Areas Commun.*, vol. 17, pp. 461–471, Mar. 1999.

[24] Y. (G.) Li, L. J. Cimini, and N. R. Sollenberger, "Robust channel estimation for OFDM systems with rapid dispersive fading channels," *IEEE Trans. Commun.*, vol. 46, pp. 902–915, July 1998.

[25] W. C. Jakes, Ed., *Microwave Mobile Communications*. New York: Wiley, 1974.

[26] C. Fragouli, N. Al-Dhahir, and W. Turin, "Training-based channel estimation for multiple-antenna broadband transmissions," *IEEE Trans. Wireless Commun.,* pp. 384–391, vol. 2, No. 2, March 2003.

[27] U. Reimers, "Digital video broadcasting," *IEEE Commun. Mag.*, vol. 36, no. 6, pp. 104–110, June 1998.

[28] M. Hsieh and C. Wei, "Channel estimation for OFDM systems based on comb-type pilot arrangement in frequency selective fading channels," *IEEE Trans. Consumer Electron.*, vol. 44, no. 1, Feb. 1998.

[29] L. J. Cimini, "Analysis and simulation of a digital mobile channel using orthogonal frequency division multiplexing," *IEEE Trans. Commun.*, vol. 33, no. 7, pp. 665–675, July 1985.

[30] Y. Zhao and A. Huang, "A novel channel estimation method for OFDM Mobile Communications Systems based on pilot signals and transform domain processing," in *Proc. IEEE 47th Vehicular Technology Conf.* Phoenix, USA, IEEE, May 1997, pp. 2089–2093.

[31] Y.-S. Shiu, S. Y. Chang, H.-C. Wu, S. C.-H. Huang, and H.-H. Chen, "Physical layer security in wireless networks: A tutorial," *IEEE Wireless Commun.,* vol no. 18, pp. 66–74, 2011.

[32] S. Buzzi, C. D'Andrea, A. Zappone, and C. D'Elia, "User-centric 5G cellular networks: Resource allocation and comparison with the cell-free massive MIMO approach," *IEEE Trans. Wireless Commun.*, vol. 19, no. 2, Feb 2020.

[33] Y. Jin, J. Zhang, S. Jin, and B. Ai, "Channel estimation for cell-free mm Wave massive MIMO through deep learning," *IEEE Trans. Veh. Technol.,* vol. 68, pp. 10325–10329, 2019.

[34] Y. Yuan, Y. Zhao, B. Zhong, and S. Parolari, "Potential key technologies for 6G mobile communications." Available online at: https://arxiv.org/ftp/arxiv/papers/1910/1910.00730.pdf

[35] M. Latva-ahom and K. Keppannen, "Key drivers and research challenges for 6G ubiquitous wireless intelligence (White Paper)," in *6G Flagship*. IEEE, Finland, 2019.

15 Industrial IoT
Challenges and Applications

S. Sangeetha, and Shweta B Thomas

15.1 INTRODUCTION TO IIOT

To enhance the manufacturing and industrial processes, industrial internet of things uses smart sensors, actuators and radio frequency tags. These devices help in collecting the data, exchanging the data and also for analysing the data. It is also defined as application of internet of things in industrial environment. The IIOT not only replaces the human in real-time data capturing and analysis, it also helps in communicating information for decision making in business. This decision-making process increases the efficiency, reduces the cost and increases the quality. It is the key component of industry 4.0 which facilitates the transformation from manual process to automated operations. It also aims at connectivity of the devices and their security in the data analysis process. IIOT is specially used for the heavy-duty tasks and handles a massive amount of data. These data should be precise, sensitive and continuous. For example, in nuclear power plant monitoring, these precise and continuous data helps to prevent the hazardous environment. The implementation of IIOT in production or industrial projects is to reduce the cost of production and maintenance; and improve the safety, product quality, stability and efficiency.

The commonly used terms of IIOT are analog data, connectivity, data streaming, data insights, depth of insights, visibility and edge computing. The sensors collect the analog data and it needs to be converted to digital for analysis as well as for storing. The connectivity helps to update and fix the software-firmware. Real-time data streaming used in safety mechanism helps to avoid hazardous events. The depth of data is an important factor depending upon the scenario where it is used. Certain monitoring of events requires immediate action to be taken and certain data requires research for many years (e.g. NASA). Visibility is the term related to the data storage in the cloud and can be accessed whenever required.

15.2 ARCHITECTURE IN IIOT

The three tier architecture (Figure 15.1) comprises of enterprise tier, platform tier and edge tier. The edge nodes are sensors, control systems and actuators. The processing of the data and control flow are involved in these three layers. The edge nodes send the data to the edge layer. It is received by the platform layer and processes the commands to the edge layer from the enterprise layer. The domain specific applications

DOI: 10.1201/9781032656021-15

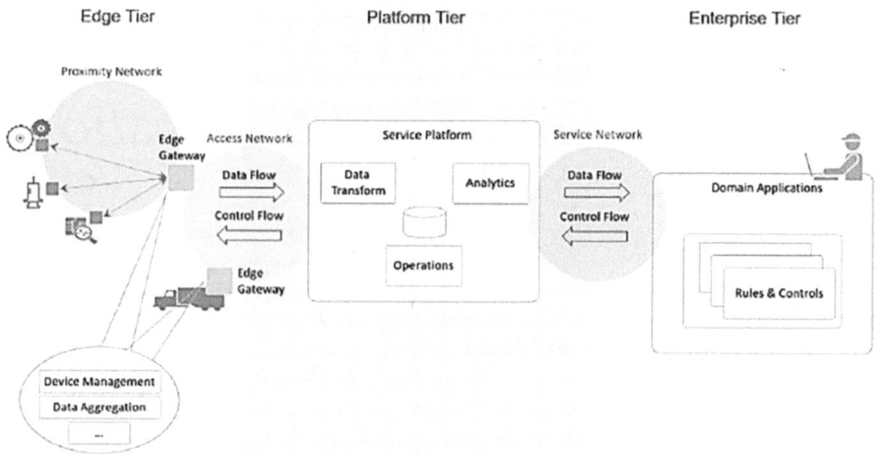

FIGURE 15.1 IIOT System Architecture.

and decisions are supported by the enterprise layer. The connectivity between the services within a tier, between the enterprise and platform layer is enabled by the service network. The advantage of this tier architecture is that the functioning partition is simple in each tier. The technology is not fully utilized until the service is fully utilized by the end users. At each level of the technology in various layers, a particular skill set is required.

15.3 COMPONENTS OF IIOT ARCHITECTURE

The various components of IIOT architecture are IIOT enabled devices, management of edge data, cloud processing, gateways and IIOT platforms.

15.3.1 IIOT Enabled Devices

These objects are present at the edge of the IoT system. They are present near to the source. In the industrial scenario, they act as the sensors and actuators. They also are processing units and edge devices such as sensors, cameras and monitoring devices.

15.3.2 Management of Edge Data

The artificial intelligence is not fully utilized when the data do not have the required quality and volume. If the immediate data are required, the data processing can be done even at the initial stage of sensor level. Hence, the edge computing processes the data at the edge level. Then, it is easy to analyse and collect the data. Instead of processing the entire data, the required features of data alone are processed. This saves the additional processing of data and amount of data to be sent also decreases.

15.3.3 Cloud Processing

The capacity of the edge computing is limited for the preprocessing. Then, the user needs to utilize the cloud for processing the data in depth. In cloud, the extensive data processing will happen and also it collects the data from different sources. It comprises of hub, storage, analytics and user interface.

a) **Hub**
It provides connectivity in the remote access. It also maintains the connection, authorization and verification of the communication channel.
b) **Analytics**
It helps in processing the data and further analysis of the data.
c) **Storage**
It helps in storing the data before and after processing it.
d) **User Interface**
The end user visualizes the data analysis using the user interface such as email, messages, calls and browsers.

15.3.4 Gateways

The data is collected and for further processing it is converted to digital channels at the gateway. After processing it is transmitted to the internet for further processing and uploaded to the cloud. The gateway can be hardware or software. The autonomous devices act as hardware gateways. It also helps in internet connectivity and linking the router. The software acts similar to the hardware gateways. It accesses the data visualization through user interface.

15.3.5 IIOT Platforms

The monitoring and controlling of the operations can be done by the IIOT systems. The IIOT platforms control the data, analytics, visualization of data and artificial intelligence. The mentioned architecture serves as the reference to develop other models. The models can be applied in various fields such as energy, health sector and governmental use.

15.4 CHALLENGES IN IIOT

More devices are connected in IIOT and also more data need to be analysed in a short span of time. The errors due to machines and humans do not have a proper solution. The poor maintenance of the devices causes failure in confidentiality, control policies and network services. Some of the challenges in industrial networks are wireless coexistence, latency, interoperability, sensor data streaming, safety, security and privacy.

a) **Wireless Coexistence**
The devices are connected safely for the purpose of communication using wireless technology. The signal interference occurs due to multipath

propagation such as refraction, scattering, diffraction and reflection. The interference causes loss of data and jitter and affects the synchronization between the communication devices, packet loss and delay. Hence, wireless coexistence has to be achieved by giving suitable distance between the devices in IIOT.

b) **Latency**
The latency is referred to as delay. The time between the collection of data and reaction of output is called latency. Depending upon the applications, the round-trip delay varies. The inter-continental communication of IIOT can have delay less than 300 ms and intra-continental communication has delay less than 50 ms [1]. The total latency is calculated by considering the time duration required for the packets to be decoded along with the complexity of the receivers.

c) **Interoperability**
It refers to the ability to connect and exchange the data without any interruptions. The cross-layer interoperability and cross-system interoperability are the two types of interoperability. Cross-layer interoperability refers to the performance of the OSI layers. It is done by using the software-defined radio or software-defined networking. The multiple technologies are supported by the wireless interfaces which are designed with a high level of flexibility [2]. This interoperability between devices has challenges such as designing efficient hardware communication. The cross-system interoperability refers to the maintenance of the connection of different system architecture. The AI based tools are enabled to make decisions and help to analyse the sensor data [3].

d) **Sensor Data Streaming**
The data streaming is required for data analysis and it has challenges such as fault tolerance in data storage and processing. The data durability is another issue in protecting the data from bit degradation and corruption. This can be avoided by using bit redundancy in data processing.

e) **Safety**
In an industrial scenario, the sensors and other devices help to increase safety and productivity. The manufacturing process focusses on preventing accidents in everyday life and other disasters such as nuclear incidents. It utilizes monitoring systems and alarming systems, emergency systems and protection systems. Most of the production system requires continuous monitoring. The monitoring systems such as sensors collect the real-time data in analog or digital form and send to the control systems to enable the control activities. The recent trends such as computer vision, wearable sensors and robotics help to achieve the required safety.

f) **Security and Privacy**
With multiple users interfacing the system, it is required to ensure the security and privacy of the system. The security system is divided as intrusion detection and prevention. Once the intrusion happens, then it is detected using the intrusion detection system. Another method is preventing the

intrusion using authorization, authentication and access control. When the same attack happens in the system, it is easy to detect the attack. If a new attack comes, then an anomaly-based intrusion detection model is preferred. This algorithm takes the decision which differs between past and future for the same event. Similarly, prevention mechanisms such as ciphering, keying, authentication, authorization and access control are used.

15.5 AUTOMATION TRENDS IN IIOT

The focus of new industry revolution is cost reduction, new technology, delivery in short duration and customizing the end product. This approach enables new technology according to the requirement of production. Along with the enabling technologies, the current industry revolution moves towards digitalization. The first industry revolution was a mechanical system in the field of production and most of the processes were manual. Steam engines were developed which led to the growth of the chemical industry. The increase in production of steel with economical cost improved the construction of bridges in the field of civil engineering. Similarly, spinning machines were introduced in the first revolution for the textile industry. The second revolution is integrated with electricity and it acts as a replacement for steam power. This revolution starts with the mass production of materials through the integration of electricity. The third revolution aims at digitalization and automation of the system with their corresponding process.

The fourth revolution begins with the globalization of technology and automation which is supported by cyber security, cloud computing, IIOT etc. Smart manufacturing systems machine to machine communication, human to machine communication came into the picture in the fourth generation of the industrial revolution. The enabling technologies for smart manufacturing are:

a) **Cloud Computing**
The large amount of data is collected and stored in the internet for accessing the buyers' as well as sellers' information. In a real-time environment, the data related to environment, machines are collected every day for various reasons such as fault diagnosis, maintenance and to improve the production. Due to the growing amount of data, cloud service is provided by Google in its different services such as software as service, platform as service and infrastructure as service. Users can connect to the cloud app using software as service. Cloud computing has advantages such as low cost of maintenance, security, independence of hardware and scalability.
b) **Cybersecurity**
The protection measurement and risk measurement are handled by cyber security. This acts as basic enabling technologies and further improves the performance of the digital transformation of the industry. The automation architecture of the manufacturing model is shown in Figure 15.2. In the first level, all the physical devices such as sensors and actuators are connected to the control layers.

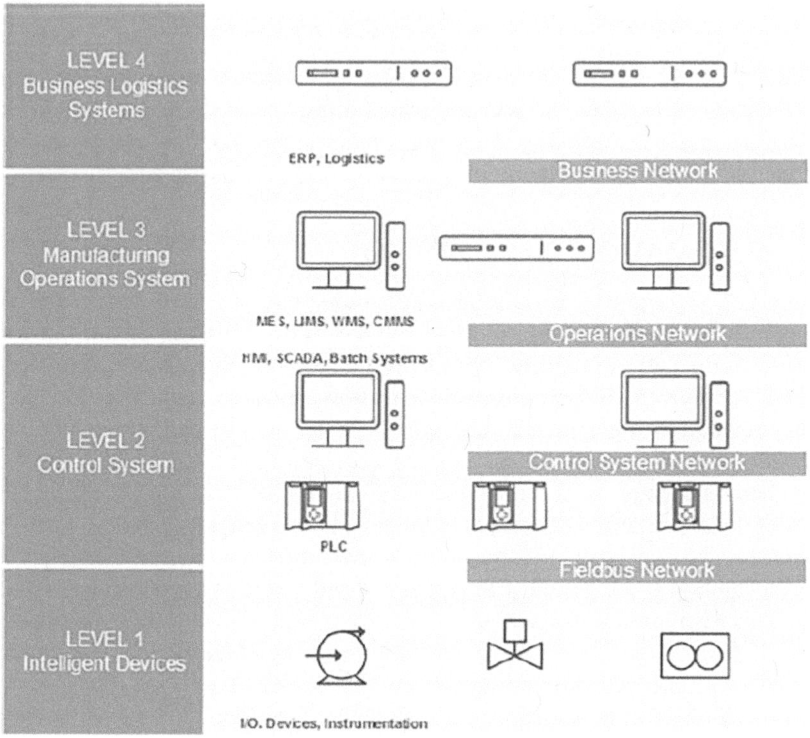

FIGURE 15.2 Architecture of the Control Level.

The control layer controls the operation of the devices, and it also supports the monitoring operation. The third level communicates between the control layer and enterprise level. The information to the end user such as quantity, orders etc is extracted by the enterprise layer.

15.6 SECURITY IN IIOT

In IIOT, decentralized computing architectures are used to subdivide the process into a number of sub-processes. This helps in hiding the malicious network to access the resources. As similar to the decentralized computing, the security mechanism also needs to be decentralized. One of the challenges in the decentralized mechanism is creation of loopholes due to the redundant processing of information. Cash payments are carried out using Bitcoins, and the concept of blockchain has come into the picture [4]. It consists of multiple blocks arranged in chronological order which are attached with a time stamp and easily accessible. Using blockchain, the data transfer between the peer group becomes efficient and the software gets upgraded automatically to identify the malicious code [5].

The blockchain is categorized as permission-based or permission-less [6–8]. The permission-less blockchain can be accessed by anyone for reading and writing the data. In the permission-based blockchain, the authorization is given to all the entities and it is controllable. It can also connect with multiple trusted users for collaboration. The challenges in IIOT such as auditability, interoperability and security issues can be solved by the permission-based block chain. Aggregate signature is generated based on different users and it is compressed for supporting public key infrastructure [9]. This improves the security in industrial internet of things. The most common attack in the IIOT is the man-in-the-middle attack and it damages the entire network. To avoid this attack, SHA-1 algorithm is proposed in [10] and also it uses location-based technology. New authentication techniques such as RF fingerprint have been proposed in [11] based on light-weight security architecture.

15.7 INTRUSION DETECTION IN DATA STREAMING, DATA ANALYTICS

The digitalization process is increasing day by day in society and the complex industrial process is being converted into smart technology. The internet of things is distributed throughout the network and it has threatening issues due to cyber-attacks. In the distributed architectures, it consists of three layers such as cloud layer, fog layer and edge layer. The data are handled and processed by the cloud layer, and it is the top end of the layer. The fog layer acts as the intermediate layer and the low-end devices are connected to this layer for the real-time experiences. The low-end devices are machines, sensors, detectors etc.

The large amount of data is collected by the data base management system using various SQL commands. The persistence of the data is removed which results in reduced latency. But the available storage is less compared to the main memory. Each of the data streams [12] is allocated with the time stamp and various operators such as aggregate, merge and stateless are used. For example, when the students are entering the process, the id is verified using the aggregate operator. It has data such as number of login attempts and it is sent to the filter. The filter sends the data to the analyst if any suspicious activity is found. It has another aggregate function which has the list of the per-user basis [13].

The commonly used stream processing engines are Apache Beam, Apache Storm, Apache Flink [14] and Aurora. The layered architecture of intrusion detection system consists of prevention, detection and mitigation. Prevention is the initial layer which protects using a firewall. The detection layer is used to identify the attacks and intrusion in the network. The mitigation system is the last layer of the system, and it has security measures when accessing the internet. The various intrusion systems are anomaly-based intrusion detection, misuse-based intrusion detection and specification-based intrusion detection systems [15]. The anomaly detection can handle the unknown attacks, it does not require frequent updates, and it is low accuracy. Misuse-based intrusion cannot handle unknown attacks, and it has high accuracy. Similarly, specification-based detection also has high accuracy and is cost effective.

15.8 REAL-TIME EXAMPLES

In the city of Australia, the automation for the water treatment facility was outsourced in the year 2000. The data acquisition and automation are initially for around 142 pumping stations. The faults are drastically increasing day by day and it is found that the employee used the wireless RTU for manipulating the commands leads to a man-in-the-middle attack and leads to spill of the sewage to the surroundings.

Similarly, a nuclear power plant was attacked by the slammer malware. It was found the attacker used the T-1 communication line by eliminating the firewall of the plant. The Black energy attack targets the interface of the machine-human. The attacker took control of the power distribution system and switched down the circuits etc. The same kind of incident happened in electrical turbines in the USA. The Stuxnet worm attacked the programmable logic controllers, and it changed the speed of the centrifuges. It completely destroyed the smart grid systems. The vampire attack took place in 2011 in the sensor nodes to destroy the communication abilities of the network. It depleted the battery of the targeted sensor nodes and increased the partition of the network. This attack is hard to detect, and it is very dangerous to the network. Some of the attacks are very difficult to identify due to the integrity behaviour of the system. The solution to these kinds of the attacks is continuous monitoring of the measurements for any changes in the networks. At the initial stage, the security risk analysis of the software, hardware and communication channel has to be handled. Based on the shortcomings, the mitigation solutions have to be analysed for resisting the cyber-attacks. It also requires educating the people working in the field and early action has to be taken for the threats.

REFERENCES

[1] Kumar, P., Lin, Y., Bai, G., Paverd, A., Dong, J.S., Martin, A.: Smart grid metering networks: A survey on security, privacy and open research issues. *IEEE Commun. Surv. Tutorials,* 21, 2886–2927 (2019).

[2] Stonebraker, M., Çetintemel, U., Zdonik, S.: The 8 requirements of real-time stream processing. *ACM Sigmod Rec.*, 34(4), 42–47 (2005).

[3] Gulisano, V., Jimenez-Peris, R., Patino-Martinez, M., Soriente, C., Valduriez, P.: Streamcloud: An elastic and scalable data streaming system. *IEEE Trans. Parallel Distrib. Syst.,* 23(12), 2351–2365 (2012).

[4] Nakamoto, S.: *Bitcoin: A Peer-to-Peer Electronic Cash System* (2009). [Online]. Available: https://bitcoin.org/bitcoin.pdf.

[5] He, S., Ren, W., Zhu, T., Choo, K.R.: BoSMoS: A blockchain-based status monitoring system for defending against unauthorized software updating in industrial Internet of Things. *IEEE Internet Things J.,* 7(2), 948–959 (Feb. 2020).

[6] Buterin, V.: A next generation smart contract & decentralized application platform. Ethereum, Zug, Switzerland, White Paper (2014).

[7] Hearn, M.: *CORDA: A Distributed Ledger* (Nov. 2016). [Online]. Available: https://docs.corda.net/static/corda-technical-whitepaper.pdf

[8] Dhillon, V., Metcalf, D., Hooper, M.: *The Hyperledger Project* (2019). [Online]. Available: www.Hyperledger.org.

[9] Qiao, Z. et al.: An efficient certificate-based aggregate signature scheme with provable security for industrial Internet of Things. *IEEE Syst. J.*, 17(1), 72–82 (Mar. 2023).

[10] Patil, P., Marathe, N., Jethani, V.: Improved ALERT protocol in MANET with strategies to prevent DOS & MitM attacks. In *Proc. Int. Conf. Autom. Control Dyn. Optim. Techn. (ICACDOT)* IEEE, Pune, India, Sep. 2016, pp. 372–377.

[11] Tian, Q. et al.: New security mechanisms of high-reliability IoT communication based on radio frequency fingerprint. *IEEE Internet Things J.*, 6(5), 7980–7987 (Oct. 2019).

[12] Stonebraker, M., Çetintemel, U., Zdonik, S.: The 8 requirements of real-time stream processing. *ACM Sigmod Rec.*, 34(4), 42–47 (2005).
[13] Gulisano, V.: Streamcloud: An elastic parallel-distributed stream processing engine. Ph.D. Thesis, Universidad Politécnica de Madrid (2012).
[14] Carbone, P., Katsifodimos, A., Ewen, S., Markl, V., Haridi, S., Tzoumas, K.: Apache flink: Stream and batch processing in a single engine. *Bull. IEEE Comput. Soc. Tech. Committee Data Eng.*, 36(4), 28–38 (2015).
[15] Fauri, D., Dos Santos, D.R., Costante, E., den Hartog, J., Etalle, S., Tonetta, S.: From system specification to anomaly detection (and back). In *Proceedings of the 2017Workshop on Cyber-Physical Systems Security and PrivaCy*. ACM, New York (2017), pp. 13–24.

16 Establishing Seamless Connectivity through UAV Communication Systems

Saif Ahmad, Mohd Javed Khan and Indrasen Singh

16.1 INTRODUCTION

Drones, often referred to as unmanned aerial vehicles (UAVs) [1], have quickly changed a number of sectors and uses, including aerial surveillance, package delivery, agriculture, and emergency management, to mention a few. These flying vehicles that operate autonomously or remotely have developed into essential tools that open up new possibilities and transform how work is done. UAV communication systems give the unmanned aircraft and the ground control station the tools for seamless connectivity, enabling real-time data transfer, command execution, and situational awareness. These systems are essential for promising the efficient and safe functioning of UAVs [2].

UAV communication requirements are unique and difficult. Unlike typical communication systems, UAV communication must take into consideration the aircraft's mobility, the possibility of long-range operations, and the necessity for dependable and secure data transfer. Furthermore, UAV communication systems must be lightweight, energy-efficient, and capable of functioning in a variety of environments [3]. Seamless connection is a vital necessity for UAVs because it allows critical information such as flight telemetry, sensor data, and video feeds to be sent between the aircraft and the ground control centre [4]. This connection provides operators with real-time situational information and allows them to make educated decisions during flight operations [5]. To provide continuous connectivity, UAV communication systems utilize a variety of technologies, including wireless communication protocols, satellite linkages, cellular networks, and ad hoc networking. To solve the obstacles faced by UAV operations, these systems make use of developments in wireless communication technology [6]. The selection of a suitable frequency range for data transmission is an important aspect in UAV communication systems. UAV communication systems can operate in licensed, unlicensed, or specialized frequency bands designated expressly for UAV operations, depending on the application and legal limitations. The frequency band selected affects the communication system's range, data throughput, and interference resilience [7]. Additionally, UAV communication systems must meet security and privacy considerations. As drones become

DOI: 10.1201/9781032656021-16

increasingly common in society, it is critical to ensure the integrity and security of data exchanged between the UAV and the ground control station. To safeguard the communication channel from unauthorized access and data manipulation, encryption, authentication measures, and secure protocols are used [8]. Finally, UAV communication technologies are the foundation of current drone operations, allowing for smooth interaction between the unmanned aircraft and the ground control centre. These systems use advanced wireless communication technologies to provide real-time data transfer, command execution, and situational awareness. As UAVs advance and find uses in a variety of sectors, the development of strong and efficient communication systems will be critical to realizing their full potential, as well as data manipulation. Communication systems are a crucial component that enables UAVs to execute their activities effectively and dependably.

16.1.1 UAV Communication Systems Overview

UAV communication systems are critical for providing dependable and effective connectivity between UAVs and their ground control stations (GCS). Wireless communication protocols, frequency ranges, and antenna systems are used in these systems to provide continuous data flow and command execution. They improve UAVs' overall operating capability by enabling real-time situational awareness [9].

UAV communication systems may use satellite connections and cellular networks in addition to typical wireless protocols like Wi-Fi and Bluetooth to enhance the range and coverage of UAV operations. Satellite links enable worldwide connectivity, making them excellent for long-distance or out-of-band operations. Cellular

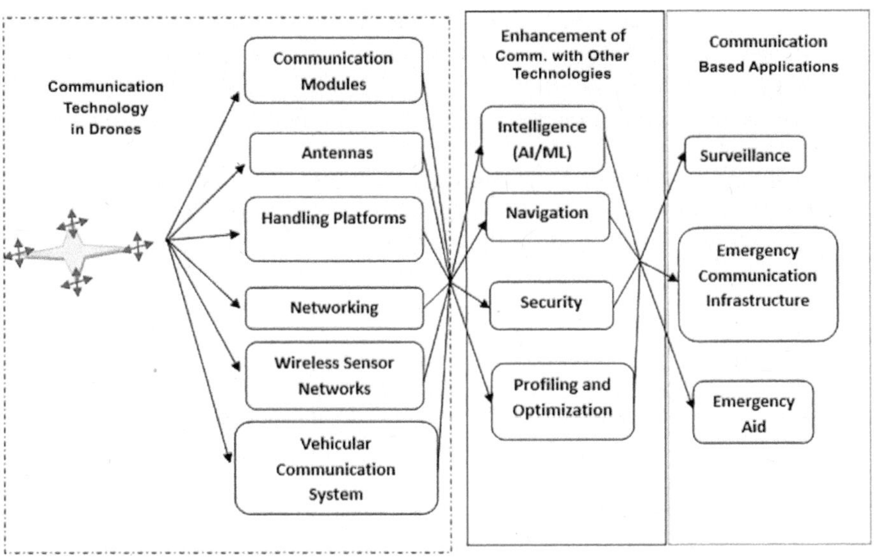

FIGURE 16.1 Drone communication technology scope.

networks, on the other hand, make use of existing infrastructure to build communication linkages, allowing UAVs to travel across greater distances [10].

The security of UAV communication systems is critical. To preserve the integrity and confidentiality of transferred data, encryption, authentication methods, and secure protocols are used, guaranteeing that the communication channel stays secure from unauthorized access and alteration [11]. Ongoing research and development activities aim to increase UAV communication systems' communication range, data rate, energy efficiency, and durability. Advances in these areas will improve UAV capabilities, allowing them to operate in a variety of situations and perform a wide range of tasks with better efficiency and dependability [12].

16.1.2 Essential Aspect of Seamless Connectivity

UAVs [13] require consistent connectivity since it allows them to:

1. **Data Transfer in Real-Time:** Seamless connectivity enables the instant transfer of crucial data such as telemetry, sensor readings, and video feeds, allowing operators to make educated decisions and successfully oversee operations.
2. **Situational Awareness:** It gives operators current information about the UAV's surroundings, assisting them in assessing the operational environment, identifying impediments, and responding rapidly to dynamic circumstances.
3. **Command and Control Execution:** Reliable and instant communication provides exact control over UAV movements, manoeuvres, and command execution in real-time.
4. **Risk and Safety Mitigation:** It enables operators to closely monitor UAV activities, detect possible problems, and respond quickly to alleviate safety concerns, minimizing the likelihood of accidents or unauthorized intrusions.
5. **Efficiency and Effectiveness of the Mission:** Improved mission outcomes and better productivity are the results of seamless connection, which optimizes resource allocation, flight routes, and adaptive decision-making based on real-time data.
6. **Combination with Other Systems:** It makes it possible for UAVs to be integrated with other technologies, allowing interoperability, data sharing, and cooperative operations and opening up fresh opportunities and uses for UAVs.

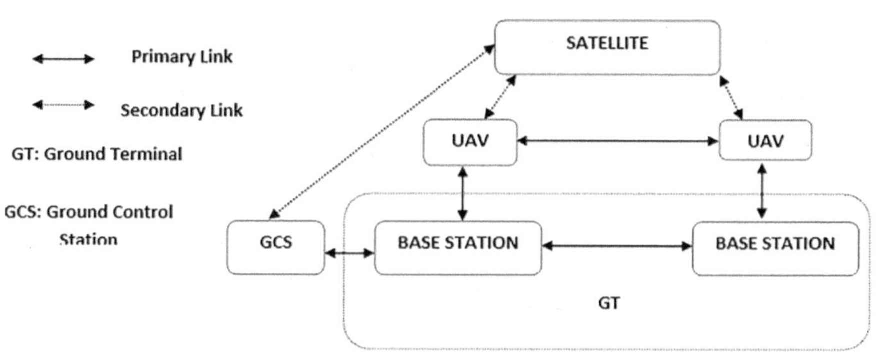

FIGURE 16.2 Communication links between UAVs and ground control station.

To summarize, seamless connection improves real-time data transfer, awareness of the situation, control, protection, operation effectiveness, and integration, allowing UAVs to operate efficiently and successfully in a variety of sectors and applications.

16.1.3 UAV Communication System Components

1. **Wireless Communication Protocols:** These networks, which include Wi-Fi, Bluetooth, and proprietary solutions, allow UAVs to communicate with ground control stations and other equipment.
2. **Frequency Bands:** UAV communication systems use licensed, unlicensed, or specialized frequency bands to determine range, data rate, and interference resilience.
3. **Antenna Systems:** To broadcast and receive signals, UAVs employ specialized antennas such as omnidirectional or directional antennas.
4. **Satellite Links:** Satellite communication systems provide worldwide coverage and allow UAVs in remote or inaccessible places to operate over long distances or beyond line-of-sight.
5. **Cellular Networks:** UAVs may establish communication links using current cellular networks, expanding their range and coverage beyond traditional line-of-sight limits.
6. **Ad Hoc Networking:** In some cases, UAVs can build ad hoc networks to communicate with one another or with the ground control station, enhancing collaboration and data exchange.
7. **Security and Encryption:** To maintain the integrity and confidentiality of sent data, UAV communication systems use encryption, authentication procedures, and secure protocols.
8. **Quality of Service (QoS):** To maintain stable and high-quality communication lines, QoS algorithms prioritize vital data, use error correction techniques, and modify transmission tactics.

In UAV communication systems, these components work together to provide seamless connectivity, real-time data transfer, situational awareness, and accurate control.

16.1.4 UAVs and Onboard Communication Systems

UAVs include onboard communication systems that are critical in providing smooth contact between the UAV and the GCS. These systems support two-way communication, allowing the flow of data, instructions, and information. Onboard communication systems provide real-time data transfer, supplying critical telemetry data, sensor readings, images, and video feeds to the GCS for monitoring and decision-making. They also allow the GCS to deliver orders to the UAV, guaranteeing accurate control and manoeuvrability. Wireless protocols and antenna systems are used to build dependable communication links, while security measures ensure the integrity and confidentiality of sent data. Overall, onboard communication systems are critical components of UAVs, allowing for effective control, data transfer, and situational awareness throughout flight operations [14, 15].

16.1.5 Ground Control Station (GCS)

The GCS is an essential component of UAV operations, functioning as the UAV's centralized command and control centre. It is a system or physical station where operators monitor and control UAV flights, get real-time data and video feeds, and make mission-critical choices. The GCS is often made up of hardware and software components including displays, control interfaces, communication systems, and data processing capabilities. Through continuous communication with the unmanned aircraft, GCS operators maintain situational awareness, execute directives, and assure the safe and efficient operating of UAVs [16].

16.1.6 Communication Links

Communication links are required to provide smooth interaction between UAVs and their GCS. These linkages include the UAV-to-GCS link, which allows for the transfer of telemetry data, sensor readings, video feeds, and instructions between the UAV and the GCS. Furthermore, satellite links enable worldwide coverage for long-distance or beyond-line-of-sight operations, whilst cellular network linkages increase communication capabilities by leveraging existing infrastructure. These communication channels enable successful and efficient UAV operations by ensuring real-time data transfer, situational awareness, and accurate control [17].

- Line-of-Sight (LOS) Communication
- Beyond Line-of-Sight (BLOS) Communication
- Inter-UAV Communication

16.2 CHALLENGES WITH UAV COMMUNICATION SYSTEMS

16.2.1 Limited Bandwidth

The term "limited bandwidth" refers to the limitation on the quantity of data that may be transferred at any one moment inside a certain communication channel or network. When communication systems have restricted bandwidth, the volume and speed of data that can be sent are constrained, which can have an influence on the system's efficiency and performance. Due to limited bandwidth, delays, poor data quality, or the need to prioritize certain types of data over others may occur. Efficient data compression algorithms, prioritization mechanisms, and adaptive transmission strategies are frequently used to maximize available bandwidth utilization and optimize communication system performance [18].

16.2.2 Signal Interference

Signal transmission interference is the disruption or degradation of a communication transmission caused by the presence of unwanted electromagnetic signals or noise. Interference can be created by a number of sources, including other electrical equipment, physical obstacles, and overlapping frequency bands. It can result in data loss, poor signal quality, or full signal interruption. Signal interference mitigation frequently incorporates techniques such as signal filtering, frequency synchronization, shielding, or the use of error correction codes to reduce the impact of unwanted signals and preserve reliable communication [19].

16.2.3 Range and Endurance

Range and endurance are critical performance variables for UAVs. Range is the greatest distance a UAV can travel or operate from its starting location, whereas endurance is the amount of time a UAV can remain in flight without needing to be refueled or recharged. Range and endurance are affected by a variety of factors, including the architecture of the UAV, its propulsion system, energy source, payload weight, and aerodynamics. Improving UAV range and endurance is critical for expanding their operational capabilities, allowing them to cover wider regions, undertake longer flights, and complete jobs more effectively. Technological advances, such as increased battery capacity, energy-efficient propulsion systems, and lightweight materials, help to increase the range and endurance of UAVs, expanding their potential applications in industries like aerial surveillance, delivery, agriculture, and search and rescue.

16.2.4 Security and Privacy

Security and privacy are essential factors in UAV operations. Security precautions are required to safeguard UAV communication systems against unauthorized access, data alteration, and interception. Encryption methods, authentication systems, and secure protocols are used to protect the integrity and confidentiality of sent data. Furthermore, UAVs may be exposed to cyber threats such as hacking or GPS spoofing, which might jeopardize their control and navigation. Privacy is also an issue since UAVs outfitted with cameras or sensors might gather and transmit sensitive data. Regulations and standards have been put in place to protect private rights and guarantee that UAV operations follow privacy principles. To encourage public trust, preserve sensitive data, and promote responsible UAV usage in accordance with legal and ethical frameworks, the correct balance of security and privacy must be struck [20].

16.3 ENHANCEMENT IN UAV COMMUNICATION SYSTEMS

16.3.1 Cognitive Radio Technology

Cognitive radio technology is a wireless communication paradigm that allows for the intelligent and dynamic use of spectrum. It enables radio equipment like UAVs to perceive and adapt to their operational environment, optimizing spectrum consumption and reducing interference. Cognitive radios can identify available frequencies, analyse spectrum quality, and dynamically alter transmission settings to avoid congestion and maximize efficiency by utilizing complex algorithms and machine learning techniques [21]. This strategy is particularly beneficial when the radio frequency spectrum is constrained or congested. Cognitive radio technology can increase the overall performance and capabilities of UAV communication systems by enhancing communication reliability and data throughput, and enabling better coordination and coexistence with other wireless devices.

16.3.2 Software-Defined Radio (SDR)

SDR is a wireless communication technology that uses software-based control and signal processing to allow flexible and variable radio capabilities. SDR uses software-based modules to replace or enhance standard radio system hardware components, providing for more versatility and reconfigurability. SDR allows for the modification and modulation of radio signals using software, allowing for the support of a diverse set of communication protocols, frequency bands, and waveform kinds on a single hardware platform. SDR's adaptability makes it especially helpful in UAV applications, as it allows for smooth integration with various communication protocols, remote setup and updates, and optimal spectrum utilization. SDR technology improves the capability and flexibility of UAV communication systems, allowing them to serve a wide range of operational needs while also simplifying interoperability with multiple devices and networks [22].

16.3.3 5G and Beyond

5G and beyond refers to the next generation of wireless communication networks that outperform current 5G technology. 5G is intended to provide faster data rates, ultra-low latency, increased device connection, and enormous machine-to-machine communication. However, "beyond 5G" refers to future developments in wireless communication, such as terahertz frequencies, huge MIMO (Multiple-Input Multiple-Output), network slicing, and enhanced beam shaping. These developments are intended to give even faster data rates, lower latency, more network capacity, and enhanced dependability. Beyond 5G networks have the potential to allow disruptive applications such as augmented reality, self-driving cars, smart cities, and the Internet of Things (IoT), further increasing the possibilities for UAV communication systems and improving the whole connection ecosystem [23].

16.3.4 Mesh Networking

Mesh networking is a decentralized communication architecture in which several nodes or devices build a network by connecting directly to one another. Mesh networking, as opposed to traditional communication methods that rely on a centralized infrastructure, enables durable and adaptable connectivity, particularly in demanding or dynamic contexts [24, 25]. Each node in the network functions as a relay, delivering and receiving data, ensuring that information may be effectively shared between devices even when direct communication pathways fail or are unavailable. Mesh networks' self-configuring and self-healing capabilities provide enhanced coverage, dependability, and scalability. Mesh networking can improve communication capabilities in the context of UAVs by allowing UAVs to build ad hoc networks, collaborate, exchange data, and relay information, allowing for more efficient coordination and increased operational options [26].

16.4 UAV COMMUNICATION PROTOCOLS

UAV communication protocols are the collection of rules and standards that regulate the transmission of data and orders between UAVs and GCS. These protocols guarantee smooth and effective communication during UAV operations. Wi-Fi, Bluetooth, Zigbee, and proprietary protocols are examples of common UAV communication methods. Wi-Fi is extensively utilized for short-range UAV communication because it provides dependable and high-speed data transfer. Bluetooth is often used for short-range communication, allowing UAVs to communicate with peripheral devices or GCS. Zigbee is well-suited to low-power and low-data-rate applications, and it is frequently used in UAVs for control and telemetry. Private protocols provide unique solutions for certain UAV systems or applications. The communication protocol selected is determined by considerations such as required range, data transfer speed, power consumption, and compatibility with current systems. UAV communication protocols are constantly changing to satisfy the growing needs of UAV applications as well as the necessity for dependable and secure connection [27–30].

16.4.1 Wireless Communication Protocols

Wireless communication protocols are sets of rules and standards that control the transmission of data between devices across wireless networks. These protocols specify how devices connect, transmit and receive data, handle faults, and manage network resources. Wi-Fi (IEEE 802.11), Bluetooth (IEEE 802.15.1), Zigbee (IEEE 802.15.4), and cellular network protocols such as 4G LTE and 5G are examples of wireless communication protocols used in many applications. Each protocol has unique features such as range, data transmission rate, power consumption, and applicability for certain use cases. These wireless communication protocols are critical for allowing seamless connectivity, real-time data transmission, and dependable communication across a wide range of wireless contexts, including UAVs and their ground control stations [31].

16.4.2 Protocol Stack for UAV Communication

The protocol stack for UAV communication refers to the layered structure of protocols that enables smooth and efficient communication between UAVs and their GCS. The stack is often made up of several levels, including the physical layer, data connection layer, network layer, transport layer, and application layer. Protocols define the transmission and receipt of raw data over a wireless media at the physical layer. The data connection layer is in charge of guaranteeing the reliable transfer of data frames between the UAV and the GCS, as well as error detection and repair. The network layer is in charge of addressing, routing, and delivering data packets across networks. The transport layer ensures end-to-end communication dependability and flow management. Finally, the application layer includes the unique protocols and interfaces necessary for UAV operations, such as telemetry, command and control, video streaming, and data interchange. The protocol stack for UAV communication, when used together, allows for the smooth transmission of data and commands, enabling efficient and reliable communication throughout UAV operations [32].

TABLE 16.1
Technology Comparison for Communication Links

Technology	Theoretical rate	Range	Advantages	Limitations	Data traffic
Satellite [1, 34]	10 Mbps uplink one Gbps downlink	Up to the entire earth	Wide coverage Provides connection in all areas	Limited bandwidth High delays Low data rates High cost	Live and real-time operations
IEEE 802.11(a,g) IEEE 802.11(b) IEEE 802.11(n) IEEE 802.11(ac) Ref. [1, 35, 36]	54 Mbps 11 Mbps Up to 300 Mbps Up to 866.7 Mbps	30 m 75–100 m 75 m 35 m	High speed Cheap	Limited range Limited support for users	Control commands Telemetry Downlink sensor data
GSM GPRS UMTS LTE 5G Ref. [1, 37, 38]	Up to 9.6 kbps Up to 144 kbps 2 Mbps Up to 1 Gbps 10 Gbps and more	Depends on BS (1 km to several Kms)	Extended coverage Redundancy One infrastructure for multiple UAVs High bandwidth Mobility support	Expensive	Control commands Telemetry Downlink sensor data (Images, Videos)
WiMAX Ref. [37, 39]	70 Mbps	Up to 50 km	High throughput Low cost Vaste coverage Easy installation Mobility support	Interference issues	Control commands Telemetry Downlink sensor data (Images, Videos)
Zigbee Ref. [40]	250 Kbps	Up to 30 m	Low cost	Low data rate Short range	Control commands
Bluetooth IEEE 802.15 Ref. [1]	1 Mbps	Up to 30 m	Low cost	Low data rate Short range Low security	Data communication Inside UAV

16.4.3 Adaptive Protocols for UAV Communication

Adaptive protocols for UAV communication are communication protocols that alter their parameters and behaviour dynamically in response to changing communication situations. These protocols are intended to improve performance, adapt to changing network circumstances, and enable dependable communication between UAVs and GCS. Adaptive protocols may use adaptive modulation and coding, dynamic spectrum access, adaptive power control, or adaptive routing algorithms. Adaptive protocols can make real-time adjustments to optimize data transfer rates, maximize link quality, and maintain a robust and efficient communication link in UAV operations by continuously monitoring and analysing factors such as signal strength, interference, bandwidth availability, and network congestion. These protocols' flexibility improves the reliability, efficiency, and overall performance of UAV communication systems in dynamic and difficult contexts [33].

16.5 ANTENNA SYSTEMS FOR UAV COMMUNICATION

Antenna systems for UAV communication are critical components that allow wireless signals to be transmitted and received between UAVs and their GCS. These systems use specialized antennas to improve signal strength, range, coverage, and directionality. UAVs may be fitted with omnidirectional antennas that offer coverage in all directions or directional antennas that focus the signal in a single direction, depending on the requirements and operating conditions. Antenna system design and location are key factors for ensuring successful communication, minimizing signal loss, and maximizing the UAV's communication range. UAV communication may accomplish dependable and efficient connectivity by selecting and deploying the proper antenna systems, enabling real-time data transfer, command execution, and situational awareness during UAV operations.

16.5.1 Types of Antennas for UAVs

For communication, UAVs can be outfitted with a variety of antennas. Omnidirectional antennas provide 360-degree coverage by emitting and receiving signals in all directions. These antennas are appropriate for applications requiring broad coverage and that do not rely primarily on long-range or directional communication. In contrast, directional antennas focus the signal in a certain direction, providing for greater range and signal intensity in a given region. When the UAV has to connect with a specific target or when flying in locations with signal blockages, these antennas come in handy. Furthermore, depending on the communication system and operational environment, UAVs may use specialized antennas such as patch antennas, helical antennas, or phased array antennas. The proper antenna type must be chosen to optimize the UAV's communication range, signal quality, and overall performance during flights.

16.5.2 Antenna Placement and Design Considerations

For UAV communication systems, antenna location and design considerations are critical for optimizing signal strength, coverage, and overall performance. Antenna installation should include aspects such as UAV aerodynamics, weight distribution, and interference minimization. Antennas are frequently purposefully placed to eliminate obstacles and maximize line-of-sight communication. Furthermore, the antenna's design, including its size, shape, and radiation pattern, is critical for obtaining the appropriate coverage and directional qualities. Impedance matching, polarization, and radiation efficiency are additional important antenna design factors to guarantee optimal signal transmission and reception. UAV communication systems may achieve dependable and efficient connection by carefully considering both antenna location and design, boosting real-time data transfer, command execution, and situational awareness during UAV operations.

16.6 UAV COMMUNICATION NETWORKS

UAV communication networks are complicated systems that allow UAVs and their GCS to communicate, share data, and collaborate. These networks are made up of a number of UAVs and GCS units that are linked together via wireless communication lines to establish a dynamic network structure. Protocols such as ad hoc networking, mesh networking, or cellular networks are used in UAV communication networks to enable reliable and efficient communication between UAVs and GCS units. These networks allow for real-time data transfer, command execution, and situational awareness, enabling collaborative operations, coordinated missions, and information sharing across numerous UAVs and GCS units. UAV communication networks are critical to improving the capabilities, efficiency, and safety of UAV operations in a variety of applications ranging from airborne surveillance and search and rescue missions to precision agriculture and package delivery [41].

16.6.1 Ad Hoc Networks for UAV Communication

Ad hoc networks are a sort of decentralized network architecture that is frequently utilized in UAV communication. UAVs may build dynamic, self-configuring networks using ad hoc networks instead of depending on a fixed infrastructure. To share data, orders, and information, UAVs can establish direct communication links with neighbouring UAVs or GCS. These networks are extremely adaptable, as UAVs may join and leave the network as required, enabling adaptive and collaborative operations. Ad hoc networks allow UAVs to communicate in situations when standard infrastructure-based networks are unavailable or unfeasible, such as in rural places or disaster-stricken areas. They improve the overall performance and capacities of UAVs in complex and dynamic operating settings by enabling efficient data exchange, coordination, and collaborative decision-making.

16.6.2 Hybrid Networks for UAV Communication

Hybrid networks for UAV communication use the capabilities of several network topologies, such as centralized and decentralized techniques, to build a resilient and adaptable communication system. These networks make use of both infrastructure-based components, such as cellular or Wi-Fi networks, and decentralized aspects, such as ad hoc or mesh networks. Hybrid networks allow UAVs to benefit from the coverage, dependability, and security afforded by infrastructure-based networks while simultaneously exploiting the autonomy, agility, and resilience of decentralized networks. Depending on the availability of infrastructure, proximity to other UAVs or GCS units, and operating requirements, UAVs can switch between several communication protocols. This hybrid solution improves the communication range, efficiency, and reliability of UAVs, allowing for continuous connectivity, real-time data transfer, and collaborative operations in a variety of demanding conditions [42].

16.7 SECURITY AND PRIVACY IN UAV COMMUNICATION SYSTEMS

In UAV communication systems, security and privacy are critical factors. To ensure the integrity and confidentiality of data exchanged between UAVs and GCS, security techniques such as encryption, authentication, and secure protocols are used. These safeguards guard against unauthorized access, manipulation, and interception, guaranteeing the secure sharing of sensitive data. Furthermore, privacy problems emerge as a result of the usage of UAVs outfitted with cameras or sensors that record and send data. Regulations and standards have been put in place to safeguard privacy rights and guarantee safe UAV operations. To build public confidence, secure sensitive data, and comply with legal and ethical frameworks, a proper balance of security and privacy is required, allowing for the safe and responsible usage of UAV communication systems [43].

16.7.1 Encryption and Authentication

Encryption and authentication are two key security technologies used in communication systems to safeguard data integrity, secrecy, and authenticity. Encryption is the process of converting data into ciphertext using cryptographic techniques in order to ensure that only authorized receivers may decipher the information. It provides a secure communication channel by preventing unauthorized access and manipulation during data transmission. The process of authentication, on the other hand, checks the identity of communication entities, guaranteeing that the transmitter and recipient of data are authentic and trustworthy. It prevents impersonation and unauthorized access by confirming the authenticity of participants using credentials, digital signatures, or biometric methods. The use of encryption and authentication together improves the security of communication systems by protecting sensitive data and limiting the dangers associated with unauthorized access, interception, or modification of information.

16.7.2 Key Management in UAV Communication

The methods and procedures associated in the generation, distribution, storage, and revocation of cryptographic keys used for encryption, decryption, and authentication are referred to as key management in UAV communication. It facilitates the safe and efficient exchange of keys between UAVs and GCS in order to create a trustworthy communication channel. Key generation techniques, secure key distribution protocols, and procedures for key storage and revocation are all part of key management. To secure the integrity and security of data exchanged between UAVs and GCS, proper key management practices are required. Secure key creation, secure key exchange, key lifespan management, key storage methods, and secure key revocation processes are all part of it. Effective key management is critical for keeping UAV communication systems secure and trustworthy, preventing unauthorized access, and protecting the confidentiality and integrity of sensitive data [44].

16.7.3 Countermeasures against Cyber Attacks

Countermeasures against cyber assaults in UAV communication systems include a variety of tactics and approaches designed to reduce the dangers provided by hostile actors. Strong network security measures such as firewalls, intrusion detection and prevention systems, and secure communication protocols are among these countermeasures. Patching and updating software on a regular basis helps to resolve vulnerabilities and guard against known attacks. Strong access restrictions, authentication procedures, and encryption are used to guarantee that only authorized organizations may access and operate UAV communication systems. Security audits, penetration testing, and continuous monitoring all aid in the detection and remediation of any vulnerabilities or unauthorized activity. Furthermore, user education and awareness programs play an important role in encouraging appropriate cyber security practices among UAV operators and staff. Organizations may improve the resilience and security of UAV communication systems by combining various countermeasures, decreasing the risk of cyber attacks and preserving crucial data and operations.

16.8 CASE STUDIES: APPLICATIONS OF UAV COMMUNICATION SYSTEMS

UAV communication systems have a wide range of applications in several sectors. UAVs equipped with communication systems have been used for precision agriculture in the agricultural sector, allowing farmers to monitor crop health, optimize irrigation, and apply fertilizers more efficiently. UAVs with communication capabilities have played an important part in disaster response, giving real-time situational awareness to emergency responders, assisting in search and rescue operations, and facilitating communication in remote or devastated locations. In the field of infrastructure inspection, UAV communication systems have been used to assess the status of bridges, power lines, and pipelines, eliminating the need for manual inspections and increasing efficiency. Furthermore, in the realm of aerial photography and cinematography, UAVs connected with communication networks have revolutionized the recording of breathtaking aerial video for movies, advertisements, and documentaries. These case studies show the wide range of applications where UAV communication systems have proved useful in improving operational efficiency, data collecting, and decision-making processes across many sectors.

16.8.1 Military and Defense

UAV communication systems are essential in military and defense applications, providing considerable benefits in intelligence, surveillance, and reconnaissance (ISR) activities as well as combat scenarios. Military-grade UAVs outfitted with modern communication systems offer military troops real-time situational awareness, allowing them to monitor enemy movements, acquire intelligence, and analyse combat conditions. These communication technologies allow high-resolution pictures, video feeds, and sensor data to be sent to ground control stations, allowing commanders to make educated judgments and alter strategy on the battlefield. UAV communication systems also aid in the communication and coordination of various UAVs, ground forces, and command centres, hence increasing mission effectiveness and situational awareness. Furthermore, UAVs outfitted with communication equipment can act as relays, increasing the reach and coverage of existing military communication networks in distant or difficult settings. UAV communication systems are critical instruments in military and defense applications because they offer efficient, secure, and reliable communication, boosting operational capabilities and delivering a tactical edge in the field.

16.8.2 Search and Rescue Operations

UAV communication systems have shown to be important assets in search and rescue missions, assisting in the location and rescue of those in need. UAVs with enhanced communication capabilities can cover huge regions quickly, enabling real-time airborne surveillance and acquiring important information about the search area. The communication technologies aboard these UAVs allow rescue crews to assess the situation, detect possible threats, and make educated decisions by transmitting live video feeds, thermal imaging, and sensor data to ground control centres. The real-time connection between the UAV and the ground control station also allows for good coordination among UAV operators, search teams, and emergency responders, which improves the efficiency and safety of the search and rescue operation. Furthermore, UAV communication systems can be outfitted with geolocation technology, allowing them to send their own locations and aid in directing ground forces to the location of missing or injured persons. UAV communication systems play a critical role in search and rescue operations, reducing reaction times, boosting situational awareness, and ultimately saving lives.

16.8.3 Disaster Management

UAV communication systems have become essential instruments in disaster management, assisting with preparedness, response, and recovery activities. During natural disasters or catastrophes, UAVs outfitted with modern communication systems may swiftly scan impacted regions, give real-time situational awareness, and aid in damage assessment. The communication technologies aboard these UAVs allow the transfer of live video feeds, aerial pictures, and sensor data to ground control stations, allowing emergency responders to make educated choices, distribute resources effectively, and coordinate rescue and relief activities. In locations where traditional communication infrastructure has been damaged or interrupted, UAVs can also function as communication relays, restoring connectivity and improving communication among emergency responders and affected people. Furthermore, UAV communication systems enable the deployment of autonomous or remotely controlled UAVs for aerial surveillance and the delivery of critical supplies to inaccessible or dangerous regions. UAV communication systems contribute considerably to disaster management by improving communication, coordination, and situational awareness, allowing for more effective response operations and helping to save lives and lessen the effects of disasters [43].

16.9 FUTURE TRENDS AND CHALLENGES

Several fascinating developments and difficulties await UAV communication systems in the future. Technological advancements will very certainly lead to more automation and autonomy, allowing UAVs to perform complicated jobs and collaborate more efficiently. The incorporation of artificial intelligence and machine learning algorithms is expected to improve UAV communication by enabling adaptive decision-making, intelligent routing, and self-optimization of network parameters. Furthermore, the advent of 5G and beyond, as well as developments in satellite communication, will increase the range, bandwidth, and reliability of UAV communication. However, with these developments come new problems, such as maintaining adequate cyber security measures to guard against cyber attacks, dealing with spectrum congestion concerns, and implementing standardized protocols for easy interoperability across different UAV systems. Furthermore, as UAVs grow more common, regulatory frameworks will need to adjust to meet safety, privacy, and airspace management concerns. Overcoming these obstacles is critical to realizing the full potential of UAV communication systems and paving the road for their widespread adoption in a variety of sectors and applications.

16.9.1 Integration with Internet of Things (IoT)

Integration of UAV communication systems with the Internet of Things (IoT) has the potential to significantly improve connection and enable new applications. A huge network of interconnected sensors, actuators, and data sources may be constructed by merging UAVs with IoT devices. UAVs outfitted with communication systems can act as mobile IoT gateways, gathering data from remote or inaccessible IoT devices and transmitting it to centralized systems. This interface provides large-scale real-time monitoring of environmental conditions, infrastructure health, and asset tracking. It also allows UAVs to accept commands from IoT systems, enabling automated operations such as targeted inspections, precision agriculture, and infrastructure maintenance. To enable smooth integration and reliable communication between UAVs and IoT devices, concerns such as interoperability, data management, and security must be solved. The integration of UAV communication systems with IoT offers up a new world of possibilities for data-driven decision-making, automation, and improved operational efficiency across a wide range of disciplines, from smart cities and industrial applications to environmental monitoring and disaster management [43].

16.9.2 Artificial Intelligence (AI) in UAV Communication

AI is playing an increasingly important role in UAV communication systems, revolutionizing their capabilities and efficacy. UAV communication systems may improve situational awareness, optimize network performance, and allow intelligent decision-making by employing AI algorithms and methodologies. AI algorithms can extract important insights and patterns from vast amounts of data acquired by UAVs, such as sensor readings, images, and real-time communication parameters. This allows for communication performance prediction, dynamic network optimization, and adaptive routing depending on changing environmental circumstances. AI also enables UAVs to make autonomous judgments, such as route planning, obstacle avoidance, and communication resource prioritization. By incorporating artificial intelligence (AI) into UAV communication systems, they become more intelligent, efficient, and responsive, allowing for enhanced mission results and extending the possibilities of UAV applications in domains such as surveillance, delivery, and disaster response.

16.9.3 Spectrum Management and Regulations

Spectrum management and regulations are critical for guaranteeing the efficient and orderly use of radio frequency spectrum in UAV communication systems. Spectrum management is designating certain frequency bands to various communication services, including UAV operations, in order to minimize interference and maximize spectrum utilization. Regulatory frameworks provide standards, licensing criteria, and operational parameters to manage UAV communication systems and assure compliance with safety, security, and privacy regulations. Spectrum management and regulations also handle issues such as spectrum congestion, interference reduction, and coordination of communication operations in common frequency bands. These regulations are intended to promote fair and equitable spectrum access, prevent against unauthorized and harmful interference, and ensure the safe and responsible operating of UAVs. Effective spectrum management and regulations are critical for stimulating innovation, supporting the expansion of UAV communication systems, and ensuring a harmonious and dependable wireless communication environment [44].

16.9.4 Ethical and Legal Considerations

In the design, implementation, and operation of UAV communication systems, ethical and legal issues are critical. UAVs must be utilized in an ethical way that protects privacy, minimizes harm, and upholds basic human rights. Operators of UAVs should prioritize safety, minimize unwanted intrusions, and safeguard sensitive data obtained during missions. Furthermore, compliance with airspace limits, licensing requirements, and privacy laws requires respect to legal frameworks and regulations.

UAV activities should not violate individuals' rights or privacy, nor should they jeopardize public safety. Data collection and use must be transparent, and proper consent must be acquired wherever possible. UAV communication systems may be implemented responsibly by addressing ethical and legal issues, encouraging public trust, preserving privacy rights, and supporting the safe and ethical integration of UAVs into society.

16.10 CONCLUSION

Finally, UAV communication systems have emerged as key components in a wide range of applications and sectors, allowing for seamless connectivity, real-time data transfer, and better operating capabilities. Communication protocol advances, antenna technology, and network design advancements have all contributed considerably to the dependability, efficiency, and security of UAV communications. UAV communication systems have proved their promise in revolutionizing sectors and boosting outcomes across the board, from precision agriculture and delivery logistics to military operations and disaster management. Future improvements in AI integration, IoT connection, and spectrum management, on the other hand, will affect the growth of UAV communication systems, bringing both possibilities and problems. To guarantee responsible and safe UAV operations, it is critical to address ethical, legal, and security concerns. UAV communication systems will continue to play a significant part in revolutionizing industries, boosting connectivity, and supporting efficient and effective operations in the years ahead with sustained innovation, collaboration, and adherence to laws.

REFERENCES

[1] Jawhar, I., Mohamed, N., Al-Jaroodi, J., Agrawal, D.P., Zhang, S.: Communication and networking of UAV-based systems: Classification and associated architectures. *J. Netw. Comput. Appl.* 5(84), 93–108 (2017).

[2] Hayat, S., Yanmaz, E., Muzaffar, R.: Survey on unmanned aerial vehicle networks for civil applications: A communications viewpoint. *IEEE Commun. Surv. Tutor.* 18(4), 2624–2661 (2016)

[3] Austin, R.: *Unmanned Aircraft Systems: UAVS Design, Development and Deployment*, vol. 54. Wiley, Hoboken (2011)

[4] www.aeronewstv.com/en/lifestyle/sports-leisure/2161-when-using-leisure-drones.html. Accessed 1 April 2018.

[5] Coffey, T., Montgomery, J.A.: The emergence of mini UAVs for military applications. *Def. Horiz.* 22, 1 (2002).

[6] Casbeer, D.W., Beard, R.W., McLain, T.W., Li, S.M., Mehra, R.K.: Forest fire monitoring with multiple small UAVs. In: *American Control Conference*, pp. 3530–3535, IEEE, Portland (2005).

[7] Eisenbeiss, H.: The autonomous mini helicopter: A powerful platform for mobile mapping. *Int. Arch. Photogram. Remote Sens. Spat. Inf. Sci.* 27, 977–983 (2008)

[8] Steffen, R., Frstner, W.: On visual real-time mapping for unmanned aerial vehicles. In: *21st Congress of the International Society for Photogrammetry and Remote Sensing (ISPRS)*, pp. 57–62. Gannett, Beijing, China (July 2008).

[9] Barr, A.: Amazon testing delivery by drone, CEO Bezos says. *USA Today*. www.usatoday.com/story/tech/2013/12/01/amazon-bezos-drone-delivery/3799021/. Accessed 2 April 2018.

[10] Johnson, T.: FAA may approve use of drones for Hollywood film-making. *Variety* 1 (2 June 2014).
[11] Costa, F.G., Ueyama, J., Braun, T., Pessin, G., Osrio, F.S., Vargas, P.A.: The use of unmanned aerial vehicles and wireless sensor network in agricultural applications. In: *International Conference on Geoscience and Remote Sensing Symposium (IGARSS)*, pp. 5045–5048. IEEE, Munich (2012).
[12] Primicerio, J., et al.: A flexible unmanned aerial vehicle for precision agriculture. *Precis. Agric.* 13(4), 517–523 (2012).
[13] Dockrill, P.: Facebook is preparing its Internet-beaming drone for maiden launch, *ScienceAlert.* www.sciencealert.com/facebook-is-preparing-its-internet-beaming-drone-for-maiden-launch. Accessed 2 April 2018.
[14] Doherty, P., Rudol, P.: A UAV search and rescue scenario with human body detection and geolocalization. In: Orgun, M.A., Thornton, J. (eds.) *AI 2007. LNCS (LNAI)*, vol. 4830, pp. 1–13. Springer, Heidelberg (2007). https://doi.org/10.1007/978-3-540-76928-61
[15] Waharte, S., Trigoni, N.: Supporting search and rescue operations with UAVs. In: *International Conference on Emerging Security Technologies (EST)*, pp. 142–147. IEEE, Canterbury (2010).
[16] Berrahal, S., Kim, J.H., Rekhis, S., Boudriga, N., Wilkins, D., Acevedo, J.: Border surveillance monitoring using quadcopter UAV-aided wireless sensor networks. *J. Commun. Softw. Syst.* 12(1), 67–82 (2016).
[17] Puri, A.: *A Survey of Unmanned Aerial Vehicles (UAV) for Traffic Surveillance*, pp. 1–29. Department of Computer Science and Engineering, Grand Hyatt Atlanta, Atlanta, GA: University of South Florida (2005).
[18] Bekmezci, I., Sahingoz, O.K., Temel, S.: Flying ad-hoc networks (FANETs): A survey. *Ad Hoc Netw.* 11(3), 1254–1270 (2013)
[19] Maxa, J.A., Mahmoud, M.S.B., Larrieu, N.: Survey on UAANET routing protocols and network security challenges. *Ad Hoc Sens. Wirel. Netw.* 37, 231–320 (2017)
[20] Gupta, L., Jain, R., Vaszkun, G.: Survey of important issues in UAV communication networks. *IEEE Commun. Surv. Tutor.* 18(2), 1123–1152 (2016).
[21] Fahlstrom, P., Gleason, T.: *Introduction to UAV Systems*. Willy, Hoboken (2012).
[22] Zaheer, Z., Usmani, A., Khan, E., Qadeer, M.A.: Aerial surveillance system using UAV. In: *Thirteenth International Conference on Wireless and Optical Communications Networks (WOCN)*, pp. 1–7. IEEE, Telangana State, India (July 2016).
[23] http://ardupilot.org/copter/docs/common-choosing-a-groundstation.htmlcomparison. Accessed 2 April 2018.
[24] *Mission Planner: Ground Control Station for APM (ArduPilotMega)*. http://planner.ardupilot.com/wiki/mission-planner-overview/. Accessed 2 April 2018.
[25] Li, J., Zhou, Y., Lamont, L.: Communication architectures and protocols for networking unmanned aerial vehicles. In: *2013 IEEE Globecom Workshops (GC Wkshps)*, Institute of Electrical and Electronics Engineers (IEEE), pp. 1415–1420. Atlanta, GA, USA (December 2013).
[26] Frew, E.W., Brown, T.X.: Airborne communication networks for small unmanned aircraft systems. *Proceedings of the IEEE*, 96(12) (2008).
[27] Andre, T., et al.: Application-driven design of aerial communication networks. *IEEE Commun. Mag.* 52(5), 129–137 (2014).
[28] Santos, N., Raimundo, A., Peres, D., Sebastio, P., Souto, N.: Development of a software platform to control squads of unmanned vehicles in real-time. In: *International Conference on Unmanned Aircraft Systems (ICUAS)*, pp. 1–5. IEEE, Miami, USA (June 2017).
[29] del Arco, J.C., Alejo, D., Arrue, B.C., Cobano, J.A., Heredia, G., Ollero, A.: Multi-UAV ground control station for gliding aircraft. In: *23th Mediterranean Conference on Control and Automation (MED)*, pp. 36–43. IEEE, Torremolinos, Spain (June 2015).

[30] Rodrigues, A.V., Carapau, R.S., Marques, M.M., Lobo, V.: Unmanned aerial vehicles: System architecture and protocols. *Sci. Bull. "Mircea cel Batran" Naval Academy* 20(1), 140 (2017).

[31] Butcher, N., Stewart, A., Biaz, S.: *Securing the MAVLink Communication Protocol for Unmanned Aircraft Systems*. Appalachian State University, Windmill Ln, Henderson, NV 89074, USA (2013)

[32] Skinnemoen, H.: UAV satellite communications live mission-critical visual data. In: *International Conference on Aerospace Electronics and Remote Sensing Technology (ICARES)*, pp. 12–19. IEEE, Yogyakarta, Indonesia (November 2014).

[33] Gabriel, C.: *WiMAX: The Critical Wireless Standard 802.16 and Other Broadband Wireless Options*. Blueprint WiFi Monthly Research Report. ARC Chart Ltd. (2003)

[34] Ribeiro, C.: Bringing wireless access to the automobile: A comparison of Wi-Fi, WiMAX, MBWA, and 3G. In: *21st Computer Science Seminar*, Rensselaer, Hartford, pp. 1–7 (April 2005).

[35] Volko, T., Moucha, V., Lipovsk, P., Draganov, K.: Possibility of usage the latest GSM generations for the purpose of UAV communication. In: *New Trends in Signal Processing (NTSP)*, pp. 1–4. Demanovska dolina, Slovakia, IEEE (October 2016).

[36] Zhu, L., Yin, D., Yang, J., Shen, L.: Research of remote measurement and control technology of UAV based on mobile communication networks. In: *IEEE International Conference on Information and Automation*, pp. 2517–2522. IEEE, Gothenburg, Sweden (August 2015).

[37] Rahman, M.A.: Enabling drone communications with WiMAX Technology. In: *The 5th International Conference on Information, Intelligence, Systems and Applications, IISA*, pp. 323–328. IEEE, Chania, Crete, Greece (July 2014).

[38] Fourty, N., Val, T., Fraisse, P., Mercier, J.J.: Comparative analysis of new high data rate wireless communication technologies from Wi-Fi to WiMAX. In: *International Conference on Autonomic and Autonomous Systems and Networking and Services, ICAS-ICNS*, p. 66. Papeete, France, IEEE (2005).

[39] Rodrigues, M., Pigatto, D.F., Fontes, J.V., Pinto, A.S., Diguet, J.P., Branco, K.R.: UAV integration into IoIT: Opportunities and challenges. In: *ICAS 2017*, vol. 95. ISO 690 IARIA, Barcelona, Spain (2017).

[40] Kouba, A., Qureshi, B., Sriti, M.F., Javed, Y., Tovar, E.: A service-oriented cloud-based management system for the internet-of-drones. In: *IEEE International Conference on Autonomous Robot Systems and Competitions (ICARSC)*, pp. 329–335. IEEE, Coimbra, Portugal (2017).

[41] Mahmoud, S., Mohamed, N.: Broker architecture for collaborative UAVs cloud computing. In: *2015 International Conference on Collaboration Technologies and Systems (CTS)*, pp. 212–219. Atlanta, GA, USA, IEEE. ISO 690 (June 2015).

[42] Yuan, Z., Huang, X., Sun, L., Jin, J.: Software defined mobile sensor network for micro UAV swarm. In: *International Conference on Control and Robotics Engineering (ICCRE)*, pp. 1–4. IEEE, Singapore (April 2016).

[43] Luo, C., Nightingale, J., Asemota, E., Grecos, C.: A UAV cloud system for disaster sensing applications. In: 2015 IEEE 81st Vehicular Technology Conference (VTC Spring), pp. 1–5. Glasgow, IEEE (May 2015).

[44] Sharma, V., Song, F., You, I., Chao, H.C.: Efficient management and fast handovers in software defined wireless networks using UAVs. *IEEE Netw.* 31(6), 78–85 (2017).

Index